Cognitive Intelligence and Robotics

Volume 1

Series Editors

Cognitive Intelligence refers to the natural intelligence of humans and animals, it is considered that the brain performs intelligent activities. While establishing a hard boundary that distinguishes intelligent activities from others remains controversial, most common behaviors and activities of living organisms that cannot be fully synthesized using artificial means are regarded as intelligent. Thus the acts of sensing and perception, understanding the environment, and voluntary control of muscles, which can be performed by lower-level mammals, are indeed intelligent. Besides the above, advanced mammals can perform more sophisticated cognitive tasks, including logical reasoning, learning, recognition, and complex planning and coordination, none of which can yet be realized artificially to the level of a baby, and thus are regarded as cognitively intelligent.

This book series covers two important aspects of brain science. First, it attempts to uncover the mystery behind the biological basis of cognition, with a special emphasis on the decoding of stimulated brain signals or images. Topics in this area include the neural basis of sensory perception, motor control, sensory-motor coordination, and understanding the biological basis of higher-level cognition, including memory, learning, reasoning, and complex planning. The second objective of the series is to publish consolidated research on brain-inspired models of learning, perception, memory, and coordination, including results that can be realized on robots, enabling them to mimic the cognitive activities performed by living creatures. These brain-inspired models of machine intelligence complement the behavioral counterparts studied in traditional artificial intelligence.

The series publishes textbooks, monographs, and contributed volumes.

Jing Yan · Xinping Guan

Maritime Unmanned Systems

Integrated Design of Detection, Communication, and Control

Springer

Jing Yan
Department of Automation
Yanshan University
Qinhuangdao, Hebei, China

Xinping Guan
Department of Automation
Shanghai Jiao Tong University
Shanghai, China

ISSN 2520-1956 ISSN 2520-1964 (electronic)
Cognitive Intelligence and Robotics
ISBN 978-981-95-6906-9 ISBN 978-981-95-6907-6 (eBook)
https://doi.org/10.1007/978-981-95-6907-6

This Springer imprint is published by the registered company Springer Nature Singapore Pte Ltd.
The registered company address is: 152 Beach Road, #21-01/04 Gateway East, Singapore 189721, Singapore

Preface

With the rapid development of robotics and communication technologies, maritime unmanned systems (MUSs) have emerged as a critical research domain due to their unprecedented capabilities in oceanographic data collection, environmental monitoring, and marine ecosystem studies. Specifically, MUSs refer to interoperable robotic systems operating synergistically in aerial, surface, and underwater environments. At present, MUSs exhibit the characteristics of diversified sensing resources, decentralized architectures, and systematic coordination, making the detection-communication-control crucial for efficient operations and multi-mission adaptability. However, the traditional research predominantly adopts separate design approaches for detection, communication, and control in MUSs. The separate design creates dual constraints: the detection-communication interactions lack effective feedback mechanisms while the control commands become vulnerable to information incompleteness. To address the above issue, it is urgent to co-design the detection, communication, and control for MUSs. This research direction has profound strategic significance and long-term value for the future development and application of marine technology.

This book is devoted to present the cutting-edge results on the integrated design of detection, communication, and control for MUSs. It first elaborates on the basic concepts of detection, communication, and control in maritime environment. After that, it presents several integrated design solutions for the detection, communication, and control in MUSs, including detection-communication integration, detection-control integration, communication-control integration, and detection-communication-control integration. Finally, this book concludes with practical applications of these integrated design methodologies in modern port monitoring systems, demonstrating their effectiveness in real-world maritime scenarios.

The results presented here provide actionable insights of developing next-generation MUSs for the researchers, engineers, and practitioners in maritime robotics, autonomous systems, and networked control. A background in control theory, signal processing, or wireless communication is recommended to fully grasp the advanced techniques. Whether you are advancing academic research or optimizing industrial systems, this book equips you with cutting-edge tools to push the boundaries of maritime autonomy.

The book is organized as follows.

Chapter 1 elaborates on the basic concepts of detection, communication, and control in maritime environment. Meanwhile, the main challenges and the current research progress are systematically summarized, through which the technical substance and the research necessity for the integration design are introduced.

Chapter 2 develops an underwater detection system by combining active and passive measurements. With regard to the unknown position and power of underwater target, a chi-square test-based local decision rule with an adaptive weight law is developed. In order to fuse the local decisions from different nodes, a hybrid Bayesian fusion algorithm is designed to reduce the number of transmissions and improve the detection rate.

Chapter 3 investigates the integrated design of detection and communication systems. It presents two shared waveform designs, i.e., communication-oriented and detection-oriented shared waveforms. It can significantly enhance spectral efficiency while simultaneously reducing overall power consumption, thereby addressing critical challenges in resource-constrained underwater environments.

Chapter 4 considers the integrated design of underwater communication and control. Particularly, it develops a communication-efficient and collision-free motion planning solution for underwater vehicles in fading channel and obstacle environment. Note that the above solution focuses on the data link layer, without considering network-layer challenges such as routing protocols, or large-scale network coordination. In view of this, an optimally persistent-based formation solution for underwater vehicles is also presented. Compared to the neighboring rule-based solutions, the proposed optimally persistent topology can balance energy efficiency and topology connectivity.

Note that the above chapters mainly investigate underwater platforms, lacking comprehensive research on integrated air-sea cross-domain systems. In view of this, Chap. 5 employs unmanned aerial/surface/underwater vehicles (UAV-USV-UUV) to develop a cooperative detection and tracking solution. The measure theory is adopted to construct a heterogeneous detection mode, such that the detection probability can be maximized. Meanwhile, a deep learning algorithm called depth deterministic policy gradient (DDPG) is designed to track the target.

Chapter 6 considers the integrated design of detection, communication, and control. Different from the above chapters, the integrated detection-communication waveform in this chapter enables simultaneous obstacle detection and self-localization. After that, it also develops an underwater teleoperation system toward human on the loop (HOTL), which mainly includes an underwater vehicle, a surface buoy, a human operator, and a communication network. With the integration design of detection, communication, and control, it can extend human intelligence and ability to perform complex tasks at a remote site.

To further validate the practicality of the theoretical methods, Chap. 7 presents typical demonstration applications of MUSs in the context of port monitoring. With the wireless communication network, the cameras mounted on underwater vehicles and shore collect the ship waterline images, such that the ship attitudes can be estimated by using deep learning technology. Sonars are used to carry out water depth detection and port wall inspection. This system can improve the reliability and reduce the cost over the manual monitoring system.

Chapter 8 summarizes the unresolved key challenges in the integrated design of detection, communication, and control for MUSs, while also providing future research directions that offer valuable references for subsequent studies.

This book was supported in part by the National Natural Science Foundation of China under Grants 62222314 and U25A20472; in part by the National Science Foundation of Hebei Province under Grant F2022203001, and in part by the Education Department Foundation of Hebei Province under Grant JCZX2025027.

Qinhuangdao, China — Jing Yan
Shanghai, China — Xinping Guan
December 2025

Competing Interests The authors have no competing interests to declare that are relevant to the content of this manuscript.

Contents

1 Introduction 1
1.1 Overview of Maritime Unmanned Systems 2
1.1.1 Collaborative Connotation of MUSs 2
1.1.2 Co-Design of Detection, Communication, and Control 5
1.2 Difficulties and Challenges 7
1.2.1 Model Construction Level 7
1.2.2 Networking-Transmission Level 8
1.2.3 Detection-Communication Level 9
1.2.4 Cooperative Control Level 10
1.3 Research Progress on Integration Design 11
1.3.1 Active-Passive Detection with Multi-Source Information Fusion 11
1.3.2 Networking and Transmission 14
1.3.3 Cooperative Control of Maritime-Aerial Unmanned Platforms 19
1.4 Application Scenarios 22
1.4.1 Maritime Joint Anti-submarine Operations 22
1.4.2 Amphibious Assaults on Islands 23
1.4.3 Maritime Search and Rescue 23
1.4.4 Maritime Infrastructure Operation and Maintenance 24
References 24

2 Integrated Detection via Active and Passive Measurements 29
2.1 Introduction 29
2.2 Network Model and Problem Formulation 31
2.2.1 Kinematic Model of Target 32
2.2.2 Active and Passive Measurement Models 32
2.2.3 Communication Energy Model 33
2.2.4 Problem Formulation 34
2.3 Underwater Target Detection 34
2.3.1 Chi-Square Test Based Local Decision Rule 35

2.3.2 Hybrid Bayesian Fusion Algorithm 39
2.3.3 Detection Performance Analysis 43
2.4 Technology Practice 48
2.4.1 Case Study I: Detection with Local Measurement 49
2.4.2 Case Study II: Detection with External Fusion Measurement 51
2.5 Summary 55
References 55

3 Integrated Design of Detection and Communication 59
3.1 Introduction 59
3.2 Integrated Wave Design 61
3.2.1 Communication-Based Shared Waveform 62
3.2.2 Detection-Based Shared Waveform 64
3.3 Implementation of the Integration System 68
3.3.1 System Hardware Architecture 68
3.3.2 Operational Procedure 69
3.4 Technology Practice 71
3.4.1 Case Study I: Integrated Design with Communication Signal 71
3.4.2 Case Study II: Integrated Design with Detection Signal 74
3.5 Summary 77
References 79

4 Integrated Design of Communication and Control 81
4.1 Introduction 81
4.1.1 Task I: Communication-Efficient and Collision-Free Motion Planning 81
4.1.2 Task II: Optimally Persistent Formation with Unknown Interaction 84
4.2 Communication-Efficient and Collision-Free Motion Planning 87
4.2.1 SNR Prediction in Fading Channel 91
4.2.2 Model-Free and Collision-Free Motion Planning Algorithm 98
4.3 Optimally Persistent Formation with Unknown Interaction Topology 106
4.3.1 Topology Inference and Optimization 111
4.3.2 Formation Controller for Multiple Underwater Vehicles 123
4.4 Technology Practice 130
4.4.1 Case Study I: Collision-Free Motion Planning for a Single Underwater Vehicle 130
4.4.2 Case Study II: Topology-Aware Formation for Multiple Underwater Vehicles 136
4.5 Summary 149
References 150

5 Integrated Design of Detection and Control 155
5.1 Introduction 155
5.2 Model Construction and Problem Formulation 158
5.3 Strategy Design for Detection and Tracking Control 162
5.3.1 Heterogeneous Detection Mode 162
5.3.2 DDPG-Based Tracking Algorithm 167
5.4 Performance Analysis 174
5.4.1 Detection Probability 174
5.4.2 Topology Connection and Convergence Analysis 175
5.5 Technology Practice 178
5.5.1 Case Study I: Detection and Tracking Control for a Surface Target 179
5.5.2 Case Study II: Detection and Tracking Control for an Underwater Target 185
5.6 Summary 193
References 193

6 Integrated Design of Detection, Communication and Control 197
6.1 Introduction 197
6.2 Co-design of Detection, Communication and Control for Underwater Target Tracking 199
6.2.1 Model Construction and Problem Formulation 199
6.2.2 Integrated Waveform Design 200
6.2.3 Design of Tracking Controller 202
6.3 Co-design of Detection, Communication and Control for Underwater Teleoperation 204
6.3.1 Model Construction and Problem Formulation 204
6.3.2 Construction of Underwater Teleoperation Solution 205
6.4 Technology Practice 210
6.4.1 Case Study I: Integrated Design for Target Tracking Control 210
6.4.2 Case Study II: Integrated Design for Teleoperation Control 212
6.5 Summary 217
References 218

7 Demonstration Application in Port Monitoring System 219
7.1 Introduction 219
7.2 Integration Methods for Port Monitoring System 222
7.2.1 Self-localization and Motion Control 222
7.2.2 Ship Attitude Monitoring 224
7.2.3 Management Platform Design 227
7.3 Construction of Port Monitoring System 228
7.3.1 Hardware Design 228
7.3.2 Software Design 229

7.4 Technology Practice 231
7.4.1 Case Study I: Effectiveness of Detection, Communication and Control 231
7.4.2 3D Reconstruction of Port Basin 231
7.4.3 Case Study II: Construction of Management Platform 235
7.5 Summary 237
References 241

8 Future Research Directions 243
8.1 Open Issues for Further Research 243
8.2 Directions for Future Research 246
8.3 Summary 248

Symbols and Acronyms

2LMG	Two-Layer Multi-Group
3D	Three Dimensional
AOA	Angle of Arrival
AUVs	Autonomous Underwater Vehicles
BRF	Body-Fixed Reference Frame
CRLB	Cramer-Rao Lower Bound
DDPG	Deep Deterministic Policy Gradient
DOA	Domain of Attraction
DOF	Degrees of Freedom
DQN	Deep Q-Network
EPSDT	End-to-End Probability of Successful Data Transmission
GPS	Global Positioning System
GSFM	Generalized Sinusoidal Frequency Modulation
HOTL	Human-on-the-Loop
I/Q	In-phase and Quadrature Components
IMU	Inertial Measurement Unit
IRF	Inertial Reference Frame
IRL	Integral Reinforcement Learning
LFM	Linear Frequency Modulation
MIMO	Multiple-Input Multiple-Output
MLE	Maximum Likelihood Estimate
MSK	Minimum Shift Keying
MUSs	Maritime Unmanned Systems
NNs	Neural Networks
OCDM	Orthogonal Chirp Division Multiplexing
OFDM	Orthogonal Frequency Division Multiplexing
PD	Proportional-Derivative
PSO	Particle Swarm Optimization
RL	Reinforcement Learning
ROVs	Remotely Operated Vehicles
SDN	Software-Defined Networking

SFM	Sinusoidal Frequency Modulation
SMC	Sliding-Mode Control
SNR	Signal-to-Noise Ratio
TDOA	Time Difference of Arrival
TOA	Time of Arrival
TOF	Time of Flight
UAVs	Unmanned Aerial Vehicles
USVs	Unmanned Surface Vehicles
UWB	Ultra-Wideband
VLC	Visible Light Communication
YOLO-T	You Only Look Twice
$\mathcal{R}$	Field of real numbers
$\mathcal{R}^n$	n-dimensional real Euclidean space
$\mathcal{R}^{n\times m}$	Space of $n \times m$ real matrices
$\mathbf{I}$	Identity matrix
$\mathbf{0}$	Zero matrix
$\mathbf{A}$	System matrix
$\mathbf{A}^{-1}$	Inverse of matrix $\mathbf{A}$
$\mathbf{A}^{\mathrm{T}}$	Transpose of matrix $\mathbf{A}$
lim	Limit
min	Minimum
max	Maximum
argmin f	Value of the variable that minimizes function f
argmax f	Value of the variable that maximizes function f
$\mathrm{tr}(\mathbf{A})$	Trace of matrix $\mathbf{A}$
$\mathrm{rank}(\mathbf{A})$	Rank of matrix $\mathbf{A}$
$\Vert \cdot \Vert$	Euclidean norm
∇f	The gradient of function f
$\forall$	For all
$\in$	Belong to
$\sum$	Sum
$\mathbf{\Pi}$	Product operation
$E\{\cdot\}$	Mathematical expectation operator
$(\mathbf{A})_s$	The sth column of matrix $\mathbf{A}$
$\mathrm{diag}\{\cdot\}$	Diagonal matrix
$\rightarrow$	Tend to, or mapping to (case sensitive)
$\lvert x \rvert$	Absolute value (or modulus) of x
$\mathrm{sgn}(\cdot)$	Signum function
$\mathrm{sign}(x)$	Sign of x
$\lambda(\mathbf{A})$	Eigenvalue of matrix $\mathbf{A}$
$\lambda_{\max}(\mathbf{A})$	Maximum eigenvalue of matrix $\mathbf{A}$
$\mathbf{0}_{n\times m}$	Zero matrix of dimension $n \times m$

Chapter 1
Introduction

The ocean covers approximately 71% of the Earth's surface and serves as a vital resource and strategic domain for human survival and development. With intensifying global climate change, increasing demands for resource exploitation, and increasingly complex national security concerns, the exploration, utilization, and protection of the ocean have become central issues of global attention. To enhance marine sensing and operational capabilities, aerial-surface-underwater unmanned platforms equipped with computation, communication, and control functions can be deployed within designated sea areas, forming the maritime unmanned systems (MUSs) characterized by broad spatial coverage, high security, and strong real-time responsiveness. Specifically, MUSs refer to a broad class of unmanned platforms equipped with autonomous navigation and mission execution capabilities, capable of performing tasks such as information acquisition, detection and identification of fixed or moving targets, localization, tracking, and encirclement or interception. These platforms include Unmanned Aerial Vehicles (UAVs), Unmanned Surface Vehicles (USVs), Autonomous Underwater Vehicles (AUVs), and Remotely Operated Vehicles (ROVs). In the context of ocean cognition and exploration, MUSs have become a crucial foundation for securing technological advantage in marine science and defense.

With the rapid developments of intelligent control and unmanned autonomous systems, the collaborative operation of MUSs has progressively transitioned from conceptual exploration to practical implementation. A notable example occurred in March 2014, when 26 countries and regions jointly deployed aerial, surface, and underwater unmanned platforms in a large-scale search operation after the disappearance of Malaysia Airlines Flight MH370 over the Indian Ocean. Although the mission ultimately did not succeed, it highlighted the significant value of swarm collaboration in maritime search and rescue. In another instance, in October 2022, Ukraine utilized a coordinated swarm of USVs and UAVs to conduct a surprise attack on Russia's Black Sea Fleet, garnering widespread international attention. Compared with traditional single-domain unmanned systems, the collaborative operation

J. Yan and X. Guan, *Maritime Unmanned Systems*, Cognitive Intelligence and Robotics 1, https://doi.org/10.1007/978-981-95-6907-6_1

of MUSs can fully leverage functional redundancy and capability complementarity across heterogeneous platforms, enabling integrated utilization of all constituent elements and thereby achieving synergistic effects and multiplicative gains in overall operational effectiveness. Simultaneously, MUSs have promoted the emergence of new operational paradigms, broadening the potential applications in marine resource exploration, maritime security patrols, environmental monitoring, and emergency response missions.

1.1 Overview of Maritime Unmanned Systems

1.1.1 Collaborative Connotation of MUSs

The collaborative connotation of MUSs can be understood as an organic integration operating primarily across maritime and aerial domains, consisting of a large number of heterogeneous unmanned platforms distributed across underwater, surface, and aerial layers. Through detection, communication, and coordinated control, these systems achieve functional complementarity and capability enhancement across different operational environments. Specifically, UAVs are characterized by their long endurance, wide-area coverage, distributed deployment, cost-effectiveness, and high efficiency. Equipped with remote sensors, cameras, and wireless communication modules, they are well-suited for reconnaissance, search, and mobile relay missions. USVs can perform formation navigation and surveillance in designated sea areas. Fitted with onboard cameras, sonars, and communication devices, they conduct monitoring and early warning tasks in key zones while also acting as relay nodes to enable cross-domain communication between aerial and underwater platforms. AUVs and ROVs are typically equipped with cameras, sonars, and wired/wireless communication units, allowing them to eliminate static monitoring blind spots and perform close-range exploration in critical underwater areas through collaborative swarm behaviors.

The core essence of MUSs lies in enabling autonomous collaboration and mission execution across multiple platforms and domains in complex marine environments. This is achieved through highly intelligent modeling, reliable and efficient networking, precise sensing and communication, as well as collaboratively optimized control strategies. Fundamentally, MUSs enhance the functional strengths of individual platforms by leveraging cross-domain information fusion and intelligent decision-making, thereby maximizing overall system performance and mission efficiency. In recent years, the significant progress has been made in the research and application of MUSs, with deployments expanding to hydrographic surveying, environmental monitoring, resource exploration, and military reconnaissance. However, the complexity and dynamism of the ocean environment characterized by currents, waves, and electromagnetic interference pose increasing challenges to traditional single-platform operations, which are struggling to meet the growing demand for diversified and fine-grained maritime missions. Unlike the traditional single-domain systems, MUSs exhibit several distinctive characteristics in multi-domain collaboration (He et al., 2021):

(1) *Diversified sensing and perception resources.* Unlike conventional systems that are typically confined to a single operational domain, MUSs integrate heterogeneous sensing and perception capabilities across aerial, surface, and subsurface domains to address the requirements of complex missions.
(2) *Decentralized system architecture.* Given the diversity of resources and tasks in multi-domain environments, MUSs adopt distributed and decentralized architectures. These structures support adaptive networking, dynamic adjustment, and optimized performance in response to mission requirements and environmental changes.
(3) *Systematic collaborative framework.* Effective cross-domain operation of MUSs requires a systematic approach to collaboration, which emphasizes the integrated task allocation, dynamic coordination, and cooperative execution to enhance the overall mission effectiveness.

Based on these characteristics, the co-design of detection, communication, and control has emerged as a key enabler for the efficient operation of MUSs. It provides the fundamental support to address diverse and complex mission needs, simultaneously catalyzing progress in marine science and technology.

Against this backdrop, how to achieve collaborative operations across multiple domains and heterogeneous platforms has become a critical challenge. The major technical bottlenecks can be summarized as follows:

(1) *Accuracy and real-time capability of model construction.* In a highly dynamic marine environment, the system must possess precise perception and predictive modeling capabilities to support mission decision-making and resource scheduling. This can be achieved through the integration of digital twin technologies and dynamic model generation methods, enabling data-driven environmental modeling and mission-level simulation.
(2) *Stability and anti-interference capability of networked communication.* The long propagation distance and high-attenuation characteristics of the ocean environment impose stringent requirements on networking technologies. In particular, achieving efficient information exchange in underwater environments remains a fundamental difficulty. To address the communication challenges in such complex maritime scenarios, it is essential to develop cross-domain and multi-modal networking techniques, together with anti-interference strategies for underwater acoustic communication.
(3) *Coverage and accuracy of sensing and communication.* In complex environments, multi-sensor data fusion is required to ensure reliable detection performance while balancing energy consumption and sensing precision. As a result, the multi-sensor fusion and cooperative data processing methods can significantly enhance both detection accuracy and communication efficiency.

(4) *Efficiency and robustness of cooperative control.* Given the dynamically changing mission objectives, multi-platform systems must employ collaborative optimization algorithms to achieve task allocation, trajectory planning, and adaptive state adjustment, thereby maintaining system robustness and responsiveness. To meet these requirements, multi-platform cooperative control algorithms based on optimization theory and reinforcement learning should be developed to tackle the challenges of task assignment and resource scheduling in complex maritime environments.

In view of this, it is necessary to realize collaborative operations across multiple domains and heterogeneous platforms. The major technical bottlenecks can be summarized as follows:

(1) *Accuracy and real-time capability of model construction*: In a highly dynamic marine environment, the system must possess precise perception and predictive modeling capabilities to support mission decision-making and resource scheduling. This can be achieved through the integration of digital twin technologies and dynamic model generation methods, enabling data-driven environmental modeling and mission-level simulation.
(2) *Stability and anti-interference capability of networked communication*: The long-range and high-attenuation characteristics of the ocean environment impose stringent requirements on networking technologies. In particular, achieving efficient information exchange in underwater environments remains a fundamental difficulty. To address the communication challenges in such complex maritime scenarios, it is essential to develop cross-domain and multi-modal networking techniques, together with anti-interference strategies for underwater acoustic communication.
(3) *Coverage and accuracy of sensing and communication*: In complex environments, multi-sensor data fusion is required to ensure reliable detection performance while balancing energy consumption and sensing precision. Research into multi-sensor fusion and cooperative data processing methods can significantly enhance both detection accuracy and communication efficiency.
(4) *Efficiency and robustness of cooperative control*: Given the dynamically changing mission objectives, multi-platform systems must employ collaborative optimization algorithms to achieve task allocation, trajectory planning, and adaptive state adjustment, thereby maintaining system robustness and responsiveness. To meet these requirements, multi-platform cooperative control algorithms based on optimization theory and reinforcement learning should be developed to tackle the challenges of task assignment and resource scheduling in complex maritime environments.

In response to these bottlenecks, the modeling, networking and transmission, sensing and communication, and cooperative control components of MUSs are highly interdependent, forming an integrated technological chain. Specifically, the modeling serves as the operational foundation, leveraging digital twin technologies and environmental modeling to provide critical support for mission planning and dynamic decision-making. The networking and transmission act as the communication backbone, ensuring efficient information exchange across multi-platform and multi-domain systems. The sensing and communication facilitate accurate environmental perception and data acquisition, enabling real-time mission adaptation and situational awareness. Finally, the cooperative control synthesizes models and sensory data through optimization algorithms and intelligent scheduling mechanisms to achieve efficient multi-platform coordination. Together, these four components empower MUSs to execute complex and diverse missions effectively in dynamic ocean environments, demonstrating synergistic performance that far exceeds the capabilities of individual platforms.

1.1.2 Co-Design of Detection, Communication, and Control

In conventional MUSs, the detection, communication, and control modules are typically designed independently. Specifically, the detection system is mainly concerned with acquiring target radiation noise or echo information through radar, hydrophones, or sonar arrays, and then applying advanced signal processing techniques to extract target features such as signature characteristics, bearing, and range. The communication system focuses on developing high-performance and practical communication protocols that can cope with the challenges of underwater and aerial acoustic channels-such as narrow bandwidth, severe multipath, frequency-selective attenuation, and high ambient noise to ensure reliable information transmission from the source to the destination. The control system, in turn, aims to design control algorithms, so as to drive unmanned vehicles to perform maneuvers such as forward and backward motion, pitching, turning, ascending, descending, and lateral translation. Clearly, the detection and communication systems exhibit an inherent overlap in functionality. When the data received by one subsystem is ignored by the other, the spectrum and energy resources are inefficiently utilized and may even become competitive. Moreover, the traditional control designs often assume ideal detection and communication performances. However, the complex and dynamic ocean environment makes this assumption unrealistic. Therefore, the separated design of detection, communication, and control not only leads to incomplete information constraints in control commands but also prevents detection and communication from forming an effective feedback loop, severely limiting the overall system performance. To address these limitations, it is essential to co-design the detection, communication and control modules, which can enable the cooperative target monitoring and adaptive coordination under uncertain ocean environments.

Figure 1.1 illustrates a typical integrated architecture of detection, communication, and control for MUSs. The primary mission of this system is to monitor designated maritime regions and to accomplish tasks such as detection, localization, identification, and tracking of moving targets. Within this system, various platforms and submersibles are equipped with active/passive sonars, forward-looking and side-scan sonars, acoustic vertical arrays, echo sounders, and sound velocity profilers, thereby

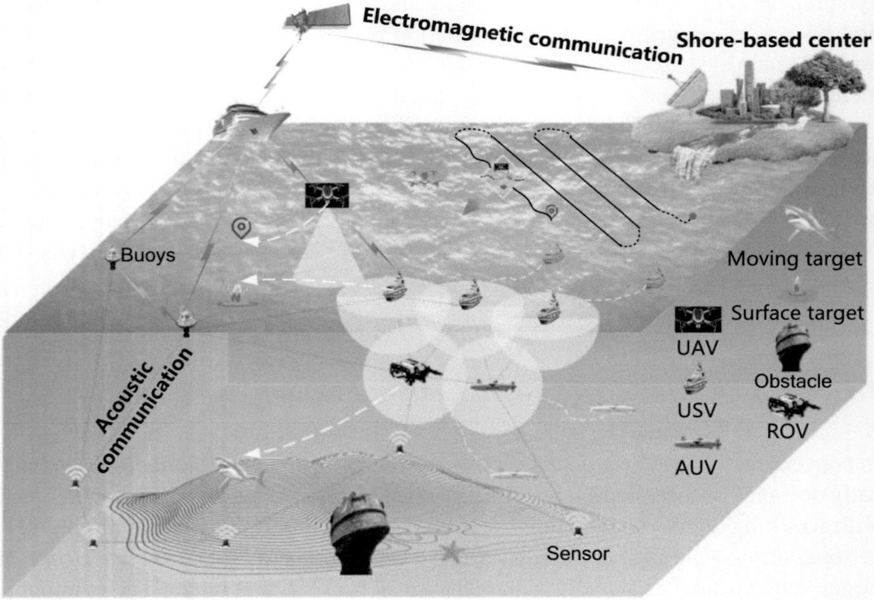

Fig. 1.1 Architecture of MUSs for underwater target monitoring

forming "aerial-surface-underwater" three-dimensional, all-weather, and all-domain monitoring network. On the one hand, the system acquires diverse dynamic information about targets-such as detection data, position, and motion state-while on the other, it also gathers auxiliary environmental information, including ocean dynamic parameters and acoustic field characteristics. The comprehensive platforms serve as the backbone nodes of the system, providing detection, communication, local data fusion and processing, as well as scheduling and control functions. Together with bottom-mounted sensors, they form clustered subnetworks that monitor different maritime regions for mid- to long-range (tens to hundreds of nautical miles) moving targets. The AUVs act as mission executors characterized by high mobility, adaptability, and controllability. They can operate in sensitive or blind-spot regions that are beyond the monitoring coverage of surface platforms. By directly detecting short- to mid-range (a few to tens of nautical miles) moving targets and cooperating with surface platforms, AUVs enable relay-based cooperative detection and continuous target tracking. The multi-source information collected by the platforms and submersibles is transmitted via underwater acoustic or optical communication links through satellites, UAVs, buoys or USVs, and may also be relayed through multi-hop links to nearshore optical fiber backbone network for delivery to the shore-based command center. The shore-based center acts as the data fusion and decision-making hub, integrating global situational awareness and performing optimized scheduling and control. The scheduling and control information is transmitted through surface vehicles and buoys, as well as the optical fiber backbone network to the AUVs and integrated platforms, forming a closed scheduling and control loop. The AUVs and platforms subsequently cooperate based on their own operational states and shared information to perform monitoring tasks. Accordingly, the system optimizes the scheduling and control of unmanned aerial, surface, and underwater vehicles by fusing multi-source dynamic sensing data, enabling efficient multi-dimensional and three-dimensional monitoring of underwater moving targets, thereby effectively enhancing the mobility and operational capability of the seabed static observation network.

1.2 Difficulties and Challenges

1.2.1 Model Construction Level

The open and dynamic properties of the ocean environment make it extremely difficult to establish fixed communication infrastructures. To meet diverse maritime monitoring requirements, the heterogeneous terminals must be deployed across different physical domains (Xing, 2019), including sonars, vertical acoustic arrays, echo sounders, sound velocity profilers, shore-based buoys, and surface monitoring platforms, among others. It is worth noting that the surface monitoring platforms can employ various broadband and narrow-band communication methods, such as satellite communication, shortwave, and global positioning system (GPS). Surface buoys typically communicate with underwater detection devices via acoustic narrow-band links, and may also establish broadband connections with surface monitoring platforms. Underwater devices, on the other hand, primarily rely on acoustic networking for narrow-band communication with buoys or other underwater nodes. Such a complex and heterogeneous communication environment, marked by a coexistence of omnidirectional and directional transmission modes, diverse frequency bands, and differentiated service priorities, results in substantial incompatibility across access technologies. This constitutes a major impediment to the realization of a self-organizing, ubiquitous architecture for detection, communication, and control collaboration in MUSs. Moreover, the intrinsic constraints of maritime communications-restricted bandwidth, significant propagation delays, and limited energy availability-along with unpredictable oceanic conditions like tides and currents, further complicate the attainment of efficient and reliable integration.

During the integration procedure, the construction of ocean current fields faces substantial challenges. Specifically, the ocean current field exhibits pronounced spatiotemporal variability and nonlinear characteristics, and its modeling must account for multi-scale dynamic factors such as tides, currents, and wind-wave interactions, as well as the multi-physics coupling effects among temperature-salinity structures, seabed topography, and geophysical morphology. However, the openness and uncertainty of the ocean environment impose severe constraints on data acquisition.

Devices such as acoustic Doppler current profilers and current meters often suffer from limited spatial coverage, measurement noise, and occasional failures, resulting in current field data with insufficient spatial-temporal resolution. The absence of robust marine communication infrastructure further aggravates the difficulty of real-time current-data transmission. In underwater scenarios, high signal attenuation and narrow bandwidth of acoustic channels lead to low transmission reliability and significant communication latency. Moreover, the energy constraints of marine platforms hinder the long-term operation of cooperative measurement and modeling tasks. Although substantial progress has been made globally in improving flow-field modeling through computational fluid dynamics, data assimilation, and artificial intelligence techniques, large-scale deployment still faces bottlenecks in multi-source heterogeneous data fusion and low-latency transmission. Constructing a robust, efficient, and dynamically adaptive cooperative network architecture thus remains a critical and challenging research focus worldwide. To this end, the establishment of a detection-communication-control architecture requires system-level coordination across overall system design, dynamic adaptability, and resource optimization. The core objective is to achieve efficient communication and coordinated control among heterogeneous nodes. This can be realized through distributed network architectures, multi-mode access, and protocol fusion strategies that integrate underwater acoustic communication, surface shortwave communication, and satellite links into a cross-domain networking framework. Developing multi-protocol-compatible access mechanisms and combining broadband and narrow-band connections can effectively meet diverse mission requirements and environmental constraints.

1.2.2 Networking-Transmission Level

The integration of detection, communication, and control in MUSs relies on a heterogeneous sensing network characterized by self-organization, ubiquity, real-time response, and high reliability. Through detection, information sharing, and cooperative control, such a network enables the transformation of information superiority into decision superiority, thereby enhancing the overall performance of MUSs. This process critically depends on the networking and transmission among physical entities. However, underwater acoustic communication inherently exhibits weak communication characteristics (Xu et al., 2016), which impose unique challenges compared with terrestrial networks: (1) Electromagnetic attenuation in water follows an exponential decay law, rendering satellite-based positioning systems. Moreover, due to high noise levels and severe multi-path interference, clock synchronization among underwater nodes becomes difficult to achieve accurately (see Fig. 1.2, on the left), where $\alpha_{P,l}$, $\alpha_{V,i}$, and $\alpha_{A,j}$ denote clock drift, and $\beta_{P,l}$, $\beta_{V,i}$, and $\beta_{A,j}$ denote clock offset respectively. (2) The non-uniform distribution of the sound velocity profile causes bending of acoustic ray paths (see Fig. 1.2, on the right), where S and R denote the sending and receiving nodes, respectively. (3) Underwater acoustic communication relies primarily on acoustic transmitters and hydrophones for data transmission and reception.

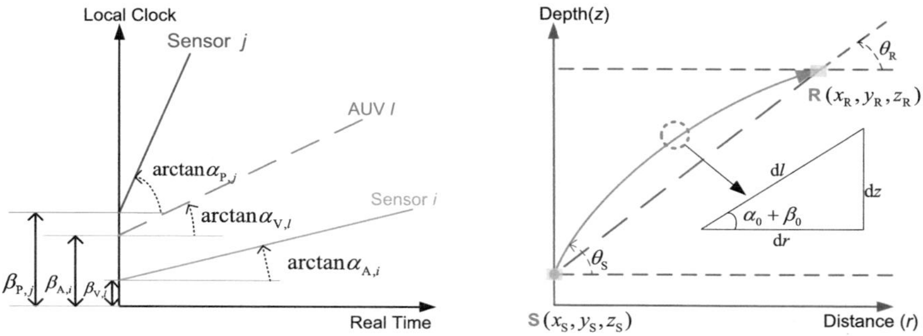

Fig. 1.2 Illustrations of underwater asynchronous clock offsets and acoustic ray bending

It suffers from long propagation delays and significantly higher energy consumption compared with radio-wave communication in terrestrial environments. Due to these weak communication characteristics of underwater acoustics, conventional networking and transmission protocols designed for terrestrial wireless systems cannot be directly applied to underwater environments.

1.2.3 Detection-Communication Level

Integrated design of detection and communication not only overcomes the limitations of conventional separated architectures-such as resource contention and low operational efficiency-but also enables shared utilization of radio-frequency resources and antenna apertures, thereby improving spectral efficiency, reducing inter-device electromagnetic interference, and enhancing system stealth. It is worth noting that the integrated design of detection and communication was first developed in the radar domain, where integration is typically achieved through time-division, frequency-division, or beam-division approaches. These systems rely on electromagnetic waves for both detection and communication, benefiting from their high propagation speed ($\approx 3 \times 10^8$ m/s), large aperture size, and broad available bandwidth (3 MHz–300 GHz). However, electromagnetic waves undergo exponential attenuation in water, rendering radar-based integration techniques inapplicable to underwater environments. At present, sonar remains the only effective means for long-range underwater sensing and communication. Compared with radar signals, sonar signals exhibit much lower propagation speed ($\approx$ 1500 m/s), smaller aperture size, narrower available bandwidth (3 Hz–97 kHz), and pronounced multipath effects (Stojanovic & Preisig, 2009). These resource constraints introduce several challenges: (1) the underwater communication signals are prone to linear and nonlinear distortions, phase jitter, and frequency offsets; (2) the out-of-band interference between underwater sensing and communication is severe and mutually suppressive, which increases the difficulty of designing shared signals. Therefore, how to overcome the inherent resource limitations of sonar systems stands as a major challenge for achieving effective underwater integrated design of detection and communication.

1.2.4 Cooperative Control Level

The detection-communication subsystem provides essential information support for the control subsystem, while the control subsystem, in turn, enhances the performance of detection and communication through the mobility of submersibles and other mobile nodes. Specifically, the mobile platforms such as UAVs, USVs and AUVs cooperate dynamically with neighboring nodes via their onboard communication modules, thereby improving detection and communication performances through trajectory planning, target tracking, and formation control. Hence, designing an effective feedback control law within the integrated detection-communication-control framework becomes of paramount importance. However, the cross-domain clusters span multiple operational dimensions-aerial, surface, and underwater, where platforms exhibit distinct kinematic and dynamic characteristics. For example, the aerial and underwater vehicles possess three-dimensional motion capabilities, whereas surface vessels are constrained to two-dimensional motion. Moreover, the influence mechanisms of environmental disturbances such as aerodynamic and hydrodynamic forces differ significantly across domains (Zhang et al., 2020). Thus, how to perform resource allocation and task decomposition for multi-domain platforms while fully accounting for kinematic disparities and environmental heterogeneity represents a major challenge in cross-domain control and decision making. Furthermore, the introduction of networking environments into cross-domain cluster systems brings additional complications. Local disturbances, communication delays, and dynamic link switching-all manifestations of the weak communication characteristics of underwater acoustics-can propagate through physical coupling and network interconnections, resulting in reduced synchronization, delayed information feedback, and ultimately degraded reliability of cross-domain cooperative control. Meanwhile, during the design of feedback control laws, the control performance optimization must not come at the expense of communication and detection performances. Therefore, in the cooperative control of cross-domain platforms, it is necessary to balance local subsystem performance with global system optimization. However, due to the limitations of underwater detection and communication technologies, certain key parameters of maritime unmanned platforms cannot be measured precisely, and the coupling relationships among heterogeneous platforms are often difficult to model accurately. These constraints directly limit the overall performance of cross-domain cooperative control systems. Consequently, how to overcome the uncertainties and uncover the coupling relationships among networking, sensing, and control information has become critical challenges in enhancing the control effectiveness of multi-domain swarm platforms.

1.3 Research Progress on Integration Design

The United States was the first country to propose the concept of maritime networking applications, and its research achievements in this field remain at the global forefront. As early as 2012, the U.S. Department of Defense (DoD) formally released the *Joint Operational Access Concept*, which for the first time articulated the theoretical notion of cross-domain maritime-aerial collaboration. In 2014, the Woods Hole Oceanographic Institution and the University of Hawaii jointly conducted a series of underwater/surface integrated localization and communication experiments using modems and hydrophones. In October 2015, the Defense Advanced Research Projects Agency launched the *Cross-Domain Maritime Surveillance and Targeting (CDMaST)* program. Subsequently, in August 2018, the DoD published the *Unmanned Systems Integrated Roadmap (2017–2042)*, outlining long-term development goals for multi-domain unmanned operations. In April 2021, the U.S. Pacific Fleet conducted the Unmanned Integrated Battle Problem 21 (UxS IBP 21) exercise, which featured joint operations between manned/unmanned aerial and maritime systems. The UK Royal Navy has also been conducting the "Unmanned Warrior" series of exercises since 2016 to evaluate the coordinated capabilities of USVs, AUVs, and UAVs for enhanced intelligence, surveillance, reconnaissance and mine countermeasure operations. In 2018, the European Union initiated the *Open Cooperation for European Maritime Awareness (OCEAN)* project, while Japan completed the *Dense Oceanfloor Network System for Earthquakes and Tsunamis (DONET)* in 2016, enabling multi-platform collaboration for real-time seismic and tsunami monitoring.

Based on the above research programs, substantial progress has been made in the detection-communication-control integration. According to Fig. 1.3, this section provides a review from the following three perspectives: (1) Integrated sensing and communication; (2) Localization and networking under weak communication conditions; (3) Collaborative control in complex and dynamic environments.

1.3.1 Active-Passive Detection with Multi-Source Information Fusion

Passive and active detections are the two most common approaches used in underwater target sensing. Passive detection relies on hydrophones or sonar arrays to receive the radiated acoustic noise generated by targets-such as flow noise induced by submarine-water friction and mechanical noise from onboard propulsion systems. By applying beamforming, direction-of-arrival (DOA) estimation, and other signal processing techniques, the passive mode extracts target-related features, including bearing, depth, and range information. In contrast, active detection operates by transmitting acoustic pulses and receiving the reflected echoes from the target. Its development has led to several major technical branches, including low-frequency high-power detection, bi-/multi-static sonar, forward-scattering sonar, and multiple-

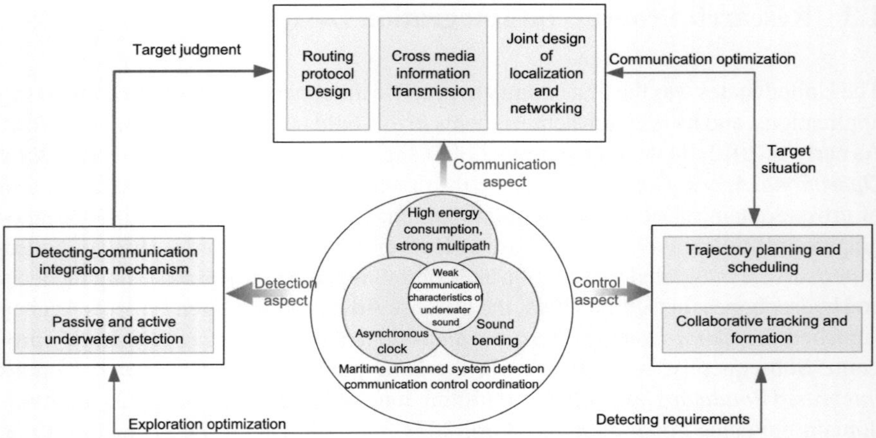

Fig. 1.3 Relationship among detection, communication, and control for MUSs

input multiple-output (MIMO) sonar architectures. Despite the maturity of both approaches, most existing passive or active sonar systems assume that sensing and communication are independent subsystems, neglecting the functional overlap and resource contention that naturally arise between them. To enable an integrated and intelligent sensing-communication system, Mealey first proposed in the 1960s the concept of modulating communication signals onto radar pulses (Mealey 1963). Inspired by this pioneering idea, researchers have developed a range of signal-processing strategies, including time-division, frequency-division, beam-division, and fully shared mechanisms. Specifically, the time-division mechanism divides the time into multiple slots, in which a switching device alternately transmits sensing or communication signals in different time slots, ensuring that sensing and communication signals cannot be transmitted simultaneously within the same slot. The frequency-division mechanism allocates distinct frequency bands for sensing and communication operations. Beam-division mechanism divides the phased-array aperture into separate sectors or beams, where some are used for sensing and others for communication. Fully shared mechanism allows the same waveform to serve both functions simultaneously and within the same frequency band, thereby enabling co-frequency, co-time sensing and communication. Considering that radar systems typically feature wide bandwidths, large apertures, and high propagation velocities, most existing radar-based integrated sensing and communication systems commonly employ time-division, frequency-division, or beam-division mechanisms. However, due to the narrow available bandwidth, small aperture, and low propagation speed of acoustic waves, underwater sonar systems are comparatively more suitable for adopting a fully shared mechanism.

In recent years, researchers have conducted preliminary studies on sonar-based integrated detection and communication under the full-shared mechanism, and have designed various types of shared waveforms from different perspectives. These shared waveforms can be broadly classified into three categories: (1) superimposed shared waveforms combining both sensing and communication signals; (2) sensing-signal-based shared waveforms; (3) communication-signal-based shared waveforms. In superimposed shared waveforms, mutually orthogonal signals are employed to perform detection and communication tasks separately (Ma et al. 2021). The advantage of this approach is that each waveform can be independently designed. However, it suffers from severe mutual interference and difficult signal separation at the receiver. In sensing-signal-based shared waveforms, communication information is modulated onto the sensing waveform (Jiang et al. 2019), thereby preserving detection performance. Nevertheless, the communication directionality is largely constrained by the sensing beam pattern, which limits its suitability for network-level transmission. In contrast, communication-signal-based shared waveforms embed sensing information into the communication signal (Lu et al. 2018), ensuring communication reliability while maintaining network connectivity. However, their relatively low source level restricts the achievable sensing performance. To illustrate these concepts more clearly, Fig. 1.4 depicts the design framework of communication-signal-based underwater shared waveforms, which can be summarized in three operational modes: (1) Active target sensing using self-emitted communication signals, where the transmitted communication waveform simultaneously performs detection and data transmission; (2) Passive target detection during communication; and (3) Using the communication signals transmitted by other nodes for detection, that is, employing the communication transmissions of other nodes as illumination sources to achieve the integration of detection and communication functions.

Research on the integration of underwater target detection and communication has gradually evolved from collaborative optimization to functional integration, forming a coherent technical development trajectory. Yan et al. (2025) proposed a heterogeneous formation-based target detection method involving UAV-USV-AUV coordination. Yan et al. (2023) designed a non-cooperative target detection algorithm that fuses active and passive measurements via a hybrid Bayesian fusion scheme, effectively improving detection accuracy while reducing communication overhead.

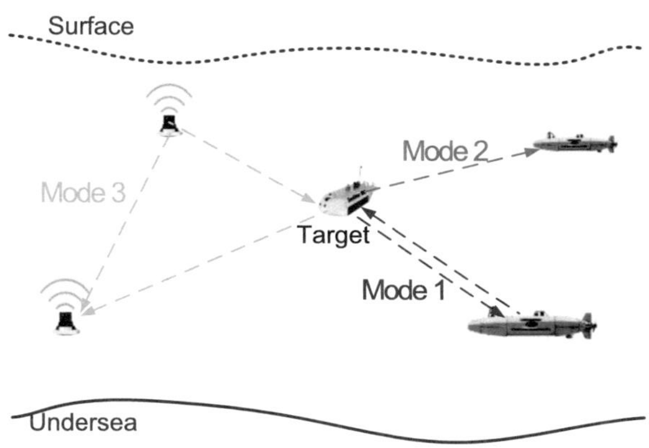

Fig. 1.4 Underwater shared waveform design model based on communication signals

Zhang et al. (2016) investigated chaotic modulation techniques, utilizing instantaneous phase and frequency features to achieve secure underwater communication. In contrast, Yan et al. (2025) introduced a time-division and multicarrier waveform design that enables joint communication and detection within a single channel, while simultaneously extracting target parameters. Collectively, these studies provide both theoretical foundations and practical pathways for optimizing underwater sensing-communication systems in complex environments. Within the context of information fusion, detection serves as an enabling function for trajectory planning and obstacle avoidance. In the field of cross-domain cooperative localization and trajectory planning, Li et al. (2024) presented a distributed message-passing framework that mitigates cumulative localization errors through AUV cooperation and depth information exchange. Li et al. (2019) formulated multi-AUV trajectory planning as a convex optimization problem, jointly optimizing detection probability and tracking accuracy through cooperative localization. Li et al. (2019) designed an Intermittent Belief Propagation-based Dead Reckoning (IBPDR) framework, which combines Cooperative Localization (CL) to reduce both communication cost and inertial drift. However, distributed frameworks demand high communication efficiency and real-time performance, while convex optimization approaches involve substantial computational cost, making real-time implementation still challenging. Although the IBPDR framework effectively reduces communication frequency, its response speed in dynamic environments still requires further optimization. To address these challenges, Yan et al. (2022) proposed an asynchronous localization algorithm that combines a model-free tracking controller with a Lyapunov function to analyze asymptotic stability. Yan et al. (2018) presented another asynchronous localization algorithm for active and passive sensor nodes, combining mobility prediction with an iterative least-squares estimator. The study also analyzes localization error convergence and derives the Cramer-Rao Lower Bound (CRLB). Although these algorithms demonstrate strong performance in simulation, their adaptability and robustness in dynamically changing environments still require further validation. In addition, Yan et al. (2019) proposed a localization algorithm based on non-feature transformation, which employs a ray-tracing method to simulate stratification effects. Yan et al. (2021), Zhao et al. (2020) designed privacy-preserving localization algorithms to prevent spoofing attacks while maintaining localization efficiency. Nonetheless, stratified-effect modeling remains computationally complex and highly dependent on environmental parameters, and the scalability of model-free tracking controllers under multi-task conditions still needs to be verified.

1.3.2 Networking and Transmission

Underwater communication media primarily include electromagnetic waves, optical waves, quantum channels, and acoustic waves (Xu et al., 2023). The characteristics and applicable scenarios of various underwater communication techniques are summarized in Table 1.1. As illustrated in Fig. 1.5, the existing studies have proposed an

Table 1.1 Comparison of cross-domain communication methods for MUSs

Communication method	Frequency band/Hz	Bandwidth/bps	Transmission Range/m	Transmission rate/bps
Cable communication	$1 \times 10^3 \sim 1 \times 10^4$	$1 \times 10^2 \sim 1 \times 10^3$	$1 \times 10^4 \sim 1 \times 10^5$	$1 \times 10^2 \sim 1 \times 10^3$
Acoustic communication	$1 \times 10^4 \sim 1.5 \times 10^4$	$1 \times 10^2 \sim 1 \times 10^3$	$10 \sim 1 \times 10^2$	$10 \sim 1 \times 10^2$
Optical communication	$\approx 10^{12}$	$\approx 10^{12}$	$1 \times 10^2 \sim 1 \times 10^3$	1×10^{11}
Electromagnetic communication	$30 \sim 300$	–	$10 \sim 10^2$	$10^2 \sim 10^6$
Quantum communication	–	–	–	–

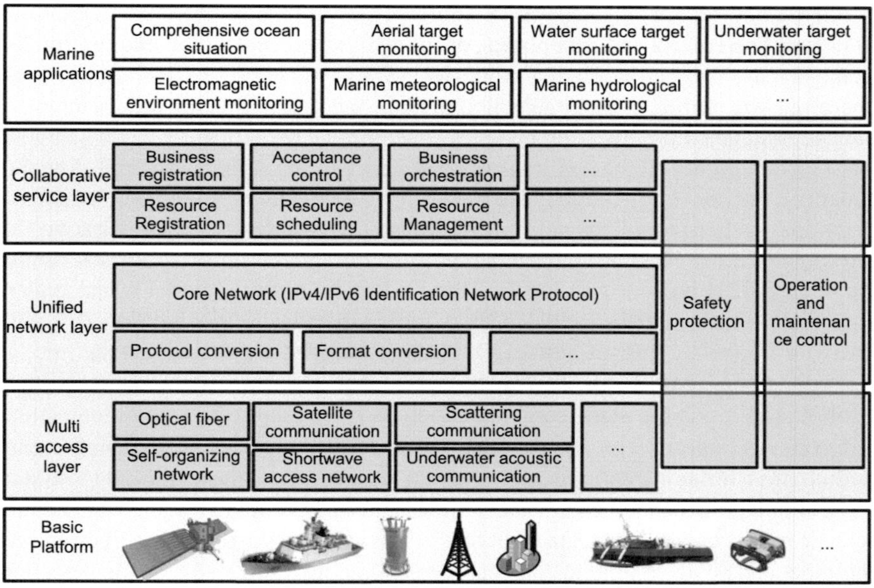

Fig. 1.5 Integrated communication network architecture for comprehensive maritime perception

omnidirectional maritime communication-networking architecture (Gao et al. 2020). However, several critical challenges remain unresolved, including the difficulty of information integration and sharing, limited real-time data transmission capability accompanied by high latency, and the lack of unified standards and protocols for cross-domain cooperative control. Moreover, the issues related to network security, anti-interference robustness, and resilience to complex marine environments still require further enhancement.

(1) Routing planning
Routing planning is critical for ensuring information reachability in maritime networks. Existing routing protocols fall into three main categories: (a) energy-based routing; (b) opportunistic routing; (c) software-defined networking (SDN). For energy-based routing, it optimizes node energy consumption to extend network lifetime. However, maritime environmental factors (sea fog, strong winds, water inhomogeneity) severely limit communication range. In view of this, Celik et al. (2022) developed a sector-based opportunistic routing protocol for acoustic-optical underwater networks. Baek et al. (2020) proposed a graph-theoretic energy-efficient routing strategy for UAV networks. Geography-based routing relies on greedy forwarding and multi-hop transmission, performing well in dense networks. However, in sparse deployments, distance-based path selection causes repeated relay usage and routing holes. For opportunistic routing, it introduces redundant packet forwarding. Data-centric routing optimizes paths based on data characteristics (type, priority, delay sensitivity). The SDN addresses traditional architecture limitations by decoupling control and data planes, enabling dynamic adaptation and centralized management. Ruby et al. (2021) proposed a SDN-enabled energy-aware protocol. Therefore, the routing strategy selection must be tailored to specific requirements to achieve reliable communication under diverse oceanic conditions.

Cross-medium information transmission between maritime and aerial domains remains a fundamental challenge. Li et al. (2020) modeled a maritime cognitive cross-domain heterogeneous network by integrating wireless, underwater acoustic, and cognitive networking characteristics to enable air-sea information exchange. Ma et al. (2021) proposed an UAV-assisted ocean monitoring model where underwater data is transmitted acoustically to surface gateways, then relayed to UAVs via radio-frequency links. Chen et al. (2022) developed wave interference mitigation approaches. Luo et al. (2022) surveyed three aerial-water interface communication types: visible light communication (VLC), relay-based communication, and non-VLC approaches. Laser-based communication offers a promising cross-medium transmission alternative. Blue-green laser wavelengths, exhibiting low seawater attenuation, penetrate the water surface and convert optical signals into acoustic waves for underwater data delivery (see Fig. 1.6). Research has focused on photo-acoustic conversion mechanisms, optical modulation, and acoustic detection schemes. Liu et al. (2023) developed an underwater wireless optical communication system using semiconductor disk lasers, improving link range and performance. Zhao et al. (2024) proposed a pulsed laser-based photo-acoustic system with fiber-optic hydro-acoustic sensors, where thermal expansion generates acoustic signals modulated to reduce bit error rates. However, laser-based cross-medium transmission remains nascent, facing challenges in signal attenuation, scattering interference, and environmental adaptability that require further research.

(2) Localization networking
Localization networking aims to determine target positions and design communication protocols for robust network formation, ensuring reliable underwater data transmission and control feedback. Underwater localization techniques are broadly

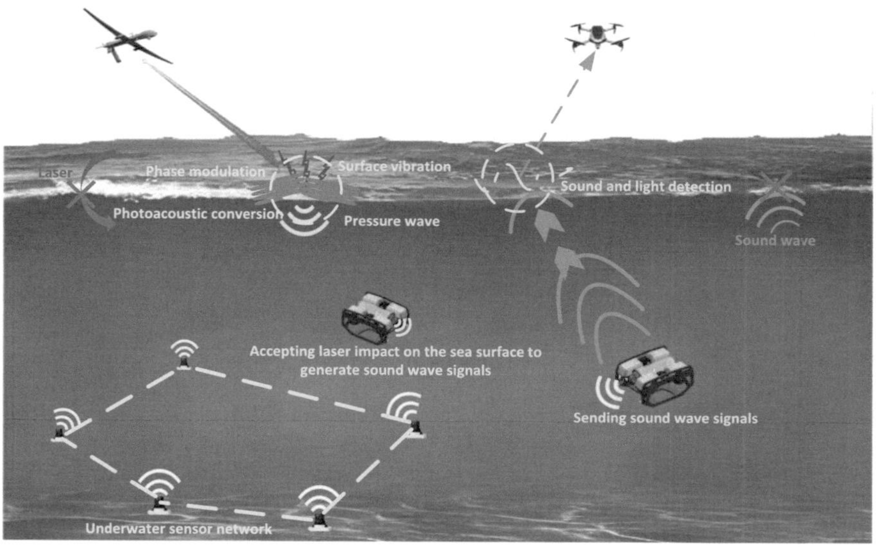

Fig. 1.6 Schematic diagram of the laser-induced acoustic principle

classified as range-based or range-free methods. Range-based methods utilize sensing devices (active/passive sonars, forward-looking/side-scan sonars, echo sounders) and offer high accuracy with minimal environmental sensitivity, making them widely adopted. Cooperative localization algorithms exploit measurements including Time Difference of Arrival (TDOA), Time of Arrival (TOA), Time of Flight (TOF), and Angle of Arrival (AOA). Liu et al. (2014) developed a TDOA-based multi-AUV cooperative localization algorithm. Zhou et al. (2011) proposed a TOA-based algorithm considering spatio-temporal correlation between AUVs and sensors to reduce energy consumption via mobility prediction. Hybrid schemes include a TOA-AOA joint algorithm, which was proposed in Chen et al. (2024) to enhance relative positioning and environmental adaptability. For clock synchronization and error optimization, a Time Difference of Communication (TDOC) was introduced in Jiang et al. (2024) as a relative metric, which optimizes nonlinear observations through Weighted Least Squares with bias mitigation and error compensation. Luo et al. (2016) analyzed passive sensor mobility and proposed a hybrid network cooperative localization algorithm. Most algorithms assume synchronized clocks and straight-line acoustic propagation. However, underwater acoustic communication limitations make clock synchronization difficult, and acoustic paths are curved. To balance topology connectivity and energy efficiency, Yan et al. (2018) pioneered using optimal rigid graphs from graph theory for energy-balanced underwater link optimization, minimizing communication complexity while maintaining 2- or 3-connectivity. Building on this, Zhao et al. (2021) proposed an optimal deployment strategy for underwater localization under weak acoustic communication and multiple constraints, using the CRLB as the accuracy metric. This rigid topology-based deployment enables optimal positioning despite asynchronous clocks, ray bending, and energy constraints. Additionally, Yan et al. (2019) proposed a localization algorithm accounting for asynchronous clocks and layered channel effects using non-feature transformation and ray-tracing simulations, providing convergence analysis and CRLB performance bounds. These advances offer theoretical and practical foundations for robust underwater networking strategies in acoustic environments.

(3) Bias correction and optimization for localization

For long-duration missions, the correction and optimization of localization are essential. In order to achieve the above objectives, Jiang et al. (2024) proposed an improved unscented Kalman filter (UKF) based on the correction range difference to solve the asynchrony problems of localization. He et al. (2018) combined spatially alternating generalized expectation (SAGE) with nonlinear weighted least squares (NWLS) to design improved algorithms for sound velocity estimation and time synchronization. Yan et al. (2020) designed an UKF-based consensus estimation algorithm to improve the localization accuracy of target. However, TOA/AOA algorithms show limited robustness to multipath interference. Yan et al. (2019) proposed a distance measurement method, and further used UKF fusion of direct and indirect measurement data to locate the target. However, the localization accuracy depends on the measurement data quality and is difficult to adapt underwater high noise environments. Hence, Yan et al. (2017) introduced a consensus-based approach with hybrid AUV-sensor architecture for self-localization and clock synchronization. This method has a high dependence on resource allocation during dynamic task switching, and its performance may decrease as task complexity increases.

Flow field modeling is crucial for target motion, communication, and sensing accuracy. In MUSs, the flow field modeling optimizes node deployment and routing to enhance network stability and guides motion planning and energy optimization. Yan et al. (2020) employed motion-compensation and ray-compensation strategies for simultaneous flow parameter and sensor position estimation, incorporating current field estimation to eliminate adverse effects of asynchronous clocks, stratification, and node mobility. Reinforcement learning (RL)-based localization offers distinct advantages in complex marine environments. Yan et al. (2022) combined Orthogonal Fractional Factorial Design (OFFD) with Deep Reinforcement Learning (DRL) to address asynchronous clock issues and optimize localization performance. Huang et al. (2025) employed the Deep Deterministic Policy Gradient (DDPG) algorithm for AUV trajectory optimization. Yan et al. (2021) proposed a DRL-based estimator that balances privacy protection and positioning accuracy. Yan et al. (2020) designed an RL-based algorithm to estimate the active/passive sensors and AUVs' positions. Yan et al. (2022) developed an asynchronous localization algorithm integrated with RL for AUV tracking control. However, DRL-based methods face several challenges, including long training times, high model complexity, limited scalability, degraded privacy-preserving accuracy under high-noise conditions, and unresolved issues in online adaptability and multi-agent coordination. To address these limitations, Yan et al. (2023) proposed a Broad Learning (BL)-based localization algorithm that uses dual neural networks to approximate policy and value functions. Nevertheless, this approach still suffers from high computational complexity, limited adaptability to large-scale distributed systems, and sensitivity to initial parameter settings.

1.3.3 Cooperative Control of Maritime-Aerial Unmanned Platforms

In cross-domain cooperative systems, the mission reliability and safety require multi-source information fusion for map construction and situational awareness, considering target position, environmental topography, and obstacle distribution. Based on accurate environmental mapping, real-time autonomous obstacle avoidance enables platforms to detect unknown obstacles and adjust navigation strategies promptly, ensuring robust path planning under dynamic conditions. Therefore, global path planning and local obstacle avoidance collaboratively construct a maritime navigation system integrating global planning with real-time avoidance mechanisms, enhancing system reliability and adaptability. Path-planning algorithms must exhibit strong adaptability to unstructured environments. Existing research relies on optical sensors (vision, LiDAR, or infrared) for detection and avoidance, often combined with artificial potential fields, map-based planning, or artificial intelligence-assisted optimization. However, these approaches generally fail to simultaneously satisfy feasibility, real-time performance, environmental complexity, and vehicle dynamics constraints in cross-domain underwater systems. RL and DRL enable unmanned systems to handle environmental uncertainties through adaptive learning. In Sutton & Barto (2018), RL allowed platforms to learn cooperative behaviors via trial-and-error environmental interactions referring. For AUV path planning, RL algorithms output continuous actions to control rudder, elevator, and propeller directly. However, the high costs and risks in real-world cross-domain experiments limit large-scale trial-based learning. The Deep Deterministic Policy Gradient (DDPG) method performs effectively in continuous action spaces. Yan et al. (2023) proposed a DDPG-based framework minimizing a cost function defined by disparity angle between AUVs and obstacles, enabling collision-free planning while maintaining communication quality. To enhance navigation capability and mission efficiency, researchers have explored integrated data and resource sharing among aerial, surface, and underwater platforms. Hu et al. (2024) proposed a distributed guidance vector field controller for heterogeneous UAV-USV systems, adopting hierarchical architecture with upper-layer guidance-speed regulation and lower-layer signal-tracking control. Li et al. (2023) developed an improved cooperative three-dimensional guidance principle enabling underwater vehicles to execute four fundamental search operations in maritime rescue. Zeng et al. (2022) comprehensively reviewed high-maneuverability cross-domain platforms. Vasilijevic et al. (2017) introduced a cooperative environmental monitoring system with underwater and surface vehicles, presenting a human-in-the-loop framework for coordinated navigation and environmental observation.

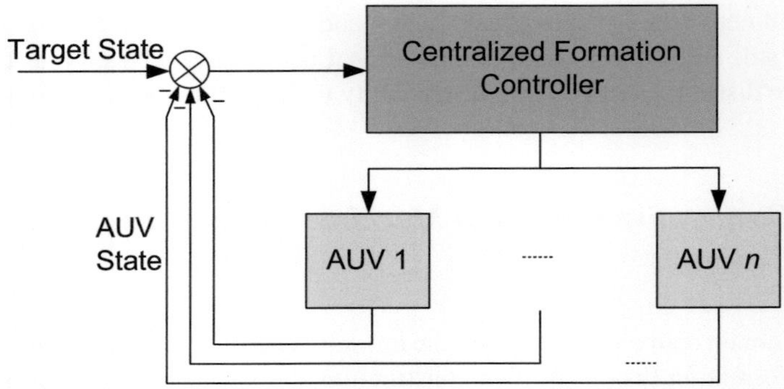

Fig. 1.7 Centralized formation structure of unmanned platforms

Building upon path planning, tracking and formation controllers are necessary for unmanned platforms to enable feedback mechanisms enhancing networking and detection performance. Tracking and formation control techniques are classified into three categories:

(1) *Architecture-based control*: The controller is categorized into centralized and decentralized schemes, with the latter divided into distributed and hierarchical architectures (see Figs. 1.7 and 1.8). The key distinction lies in decision-making processes. For example, the central controller requires complete environmental and platform knowledge (position, velocity, obstacles) collecting via sensors and communication links. And then, the controller processes data, and issues commands to each platform. In distributed architecture, each platform makes autonomous decisions based on local information, coordinating with neighbors to maintain formation.
(2) *Strategy-based control*: The control strategies include leader-follower, behavior-based, virtual structure, and artificial potential field methods. In behavior-based strategy, each platform executes predefined behavioral rules to achieve formation objectives. Virtual structure strategy treats the swarm as a rigid virtual body where each platform maintains fixed relative distances and orientations to ensure desired geometric configuration.
(3) *Controllability-based control*: Controllability assesses dynamic system stability and maneuverability. A system is controllable if a control law can drive its current state to a desired state within finite time.

(1) Tracking control
Non-cooperative target tracking and obstacle avoidance are critical problems for MUSs. Owing to their self-organizing capability, AUVs are the preferred solution in anchor-free environments. Li et al. (2020) proposed a belief-propagation (BP)-based framework for multi-AUV cooperative localization and target tracking. Yan et al. (2020) further incorporated the relationship between propagation delay and position, proposing an asynchronous localization and tracking algorithm. With respect to

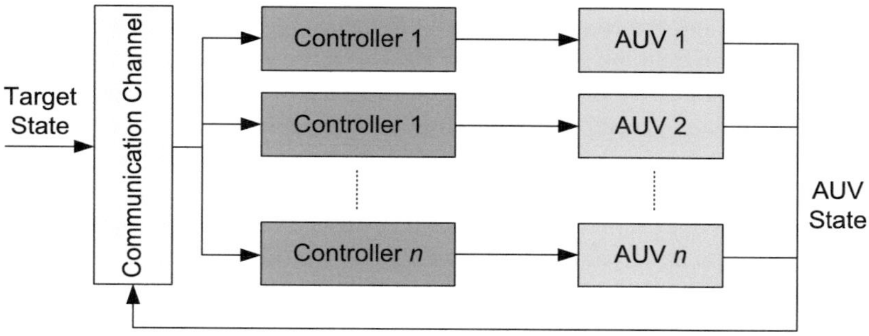

Fig. 1.8 Distributed formation structure of unmanned platforms

communication efficiency and motion planning, Yan et al. (2024) employed integral reinforcement learning (IRL) to enhance performance in fading channels, at the cost of increasing algorithmic complexity. Yan et al. (2020, 2022, 2023, 2024), Hu et al. (2024), Li et al. (2020, 2023), Zheng et al. (2022), Vasilijevic et al. (2017) combined DRL with sliding-mode control (SMC) to improve joint sensing-communication and obstacle-avoidance planning in complex environments. However, the adaptability to rapidly time-varying conditions remains insufficient. Yan et al. (2024) achieved precise localization by integrating binocular vision and acoustic communication. However, the high computational cost of multimodel data fusion limits real-time performance. Finally, Zhao et al. (2021) optimized the network topology through a minimum rigidity graph to achieve ubiquitous tracking and energy-efficiency optimization. However, its adaptability to node failures and mobile scenarios still needs improvement. In order to address environmental uncertainties and external disturbances, Yan et al. (2020, 2019) proposed methods for maintaining system stability under time-varying delay and saturation conditions. Based on this, Yan et al. (2019) optimized long-range aerial communication performance via heading control, while Yang et al. (2021, 2024) focused on improving the accuracy and robustness of path tracking for AUVs.

(2) Formation control

Beyond path planning, underactuated AUVs' formation control is of substantial importance for MUSs, particularly for bearing-based tracking under actuator constraints and unknown disturbances. Su et al. (2024, 2022) investigated bearing-based formation tracking. Specifically, Su et al. (2024) proposed a formation controller that combines adaptive robust control with heuristic gain scheduling and establishes asymptotic convergence. Su et al. (2022) developed a consensus-based bearing estimator together with adaptive laws to achieve multi-AUV formation tracking without position measurements, while optimizing control gains to balance constraints and performance criteria. Although these methods enhance the stability and robustness of control systems, their adaptability to highly dynamic environments still needs further improvement. With respect to modeling and optimization, Yan et al. (2024) implemented a minimum-time formation control approach using integral IRL, while

Yan et al. (2025) integrated game theory with backstepping techniques to enhance dynamic formation tracking performance. However, the scalability of those methods to large scale systems remains to be fully validated. In Yan et al. (2024), a rigidity-graph based formation method was proposed to reduce communication energy consumption. Yan et al. (2023), Cao et al. (2023) jointly optimized channel estimation and control, leveraging reinforcement learning and probabilistic registration techniques to improve both communication-sensing performance and formation stability. In order to address environmental uncertainty and external disturbances, Yan et al. (2023) enhanced system robustness under stochastic perturbations, while Cao et al. (2023) achieved a balance between communication efficiency and formation stability under model uncertainty. However, the increased computational complexity poses challenges for practical engineering deployment. In order to address the issues of insufficient operability under extreme conditions in underwater unmanned systems, Yan et al. (2018) developed a decentralized controller based on a remote-operated cyber-physical system framework. Zhang et al. (2024) leveraged digital twin technology to improve sensing and communication efficiency, whereas Cao et al. (2023) implemented a lightweight formation control method by using a broad learning system. However, the real-time performance of digital twins and the deployment of learning algorithms still require further investigation. Existing research has made significant progress in robustness enhancement, energy-efficient control, and optimization strategies. However, the further improvements are still required in terms of environmental adaptability, scalability to large-scale systems, and practical deployment of emerging technologies.

1.4 Application Scenarios

The integrated design of detection, communication and control for MUSs holds broad prospects across both defense and civilian domains. In order to elucidate its strategic value, we articulate representative application scenarios from military and civil perspectives.

1.4.1 Maritime Joint Anti-submarine Operations

On April 13, 2022, Ukraine deployed two Turkish-made Bayraktar TB-2 UAVs to conduct precision reconnaissance and electronic suppression against the Russian cruiser Moskva, neutralizing its aerial-defense capability. Subsequently, two Neptune anti-ship missiles were launched, striking the vessel on its port side. The Moskva soon emitted dense smoke and eventually sank into the Black Sea. Later that year, on October 29, 2022, at 04:30 a.m., Ukraine executed a cross-domain swarm operation, involving seven USVs and approximately nine to sixteen UAVs, to attack the Russian Black Sea Fleet stationed at the port of Sevastopol, Crimea.

According to reports from the UKRicon website, four Russian naval vessels were damaged during the operation, including the Admiral Makarov frigate, a Ropucha-class landing ship, another frigate, and a Project 22160 patrol vessel. Among them, the Admiral Makarov sustained the most severe damage, with its radar system critically impaired. These events demonstrate that cross-domain swarm systems composed of maritime unmanned platforms have played an irreplaceable role in the Russia-Ukraine conflict, achieving strategic effects that manned equipment alone could not realize. Therefore, the protection of critical maritime infrastructure must increasingly rely on aerial-surface-underwater integrated MUSs, operating under a networking-sensing-control co-design framework. By coordinating anti-submarine warfare missions through multi-domain collaboration, such systems can exploit the complementary advantages of heterogeneous platforms, thereby enhancing both the efficiency and effectiveness of maritime anti-submarine operations.

1.4.2 Amphibious Assaults on Islands

Maritime landings are among the most complex forms of maritime operations in both organizational and execution terms, yet they remain one of the most effective means by which naval power can influence the course of land campaigns. For example, on 6 June 1944 the Anglo-American allied forces executed a strategic amphibious landing in Normandy against German forces, committing some 45 divisions, approximately 13,000 aircraft of various types, and roughly 5,000 naval vessels; allied casualties totaled about 122,000. The German forces committed some 15 divisions, some 500 aircraft and over 500 vehicles, and suffered about 114,000 casualties and prisoners. Under the conditions of informationized warfare, modern conflict no longer necessarily requires operations on the scale of the Normandy landings. Nevertheless, conducting a maritime island assault remains exceptionally challenging: assault intentions are relatively easy to detect, and amphibious forces are exposed to the heightened risk of long-range precision fires. Against these difficulties, island operations enabled by the networking sensing control co-design of maritime unmanned systems offer new options. By integrating and coordinating military capabilities across multiple domains (land, sea, air, and cyber), unmanned system-driven campaigns can identify and precisely target the defender's weak points, enabling the seizure, control, or degradation of hostile control over islands and adjacent maritime areas.

1.4.3 Maritime Search and Rescue

In the early hours of March 8, 2014, Malaysia Airlines Flight MH370 lost contact shortly after takeoff, with 227 passengers and 12 crew on board. A multinational search was immediately launched employing assets across space, air, surface, and subsurface domains. For example, the U.S. Navy deployed the Bluefin-21 AUV for subsea search, while China's 17th naval escort task force proceeded to the Indian Ocean to support operations. The years-long search set multiple records, notably the unprecedented use of cross-domain assets spanning aerial, surface, and underwater platforms. By Day 9 (through March 17, 2014), efforts reportedly involved 40 vessels and over 58 aircraft from 26 countries and regions, covering an area exceeding 780,000 km^2. Strictly speaking, the MH370 campaign did not constitute a tightly coordinated cross-domain swarm of maritime unmanned systems, as inter-platform cooperation across nations and regions remained limited. Although the record-setting mission ultimately failed to locate the aircraft, it highlighted the significant potential of cross-domain maritime search and rescue.

1.4.4 Maritime Infrastructure Operation and Maintenance

Traditional operation and maintenance of maritime infrastructure primarily rely on physical protection networks, manual inspection, and surface patrol vessels. These approaches are typically characterized by high labor intensity, high cost, long operation cycles, and limited efficiency and reliability. Moreover, they are highly susceptible to external environmental disturbances such as complex ocean dynamics, wind, and wave forces, making it difficult to respond promptly to emergencies. For example, on September 26, 2022, the Nord Stream natural gas pipelines which transport gas from Russia through the Baltic Sea to Germany and other European countries suffered catastrophic explosions and leaks. This incident served as a stark warning about the vulnerability of subsea pipelines, underscoring the urgent need for real-time, reliable inspection and maintenance solutions. In response to such challenges, the detection-communication-control integration framework of MUSs offers a promising pathway for cross-domain infrastructure maintenance. By integrating aerial, surface, and underwater unmanned platforms, it enables comprehensive surveys of underwater topography, geology, hydrometeorology, and chemical-physical parameters, while supporting monitoring of surface and subsea transmission facilities such as cables and pipelines.

References

Baek, J., Han, S. I., & Han, Y. (Feb.2020). Energy-efficient UAV routing for wireless sensor networks. *IEEE Transactions on Vehicular Technology, 69*(2), 1741–1750.

Cao, W., Yan, J., Yang, X., et al. (Jan. 2023). Communication-aware formation control of AUVs with model uncertainty and fading channel via integral reinforcement learning. *IEEE CAA Journal of Automatica Sinica, 10*(1), 159–176.

Cao, W., Yan, J., Yang, X., et al. (May 2023). Model-free formation control of autonomous underwater vehicles: A broad learning-based solution. *IEEE CAA Journal of Automatica Sinica, 10*(5), 1325–1328.

Celik, A., Saeed, N., Shihada, B., et al. (Feb. 2022). Opportunistic routing for opto-acoustic internet of underwater things. *IEEE Internet Things Journal, 9*(3), 2165–2179.

Chen, L., Shao, Y., & Di, Y. (Mar. 2022). Underwater and water-air optical wireless communication. *The Journal of Lightwave Technology, 40*(5), 1440–1452.

Chen, Y., Yu, W., & Guan, X. (Feb. 2024). Hybrid TOA-AOA cooperative localization for multiple AUVs in the absence of anchors. *IEEE Transactions on Industrial Informatics, 20*(2), 2420–2431.

Gao, J., Xiao, S., Yu, Z., et al. (2020). An integrated ocean communication network for omni-directional comprehensive perception service. *Journal of China Academy of Electronics and Information Technology, 15*(4), 344–363.

He, Y., Qin, T., & Wang, N. (2021). Cross-domain collaboration: New trends in the development and application of unmanned systems technology. *Unmanned Systems Technology, 4*(4), 1–13.

He, C., Wang, Y., Chen, C., & Guan, X. (June 2018). Target localization for a distributed SIMO sonar with an isogradient sound speed profile. *IEEE Access, 6*(1), 2977–29783.

Hu, B., Zhang, H., Liu, B., et al. (Mar. 2024). Coordinated navigation control of cross-domain unmanned systems via guiding vector fields. *IEEE Transactions on Control Systems Technology, 32*(2), 550–563.

Huang, P., Chen, Y., Wang, Y., & Guan, X. (2025). Information entropy-based trajectory planning for AUV-aided network localization: A reinforcement learning approach. *IEEE Internet Things Journal, 12*(2), 2122–2134.

Jiang, L., Li, Y., Wang, B., et al. (Jan. 2024). Cooperative localization of asynchronous AUVs with compensation for the acoustic wave bends. *The IEEE Signal Processing Letters, 31*(1), 2240–2244.

Jiang, L., Li, Y., Yu, W., & Guan, X. (2024). Cooperative localization for asynchronous AUVs using time difference of communication in underwater anchor-free environments. *The IEEE Journal of Oceanic Engineering, 54*(11), 6531–6544.

Jiang, J., Wang, X., Duan, F., et al. (Feb. 2019). A sonar-embedded disguised communication strategy by combining sonar waveforms and whale call pulses for underwater sensor platforms. *Sensors, 145*(1), 255–266.

Li, Z., Kong, J., Liu, P., et al. (2020). Design of underwater intelligent cross domain heterogeneous network. *Ship Science and Technology, 42*(12), 137–140.

Li, Y., Li, B., Yu, W., et al. (May 2019). Cooperative localization based multi-AUV trajectory planning for target approaching in anchor-free environments. *IEEE Transactions on Vehicular Technology, 71*(3), 4707–4720.

Li, Y., Liu, L., Yu, W., et al. (Oct. 2020). Noncooperative mobile target tracking using multiple AUVs in anchor-free environments. *IEEE Internet Things Journal, 7*(10), 9819–9833.

Li, Y., Wang, Y., Yu, W., & Guan, X. (Oct. 2019). Multiple autonomous underwater vehicle cooperative localization in anchor-free environments. *The IEEE Journal of Oceanic Engineering, 44*(4), 895–911.

Li, Y., Yu, W., & Guan, X. (Apr. 2024). 3D localization for multiple AUVs in anchor-free environments by exploring the use of depth information. *Ocean Engineering, 11*(4), 1051–1053.

Li, J., Zhang, G., Jiang, C., et al. (2023). A survey of maritime unmanned search system: Theory, applications and future directions. *Ocean Engineering, 285*(1), 1–12.

Liu, J., Wang, Z., Peng, Z., et al. (2014). Suave: Swarm underwater autonomous vehicle localization. In *Proceedings of IEEE International Conference Computer Communications (INFOCOM)*, pp. 64–72. IEEE.

Liu, L., Wang, T., Tian, R., et al. (Aug. 2023). Underwater channel characteristics and wireless optical communications using a blue mode-locked semiconductor disk laser. *The Journal of Lightwave Technology,* **41**(16), 5340–5349.

Lu, J., Zhang, Q., Zhang, L., et al. (2018). Detection performance of active sonar based on underwater acoustic communication signals. In *Proceedings of IEEE International Conference on Signal Processing, Communications and Computing* (pp. 1–5). IEEE.

Luo, H., Wang, J., Bu, F., et al. (May 2022). Recent progress of air/water cross-boundary communications for underwater sensor networks: A review. *The IEEE Sensors Journal, 22*(9), 8360–8382.

Luo, H., Wu, K., Gong, Y., et al. (2016). Localization for drifting restricted floating ocean sensor networks. *IEEE Transactions on Vehicular Technology, 65*(12), 9968–9981.

Ma, J., Li, H., Zhu, J., et al. (Apr. 2021). Design and experiments of a portable seabed integrated detection sonar. *Sensors, 21*(8), 1–13.

Ma, R., Wang, R., Liu, G., et al. (Sep. 2021). UAV-aided cooperative data collection scheme for ocean monitoring networks. *IEEE Internet Things Journal, 8*(17), 13222–13236.

Mealey, R. M. (June 1963). A method for calculating error probabilities in a radar communication system. *EEE Transactions on Space Electronics and Telemetry, SET-9*(2), 37–42.

Ruby, R., Zhong, S., Elhalawany, B.M., et al. (2021). SDN-enabled energy-aware routing in underwater multi-modal communication networks. *IEEE ACM Transactions on Networking, 29*(3), 965–978.

Stojanovic, M., & Preisig, J. (Jan. 2009). Underwater acoustic communication channels: Propagation models and statistical characterization. *IEEE Communications Magazine, 47*(1), 84–89.

Su, H., Zhu, S., Chen, C., et al. (2024). Bearing-Based robust formation tracking control of underactuated AUVs with optimal parameter tuning. *IEEE Transactions on Cybernetics, 54*(7), 4049–4062.

Su, H., Zhu, S., Yang, Z., et al. (Aug. 2022). Bearing-based formation tracking control of AUVs with optimal gains tuning. *Ocean Engineering, 258*(1), 1–14.

Sutton, R., & Barto, A. (2018). *Reinforcement learning: An introduction* (pp. 1–342). MIT Press.

Vasilijevic, A., Nad, D., Mandic, F., et al. (June 2017). Coordinated navigation of surface and underwater marine robotic vehicles for ocean sampling and environmental monitoring. *IEEE ASME Transactions on Mechatronics, 22*(3), 1174–1184.

Xing, W. (Sept. 2019). A study on self-organizing ubiquitous wireless network architecture of radar communication. *Modern Rada, 41*(6), 1–7.

Xu, T., Xiao, Y., He, Y., et al. (2023). Development of cross-domain communication networking technologies for marine autonomous unmanned system. *Journal of Command and Control, 9*(6), 683–691.

Xu, W., Yan, S., Ji, F., et al. (2016). Marine information gathering, transmission, processing, and fusion: current status and future trends. *Scientia Sinica Informationis, 46*(8), 1053–1085.

Yan, J., Ban, H., Luo, X., et al. (Mar. 2022). Joint localization and tracking design for AUV with asynchronous clocks and state disturbances. *IEEE Transactions on Vehicular Technology, 68*(5), 3092–3107.

Yan, J., Cao, W., Yang, X., et al. (June 2024). Communication-efficient and collision-free motion planning of underwater vehicles via integral reinforcement learning. *IEEE Transactions on Neural Networks and Learning Systems, 35*(6), 8306–8320.

Yan, J., Cui, L., Yang, X., et al. (2025). Design of an embedded system for integrated underwater communication and detection. *EEE Embedded Systems Letters, 17*(2), 83–86.

Yan, J., Gao, J., Yang, X., et al. (July 2019). Tracking control of a remotely operated underwater vehicle with time delay and actuator saturation. *Ocean Engineering, 184*(1), 299–310.

Yan, J., Gao, J., Yang, X., et al. (July 2020). Position tracking control of remotely operated underwater vehicles with communication delay. *IEEE Transactions on Control Systems Technology, 28*(6), 2506–2514.

Yan, J., Gong, Y., Chen, C., et al. (Oct. 2020). AUV-aided localization for internet of underwater things: A reinforcement learning-based method. *IEEE Internet Things Journal, 7*(10), 9728–9746.

Yan, J., Guo, D., Luo, X., & Guan, X. (Aug. 2020). AUV-aided localization for underwater acoustic sensor networks with current field estimation. *IEEE Transactions on Vehicular Technology, 69*(8), 8855–8870.

Yan, J., Guo, Z., Yang, X., et al. (Nov. 2022). Finite-time tracking control of autonomous underwater vehicle without velocity measurements. *IEEE Systems, Man, and Cybernetics, 52*(11), 6759–6773.

Yan, J., Li, X., Yang, X., et al. (2022). Integrated localization and tracking for AUV with model uncertainties via scalable sampling-based reinforcement learning approach. *IEEE Transactions on Systems, Man, and Cybernetics: Systems, 52*(11), 6952–6967.

Yan, J., Li, X., Yang, X., et al. (2022). Joint localization and tracking for autonomous underwater vehicle: A reinforcement learning based approach. *IEEE Transactions on Systems, Man, and Cybernetics, 52*(11), 6952–6967.

Yan, J., Lin, J., Yang, X., et al. (Feb. 2025). Cooperation detection and tracking of underwater target via aerial-surface-underwater vehicle. *IEEE Transactions on Automatic Control, 70*(2), 1068–1083.

Yan, J., Meng, Y., Luo, X., & Guan, X. (Sep. 2021). To hide private position information in localization for internet of underwater things. *IEEE Internet of Things Journal, 8*(18), 14338–14354.

Yan, J., Meng, Y., Yang, X., et al. (2021). Privacy-preserving localization for underwater sensor networks via deep reinforcement learning. *The IEEE Transactions on Information Forensics and Security, 16*(1), 1880–1895.

Yan, J., Peng, S., Yang, X., et al. (Sep. 2023). Containment control of autonomous underwater vehicles with stochastic environment disturbances. *IEEE Transactions on Systems, Man, and Cybernetics: Systems, 53*(9), 5809–5820.

Yan, J., Tian, X., Luo, X., et al. (Sep. 2019). Design of an embedded communication system for underwater asynchronous localization. *IEEE Embedded Systems Letters, 11*(3), 97–100.

Yan, J., Wan, Y., Luo, X., et al. (2018). Formation control of teleoperating cyber-physical system with time delay and actuator saturation. *IEEE Transactions on Control Systems Technology, 26*(4), 1458–1467.

Yan, J., Wang, H., Yang, X., et al. (2024). Optimal rigid graph-based cooperative formation control of AUVs in anchor-free environments. *IEEE Transactions on Intelligent Vehicles, 9*(9), 5922–5938.

Yan, J., Wu, C., Yang, X., et al. (2024). Minimum time formation control of AUVs with smooth transition in communication topology. *IEEE Robotics and Automation, 9*(10), 8698–8705.

Yan, J., Xu, Z., Luo, X., et al. (Mar.2019). Feedback-based target localization in underwater sensor networks: A multisensor fusion approach. *The IEEE Transactions on Signal and Information Processing over Networks, 5*(1), 168–180.

Yan, J., Xu, Z., Wan, Y., et al. (2017). Consensus estimation-based target localization in underwater acoustic sensor networks. *International Journal of Robust and Nonlinear Control, 27*(9), 1607–1627.

Yan, J., Yang, X., Luo, X., & Chen, C. (Dec. 2018). Energy-efficient data collection over AUV assisted underwater acoustic sensor network. *IEEE System Journal, 12*(4), 3519–3530.

Yan, J., Yi, M., Yang, X., et al. (Aug. 2023). Broad learning-based localization for underwater sensor networks with stratification compensation. *IEEE Internet Things Journal, 10*(15), 13123–13137.

Yan, J., You, K., Cao, W., et al. (Aug. 2024). Binocular vision-based motion planning of an AUV: A deep reinforcement learning approach. *IEEE Transactions on Intelligent Vehicles, 9*(8), 5299–5315.

Yan, J., Zhang, X., Luo, X., et al. (Mar. 2018). Asynchronous localization with mobility prediction for underwater acoustic sensor networks. *IEEE Transactions on Vehicular Technology, 67*(3), 2543–2556.

Yan, J., Zhang, X., Yang, X., et al. (2025). Potential game-driven formation control of AUVs: An inverse reinforcement learning-based solution. *IEEE Journal of Oceanic Engineering, 50*(2), 1165–1183.

Yan, J., Zhang, L., Yang, X., et al. (Dec. 2023). Communication-aware motion planning of AUV in obstacle-dense environment: A Binocular vision-based deep learning method. *IEEE Transactions on Intelligent Transportation Systems, 24*(12), 14927–14943.

Yan, J., Zhang, Z., Yang, X., et al. (July 2023). Target detection in underwater sensor networks by fusion of active and passive measurements. *IEEE Communications Magazine, 10*(4), 2319–2333.

Yan, J., Zhao, H., Luo, X., et al. (Apr. 2019). RSSI-based heading control for robust long-range aerial communication in UAV networks. *IEEE Communications Magazine, 6*(2), 1675–1689.

Yan, J., Zhao, H., Luo, X., et al. (Oct. 2020). Asynchronous localization of underwater target using consensus-based unscented Kalman filtering. *The IEEE Journal of Oceanic Engineering, 45*(4), 1466–1481.

Yan, J., Zhao, H., Pu, B., et al. (July 2020). Energy-efficient target tracking with UASNs: A consensus-based Bayesian approach. *IEEE Transactions on Automation Sciences and Engineering, 17*(3), 1361–1375.

Yan, J., Zhao, H., Wang, Y., et al. (Apr. 2019). Asynchronous localization for UASNs: An unscented transform-based method. *The IEEE Signal Processing Letters, 26*(4), 602–606.

Yan, J., Zhou, X., Yang, X., et al. (Dec. 2023). Joint design of channel estimation and flocking control for multi-AUV-based maritime transportation systems. *IEEE Transactions on Intelligent Transportation System, 24*(12), 14520–14535.

Yang, X., Yan, J., Chen, C., et al. (Feb. 2024). Adaptive asymptotic tracking control for underactuated autonomous underwater vehicles with state constraints. *IEEE Transactions on Intelligent Transportation Systems, 25*(11), 18485–18500.

Yang, X., Yan, J., Hua, C., et al. (Feb. 2021). Trajectory tracking control of autonomous underwater vehicle with unknown parameters and external disturbances. *IEEE Communications Magazine, 51*(2), 1054–1063.

Zhang, W., Lu, X., & Han, P. (2020). Progress and challenges of overwater unmanned systems. *Acta Automatica Sinica, 46*(5), 847–857.

Zhang, C., Wang, Y., Guan , X.,et al. (2016). Chaotic modulation detection for underwater acoustic communications via instantaneous features. In *OCEANS MTS/IEEE Monterey, OCE*, pp. 1–5.

Zhang, T., Yan, J., Yang, X., et al. (2024). Digital twin-driven formation control of ROVs: An integral reinforcement learning-based solution. *IEEE Transactions on Industrial Informatics, 20*(12), 14277–14286.

Zhao, Y., Chen, Y., Huang, J., et al. (Jun. 2024). Photoacoustic communication system based on detecting laser-generated sound by optical fiber underwater acoustic sensor. *Optics and Lasers in Engineering, 177*(1), 96–105.

Zhao, H., Yan, J., Luo, X., & Guan, X. (Nov. 2020). Privacy preserving solution for the asynchronous localization of underwater sensor networks. *IEEE/CAA Journal of Automatica Sinica, 7*(6), 1511–1527.

Zhao, H., Yan, J., Luo, X., et al. (Sep. 2021). Ubiquitous tracking for autonomous underwater vehicle with IoUTs: A rigid graph-based solution. *IEEE Internet Things Journal, 8*(18), 14094–14109.

Zhao, H., Zhao, J., Luo, X., et al. (Sep. 2021). Ubiquitous tracking for autonomous underwater vehicle with IoUTs: A rigid graph-based solution. *IEEE Internet Things Journal, 8*(18), 14094–14109.

Zheng, Z., Lyu, C., Bi, Y., et al. (Mar. 2022). Review of hybrid aerial underwater vehicle: Cross-domain mobility and transitions control. *Ocean Engineering, 248*(1), 1–13.

Zhou, Z., Peng, Z., Cui, J., et al. (Mar. 2011). Scalable localization with mobility prediction for underwater sensor networks. *IEEE Transactions on Mobile Computing, 10*(3), 335–348.

Chapter 2
Integrated Detection via Active and Passive Measurements

2.1 Introduction

In MUSs, one of the critical issues is to detect the existence of underwater target before it is close enough to attack. It is a sophisticated decision procedure that conducts all available measurements from different types of sensor nodes. Compared with the traditional sonar system, the cooperation property of MUSs has the potential to greatly improve the spatiotemporal detection performance for underwater target (Ghafoor and Noh 2019).

Till now, a large variety of target detection algorithms have been developed for MUSs. For instance, an underwater passive detection algorithm by using adaptive empirical mode decomposition was developed in (Tian et al., 2022), where a single vector hydrophone was adopted to reduce the communication energy consumption. In (Hu et al., 2021), a reflection model was constructed to design a distributed generalized likelihood ratio test detector for passive sensor nodes. Also of relevance, a joint detection and tracking solution was presented in (Jiang et al., 2018), in which a Bernoulli filter associated with the path planning of mobile sensor nodes was utilized to improve the detection accuracy. In (Tian et al., 2022; Hu et al., 2021; Jiang et al., 2018), the passive measurements are employed, which means the sensor nodes detect the existence of underwater target without transmitting any messages to the networks. Although the passive measurements are energy efficient, they generally demand high signal-to-noise ratio and thus have limited anti-interference ability. Alternatively, some researchers attempt to employ the active measurements to detect the underwater target. For instance, the active measurements were adopted in (Braca et al., 2016) to provide a diffusion particle filter, whose aim was to reduce the false alarm rate of underwater target detection. In (Liang and Cheng, 2008), a maximum likelihood (ML) estimation algorithm was performed to detect the size of underwater target, where each sensor node should provide its waveform parameters to the cluster head. A machine learning-based algorithm was developed in Gong et al. (2020) to implement the continuous active detection for underwater target. In addition, some

J. Yan and X. Guan, *Maritime Unmanned Systems*, Cognitive Intelligence and Robotics 1, https://doi.org/10.1007/978-981-95-6907-6_2

other underwater active detection approaches can be found in Song et al. (2012); Diamant et al. (2019); Yan et al. (2021); Aman et al. (2018). Compared with the passive measurements, the active measurements have the advantages of improved detection area and accuracy for underwater target detection, however, they have the disadvantages of low concealment and high communication energy consumption. Through the above analyses, we know the passive and active measurements have their own advantages and shortcomings. In order to make a good use of the advantages, we believe that the detection system can benefit a lot from the united implementation of active and passive measurements, since the united implementation of all available measurements can yield improved detection robustness and accuracy in Cao et al. (2021). As far as we know, how to design an active-passive united measurement mechanism for the detection of underwater target is not well studied.

Note that an individual sensor node can only perform local detection, due to the strong noise and weak communication properties of underwater environment (Xu et al. 2021). In order to conduct a large scale detection, each sensor node requires to exchange its message with the other nodes in the sensing area. Generally, centralized and distributed communication structures are two typical setups for data exchange. In centralized structure, all sensor nodes send their observations to the control center or fusion center (Yemeni et al. 2022), which may result in the data traffic bottlenecks and the incapability to scale. In contrast, the distributed structure only exchanges the local observations to the neighboring nodes, where each node is of the same functionality and no organization exists among these *egalitarian* sensor nodes. Although the distributed structure can achieve better performance on throughput, flexibility and scalability, it is at the cost of a more time to reach consensus decision. In view of this, our previous work Wan et al. (2019) employed the two-layer multi-group (2LMG) structure to reduce the communication load and improve the convergence time, while the impact of 2LMG structure on network performance was provided in Wan et al. (2013). Specifically, the 2LMG structure is composed of sensor nodes at the bottom layer and virtual fusion centers at higher layers, which can serve as a novel mean for us to understand the trade-off between centralized and distributed structures. With the fusion of active and passive measurements, the issue of how to incorporate the 2LMG structure into the underwater target detection of MUSs is still largely unexplored.

This chapter investigates an underwater target detection issue by the fusion of active and passive measurements in MUSs. We assume that a non-cooperative target with unknown position and power is prowling in the underwater monitoring area. With regard to this, a two-phase target detection procedure is performed for MUSs, i.e., each sensor node makes a local decision based on the united implementation of active and passive measurements in the first phase, and then an external fusion procedure is conducted in the second phase to fuse the local decisions from different sensor nodes. To be particular, this chapter has the following novel contributions:

1. **Chi-square test based local decision rule with adaptive weight law.** Active and passive measurements are both employed for MUSs to develop a Chi-square test based local decision rule for each sensor node. In such rule, an adaptive

weight law is designed to maximize the contributions of reliable measurements while minimizing the contributions of those that are not reliable. Compared with the single passive measurements (e.g., Tian et al. 2022; Hu et al. 2021; Jiang et al. 2018) or passive measurements (e.g., Braca et al. 2016; Liang and Cheng 2008; Gong et al. 2020), the active-passive united measurement mechanism in this chapter can improve the detection robustness and accuracy for underwater target.

2. **Hybrid Bayesian fusion algorithm with 2LMG structure.** With regard to the unknown position and power of underwater target, a hybrid Bayesian fusion algorithm is developed to fuse the local decisions from different sensor nodes. It should be emphasized that the 2LMG structure is adopted here to make a tradeoff between centralized and distributed structures. Compared with the egalitarian distributed structure (e.g., Han et al. 2020), the proposed 2LMG structure in this chapter can reduce the number of transmissions and improve the detection rate.

2.2 Network Model and Problem Formulation

With regard to MUSs, we consider an active-passive detection system that consists of N underwater sensor nodes and a target, as depicted by Fig. 2.1. Specifically, the 2LMG structure is adopted to describe the topology relationship of sensor nodes in MUSs, which can be detailedly presented in Sect. 2.3.2. According to the definition of 2LMG structure (Wan et al. 2019, 2013), sensor nodes are divided into two categories, i.e., *ordinary sensor nodes* and *virtual fusion centers* (VFCs). Along with

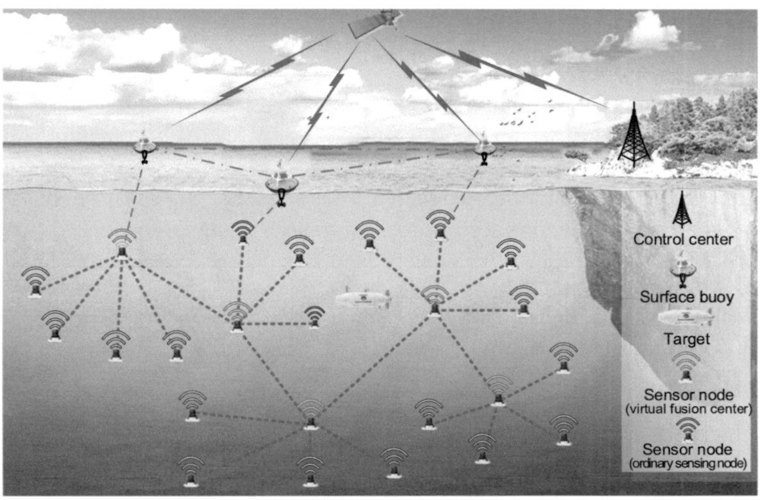

Fig. 2.1 Description of the active-passive detection system

this, each *ordinary sensor node* performs a local decision on the absence/presence of target, and then forwards it to a *VFC* for seeking a more accurate global decision. Of note, each sensor node can make direct contact with the surface buoys, such that its clock synchronization and self-localization services can be served.

2.2.1 Kinematic Model of Target

Similar to the implementation in Yan et al. (2020), we formulate the target state as a Bernoulli random finite set $\mathbf{X}_{\rm t}(k)$. Once the target is absent, $\mathbf{X}_{\rm t}(k)$ is defined as an empty set, i.e., $\mathbf{X}_{\rm t}(k) = \{\emptyset\}$. Once the target is present, $\mathbf{X}_{\rm t}(k)$ is defined as $\mathbf{X}_{\rm t}(k) = [x(k), v_x(k), y(k), v_y(k), z(k), v_z(k)]^{\rm T}$, where $x(k)$, $y(k)$ and $z(k)$ represent the positions of target on x, y and z axes, respectively. In addition, $v_x(k)$, $v_y(k)$ and $v_z(k)$ represent the velocities on x, y and z axes, respectively. Then the kinematic model of target at step k is constructed as

$$\mathbf{X}_{\rm t}(k) = \mathbf{F}(k)\mathbf{X}_{\rm t}(k-1) + \mathbf{w}(k), \tag{2.1}$$

where $\mathbf{F}(k)$ denotes the state transition function, and $\mathbf{w}(k)$ denotes the unmodeled dynamics or acceleration.

2.2.2 Active and Passive Measurement Models

As mentioned above, the active-passive united measurement mechanism is conducted by sensor nodes. For this purpose, the active and passive measurement models are presented here.

(1) *Active measurement model:* We assume that each sensor node, including ordinary sensor node and VFC, can make active and passive measurements on target. For the active measurement, sensor node $i \in \{1, 2, \ldots, N\}$ actively transmits a regular detection waveform to the monitoring area. The waveform meets the target on the path of underwater propagation, and then it is reflected back to sensor node i. When the target is present in the monitoring area, the measured signal-to-noise ratio (SNR) on sensor node $i \in \{1, 2, \ldots, N\}$ at time step k can be expressed as Zhang et al. (2021), Isbitiren and Akan (2011)

$$y_{{\rm a},i}(k) = \frac{\alpha_{\rm T} \times \alpha_{{\rm s},i}(k)}{d_{{\rm T},i}^{2n}(k) + 10^{2\alpha_f d_{{\rm T},i}(k) \times 10^{-4}}} + w_{{\rm a},i}(k), \tag{2.2}$$

with

$$\alpha_f = \frac{0.11 f^2}{1 + f^2} + \frac{44 f^2}{4100 + f^2} + 2.75 \times 10^{-4} f^2 + 0.003, \tag{2.3}$$

where α_{T} denotes the radiated power of target, following the Gaussian distribution with zero mean and variance σ_{T}^2, i.e., $\alpha_{\mathrm{T}} \sim \mathcal{N}(0, \sigma_{\mathrm{T}}^2)$. In addition, $\alpha_{\mathrm{s},i}(k)$ is the transmission power of sensor node i at time step k, following the Gaussian distribution with zero mean and variance σ_{s}^2, i.e., $\alpha_{\mathrm{s},i}(k) \sim \mathcal{N}(0, \sigma_{\mathrm{s}}^2)$. $d_{\mathrm{T},i}$ is the relative distance between target and the ith sensor node, n represents the distance attenuation coefficient, and α_f represents the absorption coefficient of underwater medium with known frequency f. Besides that, $w_{\mathrm{a},i}$ denotes the active measurement noise of sensor node i, which follows the Gaussian distribution with zero mean and variance σ_{a}^2, i.e., $w_{\mathrm{a},i} \sim \mathcal{N}(0, \sigma_{\mathrm{a}}^2)$.

We define $\mathcal{H}_0$ and $\mathcal{H}_1$ as the hypotheses of target absence and target presence, respectively. Based on this and noting with (2.2), the distribution of $y_{\mathrm{a},i}(k)$ is given as

$$y_{\mathrm{a},i}(k) \sim \begin{cases} \mathcal{N}(0, \sigma_{\mathrm{a}}^2), & \mathcal{H}_0, \\ \mathcal{N}\left(0, \dfrac{\sigma_{\mathrm{T}}^2 \times \sigma_{\mathrm{s}}^2}{(d_{\mathrm{T},i}^{2n} + 10^{2\alpha_f d_{\mathrm{T},i}(k) \times 10^{-4}})^2} + \sigma_{\mathrm{a}}^2\right), & \mathcal{H}_1. \end{cases} \tag{2.4}$$

(2) *Passive measurement model:* Different from the active measurement, the sensor nodes in passive measurement procedure detect the target without transmitting any messages to the networks. Based on the above consideration, sensor node $i \in \{1, 2, \ldots, N\}$ receives the radiated signal of target when the target is present. Particularly, the measured SNR on sensor node i at time step k can be denoted as

$$y_{\mathrm{p},i}(k) = \frac{\alpha_{\mathrm{T}}}{d_{\mathrm{T},i}^n(k) + 10^{\alpha_f d_{\mathrm{T},i}(k) \times 10^{-4}}} + w_{\mathrm{p},i}(k), \tag{2.5}$$

where $w_{\mathrm{p},i}$ denotes the passive measurement noise of sensor node i, and it follows the Gaussian distribution with zero mean and variance σ_{p}^2, i.e., $w_{\mathrm{p},i} \sim \mathcal{N}(0, \sigma_{\mathrm{p}}^2)$.

Accordingly, the distribution of $y_{\mathrm{p},i}(k)$ is given by

$$y_{\mathrm{p},i}(k) \sim \begin{cases} \mathcal{N}\left(0, \sigma_{\mathrm{p}}^2\right), & \mathcal{H}_0, \\ \mathcal{N}\left(0, \dfrac{\sigma_{\mathrm{T}}^2}{(d_{i,\mathrm{T}}^n(k) + 10^{\alpha_f d_{\mathrm{T},i}(k) \times 10^{-4}})^2} + \sigma_{\mathrm{p}}^2\right), & \mathcal{H}_1. \end{cases} \tag{2.6}$$

2.2.3 Communication Energy Model

Recall that the 2LMG structure is adopted in this chapter to reduce the communication load of MUSs. In view of this, a common communication energy model (Jiang et al. 2016) is considered in this chapter. Let r denote the relative distance from a node to its next-hop node, and hence, the energy consumption of transmitting b bits over distance r is given by

$$E_{\mathrm{tx}}(b, r) = bE_{\mathrm{elec}} + 4.2 * bT_b H e^{g(f)r} \cdot 10^{-9.5}, \tag{2.7}$$

and the receiver's energy consumption is

$$E_{\mathrm{rx}}(b, r) = bE_{\mathrm{ele}}, \tag{2.8}$$

where E_{elec} represents the unit energy consumption of processing one bit message, H denotes the depth of sensor node, and $g(f)$ denotes the absorption coefficient as defined in Jiang et al. (2016). In addition, T_b represents the bit duration.

2.2.4 *Problem Formulation*

With regard to the above system models, the objective of this chapter is to develop an underwater target detection solution by fusion of active and passive measurements in MUSs. Therefore, the following two problems are required to be solved.

Problem 2.1 (*Local decision rule design with active and passive measurements*) As noted above, the unified implementation of all available measurements can yield improved detection accuracy. In view of this, we attempt to detect the existence of underwater target with active and passive measurements. This problem can be reduced to design a local decision rule, with the consideration of (2.2), (2.3), (2.4) and (2.5).

Problem 2.2 (*External fusion algorithm design with the 2LMG structure*) Due to the strong noise and weak communication properties of underwater environment, the spatiotemporal detection ability and the installed battery power of a single sensor node is limited. Therefore, it is necessary to fuse the local detection decisions and reduce the communication energy consumption. This problem can be reduced to design an external fusion algorithm with the 2LMG structure.

2.3 Underwater Target Detection

In this section, the local decision rule employed by each sensor node is first presented. Based on this, a 2LMG structure based external fusion algorithm is designed to improve the detection performance. As last, the theory analysis for the proposed solution in this chapter are presented.

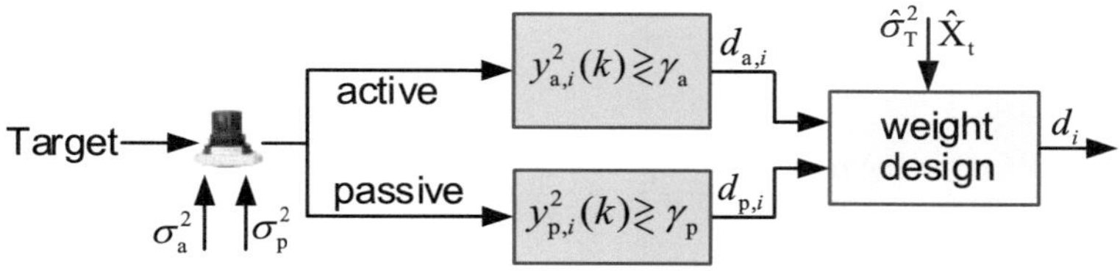

Fig. 2.2 Relationship of the three stages for local decision rule

2.3.1 Chi-Square Test Based Local Decision Rule

Considering the kinematic model of target as depicted by (2.1), we set that each sensor node can make active and passive measurements individually. According to this, the local decision procedure can be divided into the following three stages: (1) Active measurement-based local decision; (2) Passive measurement-based local decision; (3) Weight setting for local fusion. Specifically, the relationship of these stages can be depicted by Fig. 2.2, which can also be detailed as follows.

(1)*Active measurement-based local decision*
With active measurement, the following composite hypothesis test is conducted by sensor node $i \in \{1, 2, \ldots, N\}$

$$\begin{cases} \mathcal{H}_0 : y_{\mathrm{a},i}(k) = w_{\mathrm{a},i}(k), \\ \mathcal{H}_1 : y_{\mathrm{a},i}(k) = \dfrac{\alpha_{\mathrm{T}} + \alpha_{\mathrm{s},i}(k)}{2d^n_{\mathrm{T},i}(k) + 10^{2\alpha_f d_{\mathrm{T},i}(k)\times 10^{-4}}} + w_{\mathrm{a},i}(k). \end{cases} \tag{2.9}$$

Recall that the distance $d_{\mathrm{T},i}$ and the radiated power α_{T} in (2.9) are unknown to sensor node $i \in \{1, 2, \ldots, N\}$. In view of this, we utilize the energy $y^2_{\mathrm{a},i}(k)$ to make local decision, such that a uniformly most powerful (UMP) test can be conducted by sensor node $i \in \{1, 2, \ldots, N\}$. To this end, the local Log Likelihood-Ratio (LLR) of $y_{\mathrm{a},i}(k)$ can be defined as

$$\begin{aligned} \lambda_i(k) &= \ln\left[\frac{p(y_{\mathrm{a},i}(k)|\mathcal{H}_1)}{p(y_{\mathrm{a},i}(k)|\mathcal{H}_0)}\right] \\ &= \frac{1}{2}\ln\left[\frac{\sigma^2_{\mathrm{a}}}{\Phi^2}\right] + \frac{\sigma^2_{\mathrm{a}} - \Phi^2}{2\Phi^2\sigma^2_{\mathrm{a}}} y^2_{\mathrm{a},i}(k), \end{aligned} \tag{2.10}$$

with

$$p(y_{\mathrm{a},i}(k)|\mathcal{H}_0) = \frac{1}{\sqrt{2\pi\sigma^2_{\mathrm{a}}}} \exp\left(-\frac{y^2_{\mathrm{a},i}(k)}{2\sigma^2_{\mathrm{a}}}\right), \tag{2.11}$$

$$p(y_{\mathrm{a},i}(k)|\mathcal{H}_1) = \frac{1}{\sqrt{2\pi\Phi^2}} \exp\left(-\frac{y^2_{\mathrm{a},i}(k)}{2\Phi^2}\right), \tag{2.12}$$

where $\Phi^2 = \frac{\sigma_s^2 \times \sigma_T^2}{\left(d_{T,i}^{2n} + 10^{2\alpha_f d_{T,i}(k) \times 10^{-4}}\right)^2} + \sigma_a^2$.

From (2.10), (2.11) and (2.12), one can easily know that $\lambda_i(k)$ is an increasing function of $y_{a,i}^2(k)$. Besides that, $\lambda_i(k)$ does not require the values of $d_{T,i}$ and α_T in advance. The above results mean that: 1) the test of $y_{a,i}^2(k)$ is UMP by referring to the Karlin-Rubin Theorem Kay (1993); 2) one can test $y_{a,i}^2(k)$ even when the values of $d_{T,i}$ and α_T are unknown. Inspired by this, the hypothesis test in (2.9) is equivalent to

$$\begin{cases} y_{a,i}^2(k) < \gamma_a, & \text{under } \mathcal{H}_0 \\ y_{a,i}^2(k) > \gamma_a, & \text{under } \mathcal{H}_1 \end{cases} \tag{2.13}$$

where γ_a represents the local threshold, and it is associated with a given false-alarm probability $P_{f,a}$.

Referring to the Neyman-Pearson lemma Kay (1993), the relationship between γ_a and $P_{f,a}$ can be constructed as

$$P_{f,a} = 2\mathcal{Q}\left(\sqrt{\frac{\gamma_a}{\sigma_a^2}}\right) \tag{2.14}$$

where $\mathcal{Q}(x) = \int_x^\infty \frac{1}{\sqrt{2\pi}} e^{-(x^2/2)} dx$ represents the complementary cumulative distribution function.

Based on (2.13) and (2.14), the local active decision can be made by sensor node $i \in \{1, 2, \ldots, N\}$: 1) when $y_{a,i}^2(k) < \gamma_a$ is satisfied, the target can be regarded to be absent, and hence, the boolean variable $d_{a,i} = 0$; 2) when $y_{a,i}^2(k) > \gamma_a$ is satisfied, the target is present, and hence, $d_{a,i} = 1$.

(2) *Passive measurement-based local decision*
With the passive measurement as depicted by (2.5), the following composite hypothesis test is conducted by sensor node $i \in \{1, 2, \ldots, N\}$

$$\begin{cases} \mathcal{H}_0 : y_{p,i}(k) = w_{p,i}(k), \\ \mathcal{H}_1 : y_{p,i}(k) = \dfrac{\alpha_T}{d_{T,i}^n(k) + 10^{\alpha_f d_{T,i}(k) \times 10^{-4}}} + w_{p,i}(k). \end{cases} \tag{2.15}$$

Similar to the local active decision, a UMP test can be conducted by sensor node $i \in \{1, 2, \ldots, N\}$, and hence, the LLR of $y_{p,i}(k)$ is defined as

$$\begin{aligned} \lambda_i(k) &= \ln\left[\frac{p\left(y_{p,i}(k)\middle|\mathcal{H}_1\right)}{p\left(y_{p,i}(k)\middle|\mathcal{H}_0\right)}\right] \\ &= \frac{1}{2}\ln\left(\frac{\sigma_p^2}{\bar{\Phi}^2}\right) + \frac{\bar{\Phi}^2 - \sigma_p^2}{2\sigma_p^2\bar{\Phi}^2} y_{p,i}^2(k). \end{aligned} \tag{2.16}$$

with

$$P\left(y_{\mathrm{p},i}(k)\middle|\mathcal{H}_0\right) = \frac{1}{\sqrt{2\pi\sigma_{\mathrm{p}}^2}} \exp\left(-\frac{y_{\mathrm{p},i}^2(k)}{2\sigma_{\mathrm{p}}^2}\right), \tag{2.17}$$

$$P\left(y_{\mathrm{p},i}(k)\middle|\mathcal{H}_1\right) = \frac{1}{\sqrt{2\pi\bar{\Phi}^2}} \exp\left(-\frac{y_{\mathrm{p},i}^2(k)}{2\bar{\Phi}^2}\right). \tag{2.18}$$

where $\bar{\Phi}^2 = \frac{\sigma_{\mathrm{T}}^2}{\left(d_{\mathrm{T},i}^n + 10^{\alpha_f d_{\mathrm{T},i}(k)\times 10^{-4}}\right)^2} + \sigma_{\mathrm{p}}^2$.

According to (2.16), (2.17) and (2.18), the LLR is an increasing function of $y_{\mathrm{p},i}^2(k)$. Then, the UMP test is obtained as

$$\begin{cases} y_{\mathrm{p},i}^2(k) < \gamma_{\mathrm{p}}, & \text{under } \mathcal{H}_0 \\ y_{\mathrm{p},i}^2(k) > \gamma_{\mathrm{p}}, & \text{under } \mathcal{H}_1 \end{cases} \tag{2.19}$$

where γ_{p} denotes the local threshold with a given false-alarm probability $P_{\mathrm{f,p}}$, and γ_{p} is calculated as $P_{\mathrm{f,p}} = 2\mathcal{Q}\left(\sqrt{\frac{\gamma_{\mathrm{p}}}{\sigma_{\mathrm{p}}^2}}\right)$.

In view of (2.19), we get the local decision with passive detection for sensor node $i \in \{1, 2, \ldots, N\}$: 1) when $y_{\mathrm{p},i}^2(k) < \gamma_{\mathrm{p}}$ is satisfied, the target is absent, i.e., $d_{\mathrm{p},i} = 0$; 2) when $y_{\mathrm{p},i}^2(k) > \gamma_{\mathrm{p}}$ is satisfied, the target is present, i.e., $d_{\mathrm{p},i} = 1$.

(3) *Weight setting for local decision*

In order to improve the detection accuracy, a Chi-square test based weight setting strategy is performed to bias the weight of each measurement in accordance with signal reliability. To this end, we employ the *measurement error* as a criterion to evaluate the signal reliability, and hence, the active and passive measurement errors are denoted as

$$v_{\mathrm{a},i} = y_{\mathrm{a},i}(k) - f_{\mathrm{a},i}(k), \quad v_{\mathrm{p},i} = y_{\mathrm{p},i}(k) - f_{\mathrm{p},i}(k), \tag{2.20}$$

where $f_{\mathrm{a},i}(k) = \frac{\hat{\alpha}_{\mathrm{T}} \times \alpha_{\mathrm{s},i}(k)}{\hat{d}_{\mathrm{T},i}^{2n}(k) + 10^{2\alpha_f \hat{d}_{\mathrm{T},i}(k)\times 10^{-4}}}$ and $f_{\mathrm{p},i}(k) = \frac{\hat{\alpha}_{\mathrm{T}}}{\hat{d}_{\mathrm{T},i}^{n}(k) + 10^{\alpha_f \hat{d}_{\mathrm{T},i}(k)\times 10^{-4}}}$. Of note, $\hat{\alpha}_{\mathrm{T}}$ and $\hat{d}_{\mathrm{T},i}$ denote the estimations of α_{T} and $d_{\mathrm{T},i}$, respectively. It is should be emphasized that $\hat{\alpha}_{\mathrm{T}}$ and $\hat{d}_{\mathrm{T},i}$ can be calculated by the external fusion algorithm, as presented in Sect. 2.3.2.

Before the weight design, we need to standardize the active measurement error $v_{\mathrm{a},i}$ and the passive measurement error $v_{\mathrm{p},i}$. Then, the standardizations of $v_{\mathrm{a},i}$ and $v_{\mathrm{p},i}$ are given as

$$\mathbf{v}_{\mathrm{a},i} = \frac{y_{\mathrm{a},i}(k)}{\sqrt{\widehat{\Phi}^2}} - \frac{f_{\mathrm{a},i}(k)}{\sqrt{\widehat{\Phi}^2 - \sigma_{\mathrm{a}}^2}}, \quad \mathbf{v}_{\mathrm{p},i} = \frac{y_{\mathrm{p},i}(k)}{\sqrt{\widehat{\bar{\Phi}}^2}} - \frac{f_{\mathrm{p},i}(k)}{\sqrt{\widehat{\bar{\Phi}}^2 - \sigma_{\mathrm{p}}^2}}. \tag{2.21}$$

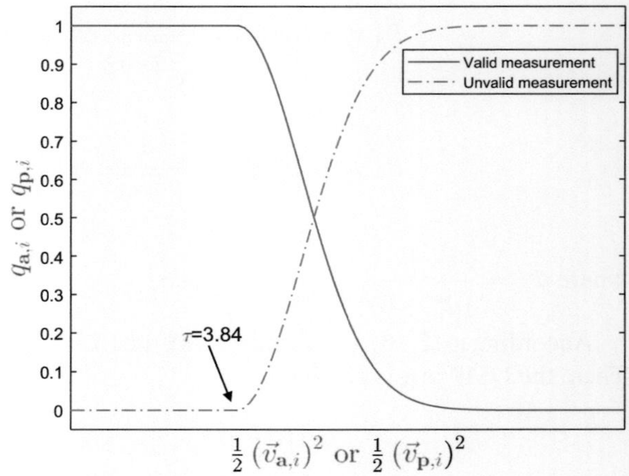

Fig. 2.3 Example of the membership functions, where the threshold τ is 3.84 by considering 95% confidence level from the χ^2 table

With (2.2), (2.5) and (2.21), one can obtain that $\mathbf{v}_{a,i}/\sqrt{2}$ and $\mathbf{v}_{p,i}/\sqrt{2}$ obey the standard normal distribution. Referring to Kay (1993), one knows the quadratic form of a standard normal random variable is distributed according to the chi-squared distribution. Based on this, one can have

$$\frac{1}{2}\left(\mathbf{v}_{a,i}\right)^2 \sim \chi_1^2,\ \frac{1}{2}\left(\mathbf{v}_{p,i}\right)^2 \sim \chi_1^2. \tag{2.22}$$

Considering a given confidence, we can get the validity domain. For example, for 95% confidence level, the threshold τ is selected as 3.84 from the χ^2 table, which means the active or passive measurement is valid if the value of $\frac{1}{2}\left(\mathbf{v}_{a,i}\right)^2$ or $\frac{1}{2}\left(\mathbf{v}_{p,i}\right)^2$ is smaller than 3.84. Define $q_{a,i}(k)$ and $q_{p,i}(k)$ as the validity probabilities of active and passive measurements, respectively. Then, the following membership functions are employed to evaluate the reliability (see Fig. 2.3), i.e.,

$$q_{a,i}(k) = \begin{cases} 1, & \text{if } \frac{1}{2}\left(\mathbf{v}_{a,i}\right)^2 \le \tau, \\ \exp\left[-\left(\frac{1}{2}\left(\mathbf{v}_{a,i}\right)^2 - \tau\right)^2\right], & \text{otherwise.} \end{cases} \tag{2.23}$$

$$q_{p,i}(k) = \begin{cases} 1, & \text{if } \frac{1}{2}\left(\mathbf{v}_{p,i}\right)^2 \le \tau, \\ \exp\left[-\left(\frac{1}{2}\left(\mathbf{v}_{p,i}\right)^2 - \tau\right)^2\right], & \text{otherwise.} \end{cases} \tag{2.24}$$

.

Based on (2.23) and (2.24), the weights of active local decision $d_{a,i}$ and passive local decision $d_{p,i}$ are designed as

Algorithm 1: Main procedure of the local fusion decision

Input: Initialize γ_{a} and γ_{p}
for $k = 1 : M$ **do**
    if $k = 1$ **then**
        Obtain active local decision $d_{\mathrm{a},i}(1)$ by (2.13);
        Obtain passive local decision $d_{\mathrm{p},i}(1)$ by (2.19);
        Set $\mu_{\mathrm{a},i}(1) = \mu_{\mathrm{p},i}(1) = 0.5$;
        Make an initial rough local decision through $d_i(1) = \mu_{\mathrm{a},i}(1)d_{\mathrm{a},i}(1) + \mu_{\mathrm{p},i}(1)d_{\mathrm{p},i}(1)$;
        Make an external decision fusion in Section 2.3.2, such that $\hat{\alpha}_{\mathrm{T}}(1)$ and $\hat{d}_{\mathrm{T},i}(1)$ are obtained;
    else
        Update active local decision $d_{\mathrm{a},i}(k)$ by (2.13);
        Update passive local decision $d_{\mathrm{p},i}(k)$ by (2.19);
        $\mu_{\mathrm{a},i}(k) \leftarrow \mu_{\mathrm{a},i}(k-1)$;
        $\mu_{\mathrm{p},i}(k) \leftarrow \mu_{\mathrm{p},i}(k-1)$;
        Acquire $q_{\mathrm{a},i}(k)$ and $q_{\mathrm{p},i}(k)$ by (2.23) and (2.24);
        Update $\mu_{\mathrm{a},i}(k)$ and $\mu_{\mathrm{p},i}(k)$ through (2.25);
        Update the local fusion decision through $d_i(k) = \mu_{\mathrm{a},i}(k)d_{\mathrm{a},i}(k) + \mu_{\mathrm{p},i}(k)d_{\mathrm{p},i}(k)$;

Output: Local fusion decision $d_i(k)$

$$\mu_{\mathrm{a},i}(k)=\frac{q_{\mathrm{a},i}(k)}{q_{\mathrm{a},i}(k)+q_{\mathrm{p},i}(k)},\ \mu_{\mathrm{p},i}(k)=\frac{q_{\mathrm{p},i}(k)}{q_{\mathrm{a},i}(k)+q_{\mathrm{p},i}(k)}, \tag{2.25}$$

and thereby, the internal fusion decision $d_i(k)$ at time step k can be provided as

$$d_i(k) = \mu_{\mathrm{a},i}(k)d_{\mathrm{a},i} + \mu_{\mathrm{p},i}(k)d_{\mathrm{p},i}, \tag{2.26}$$

where the values of $d_{\mathrm{a},i}$ and $d_{\mathrm{p},i}$ are obtained by Stage 1 and Stage 2, respectively. For clear description, the main procedure of the above stages is presented in Algorithm 1, where M denotes the total time steps of the monitoring network.

2.3.2 *Hybrid Bayesian Fusion Algorithm*

Through the above local decision rule, the local detection result d_i can be estimated by sensor node $i \in \{1, 2, \ldots, N\}$. Recall that the spatiotemporal detection ability and the installed battery power of a single sensor node are limited. In view of this, we develop a 2LMG structure-based external fusion algorithm to fuse the local detection decisions and reduce the communication energy consumption. To this end, these sensor nodes are categorized into $\mathbf{M}$ ordinary sensor nodes and M VFCs, where $\mathbf{M} + M = N$. Some existing underwater routing algorithms, e.g., the rigid-graph based routing (Yan et al. 2018) and Q-learning based routing (Shen et al. 2022), can be adopted to classify these nodes, and hence omitted here. Based on this, the topology relationship of ordinary sensor nodes and VFCs can be rearranged

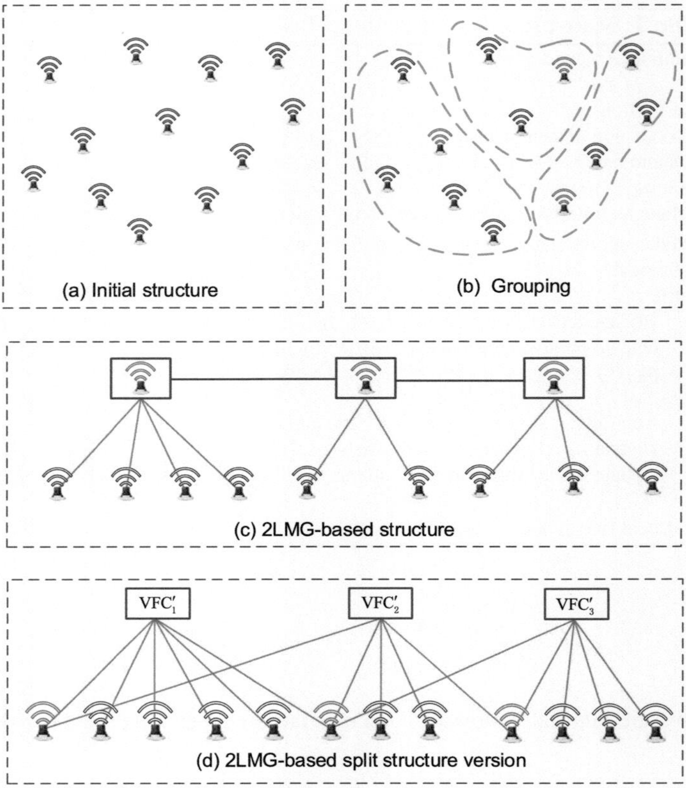

Fig. 2.4 Example of the generation of the 2LMG structure

as a 2LMG structure, as depicted by Fig. 2.4. In view of this, the following communication procedure is conducted by the 2LMG structure: (1) each node including ordinary sensor node and VFC sends its local detection result to the connected VFC; (2) each VFC updates its detection result based on the received results from the neighboring ordinary sensor nodes and VFCs; (3) each VFC sends its updated result to the neighboring ordinary sensor nodes; (4) each ordinary sensor node updates its local detection result.

Without loss of generality, we consider an arbitrary group G_j for $j \in \{1, \ldots, M\}$, which includes S_j ordinary sensor nodes labelled as $j_1, \ldots, j_{\mathrm{S}_j}$ and one VFC labelled as j_{VFC}. Of note, $\sum_{j=1}^{M} \mathrm{S}_j = \mathbf{M}$, and the neighboring VFC set for VFC j_{VFC} is denoted by $\mathcal{N}^{\#}_{j_{\text{VFC}}}$ according to the network topology. Along with this, the local detection results for ordinary sensor nodes $j_1, \ldots, j_{\mathrm{S}_j}$ are denoted by $d_{j_1}, \ldots, d_{j_{\mathrm{S}_j}}$. Meanwhile, the local detection result for VFC j_{VFC} is represented by $d_{j_{\text{VFC}}}$, and the neighboring local detection results for $\mathcal{N}^{\#}_{j_{\text{VFC}}}$ are collected in a set $D_{\mathcal{N}^{\#}_{j_{\text{VFC}}}}$. For ease of calculation, we rearrange the local decision $\mathbf{d}_j = \left\{d_{j_1}, ..., d_{j_{S_j}}, d_{j_{\rm VFC}}, D_{\mathcal{N}^{\#}_{j_{\rm VFC}}}\right\}$ of VFC $j_{\rm VFC}$ as $\mathbf{d}_j = \left\{d_{j,1^{\#}},\ d_{j,2^{\#}},\ \ldots,\ d_{j,n^{\#}}\right\}$, where $\left|D_{\mathcal{N}^{\#}_{j_{\rm VFC}}}\right|$ denotes the number of elements in $D_{\mathcal{N}^{\#}_{j_{\rm VFC}}}$, and $n^{\#} = {\rm S}_j + \left|D_{\mathcal{N}^{\#}_{j_{\rm VFC}}}\right| + 1$.

In the following, the VFC $j_{\rm VFC}$ requires to fuse the detection result $\mathbf{d}_j$ such that a more accurate result can be obtained. It is worth noting that the target position $\mathbf{X}_{\rm t}$ and target power $\sigma_{\rm T}^2$ are unknown to VFC $j_{\rm VFC}$, and hence, the test approach used in Sect. 2.3.1 cannot be directly utilized here. For that reason, we attempt to employ the Hybrid Bayesian approach to fuse the local detection result, where the prior distribution and the Maximum Likelihood Estimate (MLE) are jointly utilized to estimate the values of unknown parameters. To this end, we first calculate the *probability mass function* (pmf) of $\mathbf{d}_j$ under $\mathcal{H}_1$ hypothesis. Since the elements in $\mathbf{d}_j$ are independent of each other, then the pmf of $\mathbf{d}_j$ under hypothesis $\mathcal{H}_1$ and $\mathcal{H}_0$ can be expressed by independent Bernoulli probability mass functions $P\left(\mathbf{d}_j|\mathcal{H}_1\right)$ and $P\left(\mathbf{d}_j|\mathcal{H}_0\right)$, i.e.,

$$
\begin{aligned}
P\left(\mathbf{d}_j\middle|\mathcal{H}_1\right) &= \prod_{i=1}^{n}\left(P_{{\rm d},i^{\#}}\right)^{d_{j,i^{\#}}}\left(1-P_{{\rm d},i^{\#}}\right)^{1-d_{j,i^{\#}}},\\
P\left(\mathbf{d}_j\middle|\mathcal{H}_0\right) &= \prod_{i=1}^{n}\left(P_{{\rm f},i^{\#}}\right)^{d_{j,i^{\#}}}\left(1-P_{{\rm f},i^{\#}}\right)^{1-d_{j,i^{\#}}}.
\end{aligned}
\tag{2.27}
$$

with

$$
\begin{aligned}
P_{{\rm d},i^{\#}} &= 2\mu_{{\rm a},i^{\#}}\,\mathcal{Q}\left(\sqrt{\frac{\gamma_{\rm a}}{(\Phi^{\#})^2}}\right) + 2\mu_{{\rm p},i^{\#}}\,\mathcal{Q}\left(\sqrt{\frac{\gamma_{\rm p}}{(\bar{\Phi}^{\#})^2}}\right),\\
P_{{\rm f},i^{\#}} &= \mu_{{\rm a},i^{\#}}\,P_{\rm f,a} + \mu_{{\rm p},i^{\#}}\,P_{\rm f,p}.
\end{aligned}
\tag{2.28}
$$

Referring to the standard Hybrid Bayesian approach Ciuonzo and Rossi (2017), the Hybrid Bayesian statistic $\Lambda_{{\rm HB},j_{\rm VFC}}\left(\mathbf{d}_j\right)$ is denoted as

$$
\begin{aligned}
\Lambda_{{\rm HB},\,j_{\rm VFC}} &\triangleq \ln\left[\frac{\max\limits_{\sigma_{\rm T}^2}\int P\left(\mathbf{d}_j\middle|\mathcal{H}_1;\ \mathbf{X}_{\rm t},\sigma_{\rm T}^2\right)p(\mathbf{X}_{\rm t})\,{\rm d}\mathbf{X}_{\rm t}}{P\left(\mathbf{d}_j\middle|\mathcal{H}_0\right)}\right]\\
&= \ln\max_{\sigma_{\rm T}^2}\left\{\int\prod_{i=1}^{n}\left(\frac{P_{{\rm d},i^{\#}}\left(\mathbf{X}_{\rm t},\sigma_{\rm T}^2\right)}{P_{{\rm f},i^{\#}}}\right)^{d_{j,i^{\#}}}\right.\\
&\quad\left.\times\left(\frac{1-P_{{\rm d},i^{\#}}\left(\mathbf{X}_{\rm t},\sigma_{\rm T}^2\right)}{1-P_{{\rm f},i^{\#}}}\right)^{1-d_{j,i^{\#}}}p(\mathbf{X}_{\rm t})\,{\rm d}\mathbf{X}_{\rm t}\right\}.
\end{aligned}
\tag{2.29}
$$

where $p\left(\mathbf{X}_{\rm t}\right)$ is the given prior probability density function.

Algorithm 2: Main procedure of external fusion decision

Input: Initialize d_i for each sensor nodes

Sensor nodes are rearranged into the 2LMG structure, including M groups with **M** VFCs;

for $j = 1 : M$ **do**

  VFC j_{VFC} acquires the local decision set $\mathbf{d}_j = \left\{ d_{j1}, ..., d_{jS_j}, d_{j\text{VFC}}, D_{\mathcal{N}^{\#}_{j\text{VFC}}} \right\}$;

  VFC j_{VFC} obtains the pmf of $\mathbf{d}_j$ by (2.27);

  VFC j_{VFC} makes the Hybrid Bayesian decision test by comparing $\Lambda_{\text{HB},j_{\text{VFC}}} (\mathbf{d}_j)$ and the threshold $\tilde{\gamma}_{j\text{VFC}}$;

  VFC j_{VFC} gets external fusion decision $d_{\text{v},j}$ from (2.30);

  for $l = 1 : S_j$ **do**

    VFC j_{VFC} sends $d_{\text{v},j}$ to ordinary sensor nodes j_l;

    Ordinary sensor node j_l updates d_{j_l} by (2.31);

Output: $d_{\text{v},j}$ and d_{j_l}

Based on this, the following Hybrid Bayesian decision test is conducted by the group G_j for $j \in \{1, \ldots, M\}$

$$\begin{cases} \Lambda_{\text{HB},j_{\text{VFC}}} (\mathbf{d}_j) < \tilde{\gamma}_{j\text{VFC}}, & \text{under } \mathcal{H}_0 \\ \Lambda_{\text{HB},j_{\text{VFC}}} (\mathbf{d}_j) > \tilde{\gamma}_{j\text{VFC}}, & \text{under } \mathcal{H}_1 \end{cases} \tag{2.30}$$

where $\tilde{\gamma}_{j\text{VFC}}$ denotes the threshold, aiming to ensure that the group false alarm rate by Monte Carlo simulation.

From (2.30), the fused detection result of VFC j_{VFC} for $j \in \{1, \ldots, M\}$ can be expressed as follows: (1) when $\Lambda_{\text{HB},j_{\text{VFC}}} < \tilde{\gamma}_{j\text{VFC}}$ is satisfied, the target is absent, i.e., $d_{\text{v},j} = 0$; (2) when $\Lambda_{\text{HB},j_{\text{VFC}}} > \tilde{\gamma}_{j\text{VFC}}$ is satisfied, the target is present, i.e., $d_{\text{v},j} = 1$.

Next, VFC j_{VFC} for $j \in \{1, \ldots, M\}$ sends its updated result to the ordinary sensor nodes $j_1, ..., j_{S_j}$, as depicted by Fig. 2.4(c). With regards to this, ordinary sensor node $i \in \{1, \ldots, \text{S}_j\}$ adopts the average consensus strategy to updates its local detection result, i.e.,

$$d_{j_i}(k) = \frac{1}{2} \left(d_{\text{v},j}(k) + d_{j_i}(k) \right). \tag{2.31}$$

Correspondingly, the main procedure of the above external fusion decision is shown in Algorithm 2.

2.3.3 Detection Performance Analysis

Detection and False Alarm Probabilities

The detection and false alarm probabilities are the key indexes to evaluate the detection performance. In view of this, we first derive the detection and false alarm probabilities for the Hybrid Bayesian fusion rule (2.29).

Theorem 2.1 *Consider the 2LMG structure based Hybrid Bayesian fusion rule (2.29), if the data record is sufficiently large and the MLE can reach probability density function (PDF), the detection and false alarm probabilities are calculated as*

$$\begin{aligned} P_{D,j} &= \Pr\Big(T_{LMP} > \gamma' \,\Big|\, \mathcal{H}_1\Big) = \mathcal{Q}\Big(\sqrt{I(\sigma_0^2)}\,(\sigma_T^2 - \sigma_0^2)\Big), \\ P_{F,j} &= \Pr\Big(T_{LMP} > \gamma' \,\Big|\, \mathcal{H}_0\Big) = \mathcal{Q}(\gamma'), \end{aligned} \tag{2.32}$$

with

$$T_{LMP} = \left.\frac{\partial \ln\left[P(\mathbf{d}_j|\mathcal{H}_1, \sigma_T^2)\right]}{\partial \sigma_T^2}\right|_{\sigma_T^2=\sigma_0^2} \Big/ \sqrt{I\left(\sigma_0^2\right)}, \tag{2.33}$$

where T_{LMP} is the Local maximum power (LMP) test statistic. In addition, $P_{D,j}$ and $P_{F,j}$ are the detection and false alarm probabilities for the jth sensor node group, respectively. $I\left(\sigma_0^2\right)$ represents the Fisher information. σ_T^2 and σ_0^2 are the actual values of target power and the target power under hypothesis $\mathcal{H}_0$, respectively. Besides that, γ' is the global threshold that ensures the given false-alarm probability.

Proof The main procedure of the proof is summarized as follows: Calculate the test statistic T_{LMP} by (2.27) and (2.29), and hence, one obtains the probability distribution of Hybrid Bayesian rule in $\mathcal{H}_1$ and $\mathcal{H}_0$, such that $P_{\mathrm{D},j}$ and $P_{\mathrm{F},j}$ can be acquired. We first calculate the probability mass function (pmf) of $\mathbf{d}_j$ under $\mathcal{H}_1$ hypothesis by (2.27), i.e.,

$$\begin{aligned} &\ln\left[P\left(\mathbf{d}_j|\mathcal{H}_1, \sigma_{\mathrm{T}}^2\right)\right] \\ &= \ln\left[\int \prod_{i=1}^{n} \left(P_{\mathrm{d},i^\#}\right)^{d_{j,i^\#}} \left(1 - P_{\mathrm{d},i^\#}\right)^{\left(1-d_{j,i^\#}\right)} p\left(\mathbf{X}_{\mathrm{t}}\right) d\mathbf{X}_{\mathrm{t}}\right], \end{aligned} \tag{2.34}$$

with

$$P_{\mathrm{d},i^\#} = 2\mu_{\mathrm{a},i^\#}\mathcal{Q}\left(\sqrt{\frac{\gamma_{\mathrm{a}}}{\left(\Phi^\#\right)^2}}\right) + 2\mu_{\mathrm{p},i^\#}\mathcal{Q}\left(\sqrt{\frac{\gamma_{\mathrm{p}}}{\left(\bar{\Phi}^\#\right)^2}}\right). \tag{2.35}$$

With (2.34) and (2.35), one can have

$$\frac{\partial \ln\left[P\left(\mathbf{d}_j|\mathcal{H}_1,\sigma_{\mathrm{T}}^2\right)\right]}{\partial\sigma_{\mathrm{T}}^2}$$
$$=\frac{\int \frac{\partial P\left(\mathbf{d}_j|\mathcal{H}_1,\mathbf{X}_{\mathrm{t}},\sigma_{\mathrm{T}}^2\right)}{\partial\sigma_{\mathrm{T}}^2} p\left(\mathbf{X}_{\mathrm{t}}\right) d\mathbf{X}_{\mathrm{t}}}{\int \prod_{i=1}^{n}\left(P_{\mathrm{d},i^{\#}}\right)^{d_{j,i^{\#}}}\left(1-P_{\mathrm{d},i^{\#}}\right)^{\left(1-d_{j,i^{\#}}\right)} p\left(\mathbf{X}_{\mathrm{t}}\right) d\mathbf{X}_{\mathrm{t}}}. \tag{2.36}$$

Note that the logarithmic form of derivative at the integral inner numerator is employed to calculate (2.36). With respect to this, the derivative can be detailed as

$$\begin{aligned}
&\ln P\left(\mathbf{d}_j|\mathcal{H}_1,\mathbf{X}_{\mathrm{t}},\sigma_{\mathrm{T}}^2\right)\\
&=\ln\left[\prod_{i=1}^{n}\left(P_{\mathrm{d},i^{\#}}\right)^{d_{j,i^{\#}}}\left(1-P_{\mathrm{d},i^{\#}}\right)^{\left(1-d_{j,i^{\#}}\right)}\right]\\
&=\sum_{i=1}^{n}\left(d_{j,i^{\#}}\ln P_{\mathrm{d},i^{\#}}+(1-d_{j,i^{\#}})\ln(1-P_{\mathrm{d},i^{\#}})\right).
\end{aligned} \tag{2.37}$$

From (2.37), we can calculate that

$$\begin{aligned}
&\frac{\partial \ln P\left(\mathbf{d}_j|\mathcal{H}_1,\mathbf{X}_{\mathrm{t}},\sigma_{\mathrm{T}}^2\right)}{\partial\sigma_{\mathrm{T}}^2}\\
&=\frac{\frac{\partial P\left(\mathbf{d}_j|\mathcal{H}_1,\mathbf{X}_{\mathrm{t}},\sigma_{\mathrm{T}}^2\right)}{\partial\sigma_{\mathrm{T}}^2}}{\prod_{i=1}^{n}\left(P_{\mathrm{d},i^{\#}}\right)^{d_{j,i^{\#}}}\left(1-P_{\mathrm{d},i^{\#}}\right)^{\left(1-d_{j,i^{\#}}\right)}}\\
&=\sum\nolimits_{i=1}^{n}\frac{d_{j,i^{\#}}-P_{\mathrm{d},i^{\#}}}{P_{\mathrm{d},i^{\#}}\left(1-P_{\mathrm{d},i^{\#}}\right)}\frac{\partial P_{\mathrm{d},i^{\#}}}{\partial\sigma_{\mathrm{T}}^2}.
\end{aligned} \tag{2.38}$$

Based on (2.37) and (2.38), one can further have

$$\begin{aligned}
&\frac{\partial P\left(\mathbf{d}_j|\mathcal{H}_1,\mathbf{X}_{\mathrm{t}},\sigma_{\mathrm{T}}^2\right)}{\partial\sigma_{\mathrm{T}}^2}\\
&=\left(\prod\nolimits_{i=1}^{n}\left(P_{\mathrm{d},i^{\#}}\right)^{d_{j,i^{\#}}}\left(1-P_{\mathrm{d},i^{\#}}\right)^{\left(1-d_{j,i^{\#}}\right)}\right)\\
&\quad\times\sum\nolimits_{i=1}^{n}\frac{d_{j,i^{\#}}-P_{\mathrm{d},i^{\#}}}{P_{\mathrm{d},i^{\#}}\left(1-P_{\mathrm{d},i^{\#}}\right)}\frac{\partial P_{\mathrm{d},i^{\#}}}{\partial\sigma_{\mathrm{T}}^2}.
\end{aligned} \tag{2.39}$$

Hence, the derivative in the form at $\sigma_{\mathrm{T}}^2=\sigma_0^2$ is denoted as

$$\frac{\partial \ln\left[P\left(\mathbf{d}_j|\mathcal{H}_1,\sigma_{\mathrm{T}}^2\right)\right]}{\partial \sigma_{\mathrm{T}}^2}\Big|_{\sigma_{\mathrm{T}}^2=\sigma_0^2}$$
$$= \frac{\int \Gamma p(\mathbf{X}_{\mathrm{t}}) d\mathbf{X}_{\mathrm{t}}}{\int \prod_{i=1}^{n} (P_{\mathrm{d},i^\#})^{d_{j,i^\#}} (1-P_{\mathrm{d},i^\#})^{\left(1-d_{j,i^\#}\right)} p(\mathbf{X}_{\mathrm{t}}) d\mathbf{X}_{\mathrm{t}}}\Big|_{\sigma_{\mathrm{T}}^2=\sigma_0^2},$$

with

$$\frac{\partial P_{\mathrm{d},i^\#}}{\partial \sigma_{\mathrm{T}}^2}$$
$$= \frac{\mu_{\mathrm{a},i^\#}}{\sqrt{2\pi}} \exp\left(-\frac{\gamma_{\mathrm{a}}}{2(\Phi^\#)^2}\right) \frac{\sqrt{\gamma_{\mathrm{a}}}(\Phi^\# - \sigma_{\mathrm{a}}^2)}{(\sigma_{\mathrm{S}}^2+\sigma_{\mathrm{T}}^2)(\Phi^\#)^3}$$
$$+ \frac{\mu_{\mathrm{p},i^\#}}{\sqrt{2\pi}} \exp\left(-\frac{\gamma_{\mathrm{p}}}{2(\bar{\Phi}^\#)^2}\right) \frac{\sqrt{\gamma_{\mathrm{p}}}(\bar{\Phi}^\# - \sigma_{\mathrm{a}}^2)}{\sigma_{\mathrm{T}}^2(\bar{\Phi}^\#)^3}, \tag{2.40}$$
$$\Gamma$$
$$= \left(\prod_{i=1}^{n} (P_{\mathrm{d},i^\#})^{d_{j,i^\#}} (1-P_{\mathrm{d},i^\#})^{\left(1-d_{j,i^\#}\right)}\right)$$
$$\times \sum_{i=1}^{n} \frac{d_{j,i^\#} - P_{\mathrm{d},i^\#}}{P_{\mathrm{d},i^\#}(1-P_{\mathrm{d},i^\#})} \frac{\partial P_{\mathrm{d},i^\#}}{\partial \sigma_{\mathrm{T}}^2}. \tag{2.41}$$

Under the condition that MLE reaches PDF, one refers to Kay (1993), and hence, the Fisher information is defined as

$$I\left(\sigma_0^2\right)$$
$$= E\left[\left(\frac{\partial \ln P\left(\mathbf{d}_j|\mathcal{H}_1,\sigma_{\mathrm{T}}^2\right)}{\partial \sigma_{\mathrm{T}}^2}\Big|_{\sigma_{\mathrm{T}}^2=\sigma_0^2}\right)^2\right]$$
$$= \frac{\Psi^2+\bar{\Psi}^2}{\left(\int (P_{\mathrm{d},i^\#})^n p(\mathbf{X}_{\mathrm{t}}) d\mathbf{X}_{\mathrm{t}} \int (1-P_{\mathrm{d},i^\#})^n p(\mathbf{X}_{\mathrm{t}}) d\mathbf{X}_{\mathrm{t}}\right)^2}\Big|_{\sigma_{\mathrm{T}}^2=\sigma_0^2}, \tag{2.42}$$

with

$$\Psi = \int \left(P_{\mathrm{d},i^\#}\right)^n \sum_{i=1}^{n} \frac{\partial P_{\mathrm{d},i^\#}}{\partial \sigma_{\mathrm{T}}^2} p(\mathbf{X}_{\mathrm{t}})\, \mathrm{d}\mathbf{X}_{\mathrm{t}} \int \left(1 - P_{\mathrm{d},i^\#}\right)^n p(\mathbf{X}_{\mathrm{t}})\, \mathrm{d}\mathbf{X}_{\mathrm{t}}. \tag{2.43}$$

$$\bar{\Psi} = \int \left(P_{\mathrm{d},i^\#}\right)^n p(\mathbf{X}_{\mathrm{t}})\, \mathrm{d}\mathbf{X}_{\mathrm{t}} \int \left(1 - P_{\mathrm{d},i^\#}\right)^n \sum_{i=1}^{n} \frac{\partial P_{\mathrm{d},i^\#}}{\partial \sigma_{\mathrm{T}}^2} p(\mathbf{X}_{\mathrm{t}})\, \mathrm{d}\mathbf{X}_{\mathrm{t}}. \tag{2.44}$$

Based on the above deduction, the T_{LMP} can be denoted as

$$
\begin{aligned}
&T_{\mathrm{LMP}} \\
&= \frac{\int \Gamma p(\mathbf{X}_{\mathrm{t}}) d\mathbf{X}_{\mathrm{t}}}{\frac{\int \prod_{i=1}^{n} \left(P_{\mathrm{d},i\#}\right)^{d_{j,i\#}} \left(1-P_{\mathrm{d},i\#}\right)^{\left(1-d_{j,i\#}\right)} p(\mathbf{X}_{\mathrm{t}}) d\mathbf{X}_{\mathrm{t}}}{\sqrt{\frac{\Psi^2+\bar{\Psi}^2}{\left(\int \left(P_{\mathrm{d},i\#}\right)^{n} p(\mathbf{X}_{\mathrm{t}}) d\mathbf{X}_{\mathrm{t}} \int \left(1-P_{\mathrm{d},i\#}\right)^{n} p(\mathbf{X}_{\mathrm{t}}) d\mathbf{X}_{\mathrm{t}}\right)^2}}}} \Bigg|_{\sigma_{\mathrm{T}}^2=\sigma_0^2} .
\end{aligned} \tag{2.45}
$$

Referring to the Kay (1993), we have the probability distribution T_{LMP} of Hybrid Bayesian rule in $\mathcal{H}_1$ and $\mathcal{H}_0$ by (2.45) if the data record is sufficiently large. Accordingly, one obtains

$$
T_{\mathrm{LMP}} \overset{a}{\sim} \begin{cases} \mathcal{N}(0, 1) & \text{under } \mathcal{H}_0, \\ \mathcal{N}\left(\sqrt{I\left(\sigma_0^2\right)}\left(\sigma_{\mathrm{T}}^2 - \sigma_0^2\right), 1\right) & \text{under } \mathcal{H}_1. \end{cases} \tag{2.46}
$$

In view of (2.46), the PDFs under $\mathcal{H}_1$ and $\mathcal{H}_0$ hypotheses are

$$
\begin{aligned}
p\left(T_{\mathrm{LMP}} \middle| \mathcal{H}_1\right) &= \frac{1}{\sqrt{2\pi}} \exp\left(-\frac{\left(x - \sqrt{I(\sigma_0^2)}\,(\sigma_{\mathrm{T}}^2 - \sigma_0^2)\right)^2}{2}\right), \\
p\left(T_{\mathrm{LMP}} \middle| \mathcal{H}_0\right) &= \frac{1}{\sqrt{2\pi}} \exp\left(-\frac{x^2}{2}\right).
\end{aligned} \tag{2.47}
$$

Based on (2.47), we can get the complementary cumulative distribution function (CCDF) as

$$
\begin{aligned}
P\left(T_{\mathrm{LMP}} > x \,\middle|\, \mathcal{H}_1\right) &= \int_x^{\infty} \frac{1}{\sqrt{2\pi}} \exp\left(-\frac{\left(t - \sqrt{I(\sigma_0^2)}\,(\sigma_{\mathrm{T}}^2 - \sigma_0^2)\right)^2}{2}\right) \mathrm{d}t, \\
P\left(T_{\mathrm{LMP}} > x \,\middle|\, \mathcal{H}_0\right) &= \int_x^{\infty} \frac{1}{\sqrt{2\pi}} \exp\left(-\frac{t^2}{2}\right) \mathrm{d}t.
\end{aligned} \tag{2.48}
$$

Giving a pre-known threshold γ', we can obtain the detection rate $P_{\mathrm{D},j}$ and false alarm rate $P_{\mathrm{F},j}$ by substituting γ' into (2.48), as presented in (2.32). That completes the proof. □

In the following, we analyze the advantages of the 2LMG structure over traditional egalitarian distributed structure. We adopt the number of transmission and communication load to analyze communication performance. To this end, we consider the 2LMG structure in which each group has S_j ordinary sensor nodes. For the convenience of calculation, we separate the functions of the VFC j_{VFC} into a virtual fusion center VFC'_j and an ordinary sensor node $S_{j_{VFC}}$, as depicted by Fig. 2.4c. Meanwhile, we use $N(\cdot)$ to express the sum of transmissions. Based on this, the following theorem is presented

Convergence and Communication Overhead

Theorem 2.2 *Compared with the traditional egalitarian structure G_0, the reduced number of communication transmissions in the 2LMG structure G can be expressed as*

$$N(G_0) - N(G) = S_{M-1} + S_2 + \sum_{j=2}^{M-2}(S_{j-1} + S_{j+1} - 1) + \sum_{j=1}^{M}(S_j^2 - S_j - 1). \tag{2.49}$$

Proof In the G structure, sensor nodes S_1 and $S_{1_{VFC}}$ send one-bit result to VFC'_1. Accordingly, the number of communication transmissions is $S_1 + 1$. Meanwhile, S_M sensor nodes in the group G_M communicate with VFC'_M, whose communication number is $S_M + 1$. Beside that, the sensor nodes in the rest groups communicate with VFC'_j and the neighboring sensor nodes, and hence, the communication number is $\sum_{j=1}^{M-2}(S_j + 3)$. On the other hand, the VFC'_j communicates with $S_j + 1$ ordinary sensor nodes, i.e., the communication number $\sum_{j=1}^{M}(S_j + 1)$ is required. Hence, the number of transmission can be obtained as

$$N(G) = S_1 + S_M + 2 + \sum_{j=1}^{M-2}(S_j + 3) + \sum_{j=1}^{M}(S_j + 1). \tag{2.50}$$

In the G_0 structure, each of the $S_{M_{VFC}}$ sensor nodes via the VFCs communicates with the connected sensor nodes. Noted that sensor node $S_{1_{VFC}}$ and $S_{M_{VFC}}$ have a neighboring VFC. Sensor node $S_{1_{VFC}}$ communicates with the connected S_1+S_2+1 sensor nodes via VFCs. Sensor node $S_{M_{VFC}}$ connects to the $S_M+S_{M-1}+1$ sensor nodes through the VFCs. Each of the $S_{M_{VFC}} - 2$ sensor nodes via VFCs communicates with the connected $S_j+S_{j-1}+S_{j+1}+2$ sensor nodes, the total number of communication transmissions is $\sum_{j=2}^{M-2}(S_j + S_{j-1} + S_{j+1} + 2)$. Each of the S_j sensor nodes in each of the M groups communicates with the remaining sensor nodes in its group, a total of $\sum_{j=1}^{M} S_j^2$ transmissions are obtained. Then, the number of transmission is

$$\begin{aligned} N(G_0) &= (S_1 + S_2 + 1) + (S_M + S_{M-1} + 1) \\ &+ \sum_{j=2}^{M-2}(S_j + S_{j-1} + S_{j+1} + 2) + \sum_{j=1}^{M} S_j^2. \end{aligned} \tag{2.51}$$

From (2.50) and (2.51), we can get the reduced number of communication transmissions by using 2LMG structure, i.e., $N(\mathrm{G}_0) - N(\mathrm{G}) = \mathrm{S}_{M-1} + \mathrm{S}_2 + \sum_{j=2}^{M-2} \left(\mathrm{S}_{j-1} + \mathrm{S}_{j+1} - 1\right) + \sum_{j=1}^{M} \left(\mathrm{S}_j^2 - \mathrm{S}_j - 1\right)$ as presented in (2.49). □

2.4 Technology Practice

In this section, simulation results are conducted on MATLAB 2018a. With regard to the capabilities of the modems and the communication channels, it is assumed that each node can always efficiently communicate with the other sensor nodes. Meanwhile, the sensor nodes are randomly distributed over a region of $100 \times 100 \times 100\ \mathrm{m}^3$ and there are 20 sensor nodes, as shown in Fig. 2.5. Particularly, the position of each sensor node can be accurately obtained through the anchor nodes. The sensor network detects the target area at all times. When the received signal fluctuates weakly in the detection area, the sensor nodes are scheduled to detect the target.

In addition, we define the $\mathrm{SNR} = 10\lg \frac{\sigma_\mathrm{T}^2}{\sigma_\mathrm{w}^2}$. The hybrid Bayesian statistic Λ_{HB} is implemented via grids for σ_T^2 and $\mathbf{X}_\mathrm{t}$ in VFC. To be specific, the target can be regarded as existence, and the target power σ_T^2 interval is $[0.9 * \sigma_\mathrm{T}^2,\ 1.1 * \sigma_\mathrm{T}^2]$, and this grid of the target power σ_T^2 is given by sampling $N_{\sigma_\mathrm{T}^2} = 10$. The target position $\mathbf{X}_\mathrm{t}$ space is assumed to be the same as the detection area of the target, where the three-dimensional grid is obtained by sampling $N_{\mathbf{X}_\mathrm{t}} = 10^6$. Accordingly, the corresponding simulation results are obtained by 10^7 Monte Carlo simulation times, and some parameters are set as follows in Table 2.1.

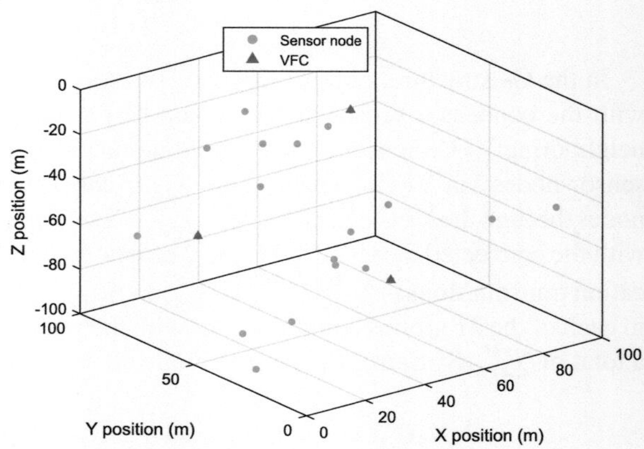

Fig. 2.5 Deployment of sensor nodes

2.4.1 Case Study I: Detection with Local Measurement

We first investigate the local detection performance, as presented in Sect. 2.3.1. The detection performance is evaluated by the false alarm probability P_{f} and the detection probability P_{d}. Typically, the receiver operating characteristic (ROC), which is defined by P_{f} versus P_{d}, can be employed to check the quality of detector. Based on this, we adopt sensor node 1 to detect the existence of target, and hence, the active measurement noise σ_{a}^2 is set as $\sigma_{\mathrm{a}}^2 = 1$ while the active transmit power is $\sigma_{\mathrm{s}}^2 = 700$. Through active measurement-based local decision (e.g., Tian et al. 2022; Hu et al. 2021; Jiang et al. 2018), the variation of active local detection probability with distance $d_{\mathrm{T},1}$ can be shown in Fig. 2.6. From Fig. 2.6, the active detection has a high detection probability when the false alarm probability $P_{\mathrm{f,a}}$ is low. Without loss of generality, the depth of target is regarded to be fixed for clear depiction. Similarly, the passive measurement noise σ_{p}^2 is set as $\sigma_{\mathrm{p}}^2 = 1$. Through passive measurement-based local decision (e.g., Braca et al. 2016; Liang and Cheng 2008; Gong et al. 2020), the variation of passive local detection probability with distance $d_{\mathrm{T},1}$ is shown in Fig. 2.7. From Fig. 2.7, we can find that the local passive detection probability is low when the false alarm probability is fixed. Besides that, the passive detection probability has lower detection probability as compared with the one with active detection mode.

Note that one cannot know the detection quality in advance, i.e., we are not sure which mode is the good. Therefore, it is necessary to fuse the local active and passive detection results, as presented in Algorithm 1. In view of this, the variation of active and passive local fusion detection probability $P_{\mathrm{d},1}$ with distance $d_{\mathrm{T},1}$ is shown in Fig. 2.8, where the active weight $\mu_{\mathrm{a},1^{\#}}$ is 0.9731 and the passive weight $\mu_{\mathrm{a},1^{\#}}$ is 0.0269. Clearly, the fused detection probability is close to the best case (i.e., $P_{\mathrm{d},1} = 0.9345$), even when the detection quality for active or passive detection is not pre-known. Meanwhile, the ROC curves for the different cases are shown in Fig. 2.9. It is clear that, the fused local detection has a high detection probability under the same false alarm probability $P_{\mathrm{f,a}}$, since the reliable measurements are maximized and the unreliable measurements are minimized. The above results indicate that the necessary and meaningful of local fusion measurements.

Table 2.1 Some parameters used for the simulation

Parameter	Value	Specification
n	0.4	Distance attenuation coefficient
σ_{T}^2	20	Target power
f	10	Frequency
b	1	Bit length of data packet
$P_{\mathrm{f,a}}$	0.05	Local active false alarm probability
$P_{\mathrm{f,p}}$	0.05	Local passive false alarm probability
E_{elec}	0.5	Unit energy consumption
T_{b}	1000	Bit duration

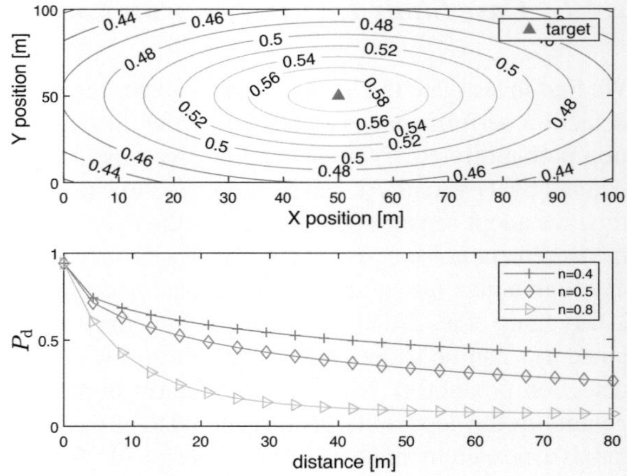

Fig. 2.6 Active detection probability field

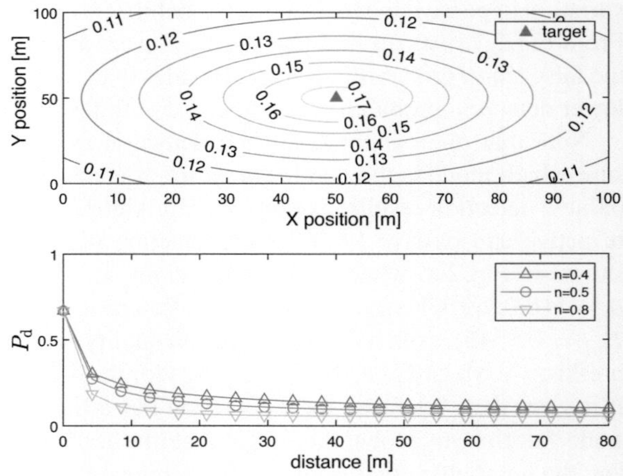

Fig. 2.7 Passive detection probability field

In the following, we consider a special scenario, i.e., the active detection is affected by underwater noise. Particularly, the active measurement noise σ_{a}^2 is set from $\sigma_{\mathrm{a}}^2 = 1$ to $\sigma_{\mathrm{a}}^2 = 5$. Then, the variation of active local detection probability with distance $d_{\mathrm{T},1}$ is shown in Fig. 2.10. Correspondingly, the variation of active and passive fusion local detection probability with distance $d_{\mathrm{T},1}$ is shown in Fig. 2.11, where the active weight $\mu_{\mathrm{a},1^{\#}}$ is 0.4729 and the passive weight $\mu_{\mathrm{p},1^{\#}}$ is 0.5271. From Figs. 2.10 and 2.11, one knows the local detection probability decreases in the fixed false alarm probability. Clearly, the fusion result cannot reach the requirement of accurate detection when the underwater noises are high. The above results reflect that the external fusion measurement is required in underwater environment, especially when the active or passive measurements are affected with external disturbance.

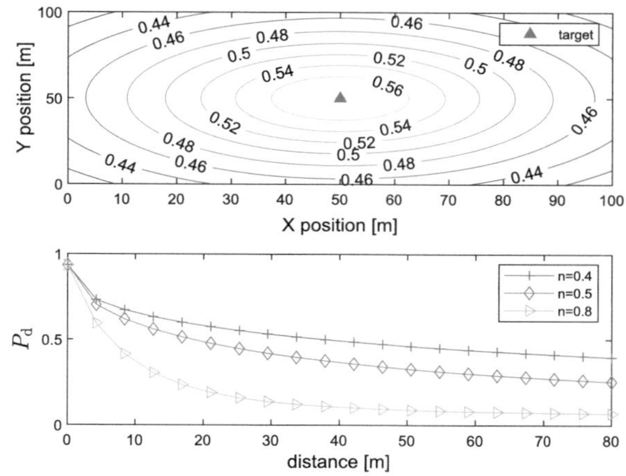

Fig. 2.8 Active-passive fusion detection probability field

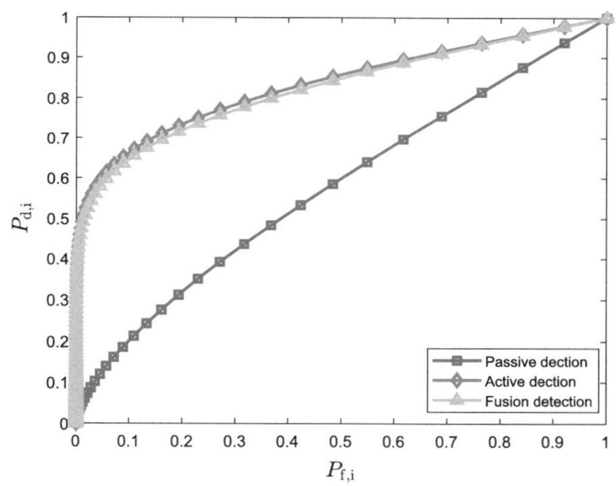

Fig. 2.9 The ROC curves in different modes

2.4.2 *Case Study II: Detection with External Fusion Measurement*

As mentioned above, the external fusion detection is required in underwater environment. In view of this, the active-passive detection result for each sensor node is fused in 2LMG structure. It is noted that the target is with unknown position $\mathbf{X}_t$ and power σ_{T}^2 in MUSs based on the 2LMG structure. For comparison, we also apply the measurement fusion detection in the egalitarian structure, e.g., the structure in Wan et al. (2013).

To be particular, we consider a scenario, i.e., target detection in the 2LMG structure as shown in Fig. 2.12. The 2LMG structure has three VFCs and three groups, where the positions of VFCs are [22.9, 78.02, −63.06], [59.86, 58.91, −11.17] and

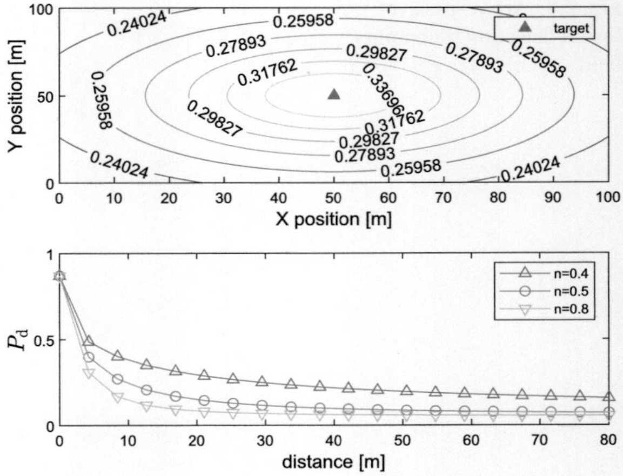

Fig. 2.10 Active detection probability field with disturbance

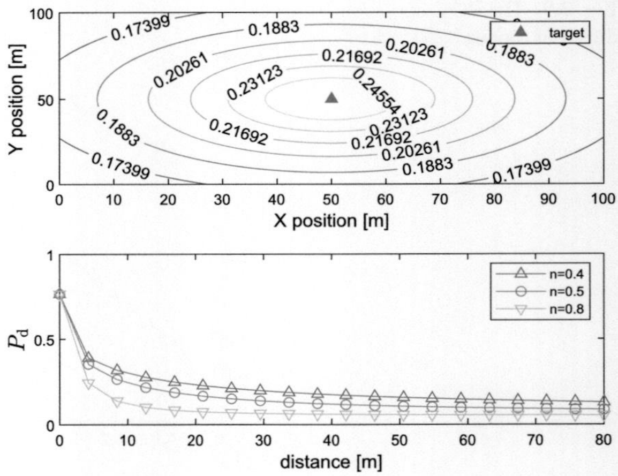

Fig. 2.11 Fusion detection probability field

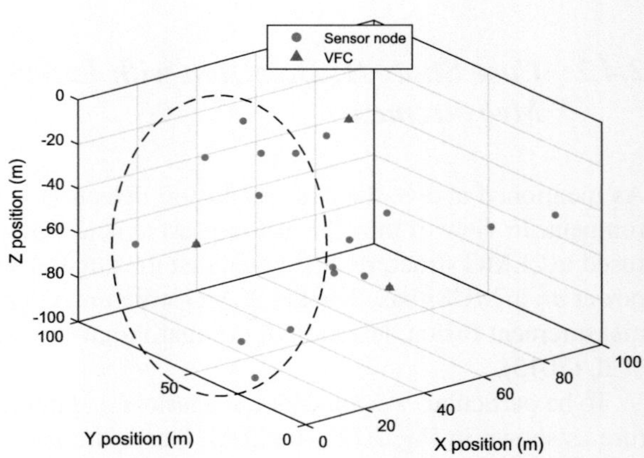

Fig. 2.12 Deployment of sensor nodes

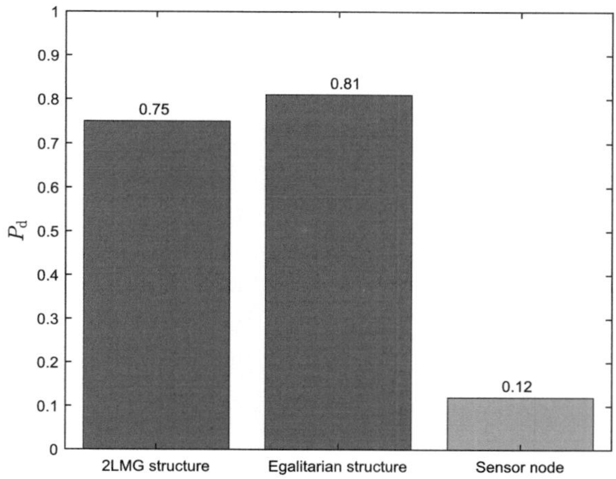

Fig. 2.13 Detection probability in different structure

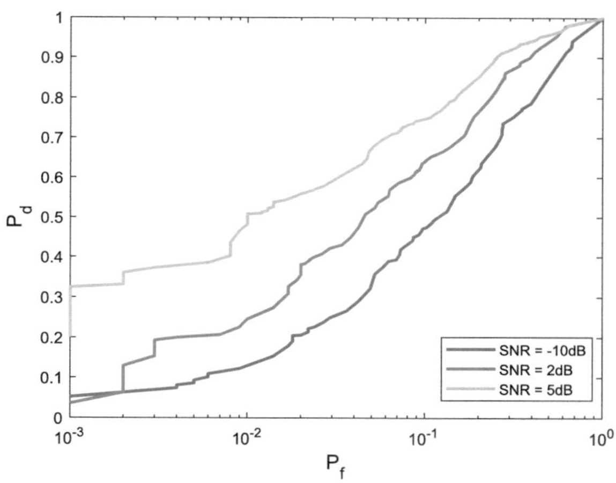

Fig. 2.14 ROC curves of fusion detection in the 2LMG structure

[43.64, 19.87, −63.33]. With the external fusion detection Algorithm 2, the global detection probability P_D for group 1 is presented in Fig. 2.13. Meanwhile, ROC curves with the different SNRs (changing the target transmit power σ_T^2) are shown in Fig. 2.14. The active detection weights are [0.50, 0.50, 0.86, 0.50, 0.46, 0.50, 0.50, 0.66, 0.50, 0.93], and the passive detection weights are [0.50, 0.50, 0.14, 0.50, 0.54, 0.50, 0.50, 0.34, 0.50, 0.07]. In the case of low SNRs, the detection performance under this structure can also obtain high performance. With the increasing of SNR, the external fusion detection probability P_D gradually increases if the false alarm rate P_F is fixed. Correspondingly, the energy consumption per sensor node in group 1 is shown in Fig. 2.15.

On the other hand, we also adopt the egalitarian structure Wan et al. (2019) to make external fusion. In the egalitarian structure, each sensor node can obtain the decision information of all the other sensor nodes in group 1. With the fusion detec-

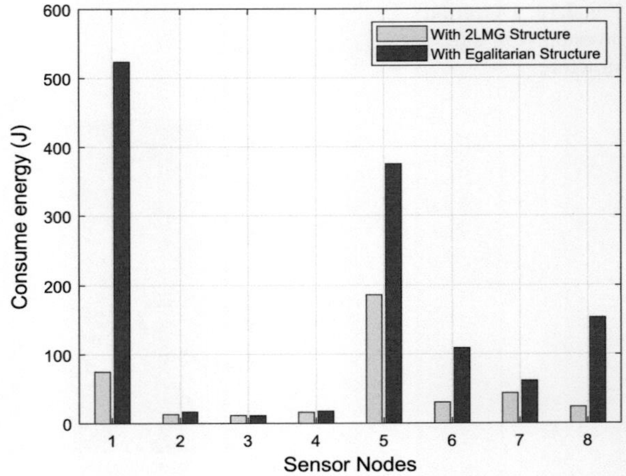

Fig. 2.15 Energy consumption of sensor nodes

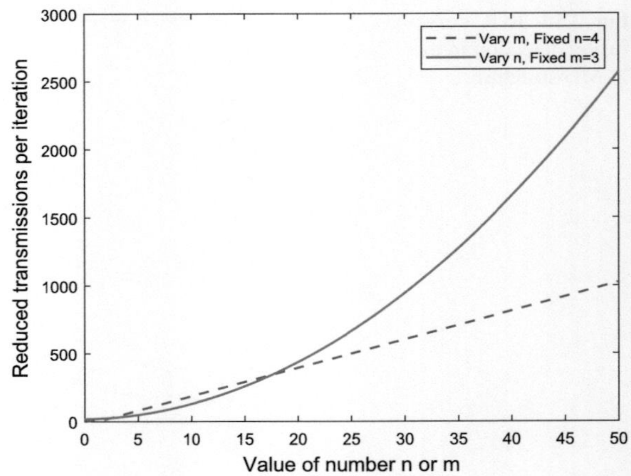

Fig. 2.16 Difference of transmission times

tion in egalitarian structure, the global detection probability P_{D} is also presented in Fig. 2.13, whose ROC curves are also depicted in Fig. 2.17. The active detection weights are [0.50, 0.19, 0.50, 0.50, 0.49, 0.50, 0.50, 0.73, 0.50, 0.50], while the passive detection weights are [0.50, 0.81, 0.50, 0.50, 0.51, 0.50, 0.50, 0.27, 0.50, 0.50]. Besides that, the energy consumption per sensor node in the egalitarian structure is shown in Fig. 2.15, and the difference of transmission times is considered about two cases. Case I: fix the number of network groups M and change the number of sensor nodes S_j in each group. Case II: fix the number of sensor nodes S_j in each group and change the number of groups M in the network. The difference of transmission times are shown in the Fig. 2.16.

From Figs. 2.13, 2.14 and 2.17, one has the following two conclusions: (1) the detection probability P_{D} with 2LMG structure and egalitarian structure is very close; (2) the detection performance with 2LMG structure and egalitarian structure is much

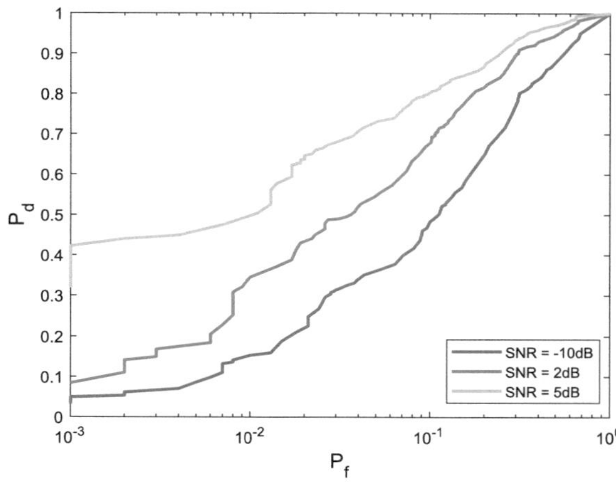

Fig. 2.17 ROC curve of fusion detection in egalitarian structure

better than the result in Figs. 2.10 and 2.11. The above conclusions verify the necessary and meaningful of external fusion detection in underwater environment. From Figs. 2.15 and 2.16, we find that the energy consumption and transmission times in 2LMG structure are smaller than the egalitarian structure, i.e., the proposed 2LMG structure in this chapter can reduce the number of transmissions.

2.5 Summary

This chapter studies target detection issue in MUSs. To reduce the influence of strong noise and weak communication in underwater environment, we design a local decision-making method with fusion of active and passive measurements. With the local decision, we propose a hybrid Bayesian fusion algorithm under 2LMG network to obtain final decision. Finally, the simulation results demonstrate that the proposed approach in this chapter not only integrates active measurement and passive measurement, but also can improve the detection performance and reduce the communication load.

References

Aman, W., Rahman, M., Qadir, J., Pervaiz, H., & Ni, Q. (2018). Impersonation detection in line of sight underwater acoustic sensor networks. *IEEE Access., 6*, 59–72.

Braca, P., Goldhahn, R., Ferri, G., & LePage, K. (2016). Distributed information fusion in multistatic sensor networks for underwater surveillance. *EEE Sensors Journal, 16*(11), 4003–4014.

Cao, B., Yang, F., Gao, L., Hu, H., Jiang, Z., Sun, B., & Luo, L. (2021). An active passive fusion strategy and accuracy evaluation for shallow water bathymetry based on icesat-2 atlas laser point cloud and satellite remote sensing imagery. *International Journal of Remote Sensing, 42*(8), 2783–2806.

Ciuonzo, D., & Rossi, P. (2017). Distributed detection of a non-cooperative target via generalized locally optimum approaches. *Information Fusion., 36*(1), 261–274.

Diamant, R., Kipnis, D., Bigal, E., Scheinin, A., Tchernov, D., & Pinchasi, A. (2019). An active acoustic track before detect approach for finding underwater mobile targets. *IEEE Journal on Selected Areas in Communications, 13*(1), 104–119.

Ghafoor, H., & Noh, Y. (2019). An overview of next generation underwater target detection and tracking: An integrated underwater architecture. *IEEE Access, 7*(1), 98841–98853.

Gong, Z., Li, C., & Jiang, F. (2020). A machine learning based approach for auto detection and localization of targets in underwater acoustic array networks. *IEEE Transactions on Vehicular Technology, 69*(12), 15857–15866.

Han, G., Du, J., Lin, C., Wu, H., & Guizani, M. (2020). An energy balanced trust cloud migration scheme for underwater acoustic sensor networks. *IEEE Transactions on Wireless Communications, 19*(3), 1636–1649.

Hu, L., Wang, X., & Wang, S. (2021). Decentralized underwater target detection and localization. *EEE Sensors Journal, 21*(2), 2385–2399.

Isbitiren, G., & Akan, O. (2011). Three dimensional underwater target tracking with acoustic sensor networks. *IEEE Transactions on Vehicular Technology, 60*(8), 3897–3906.

Jiang, C., Li, J., & Xu, W. (2018). Joint detection and tracking via path planning in the mobile underwater sensing network. In *International Conference on Wireless Communications and Signal Processing (WCSP)* (pp. 1–6).

Jiang, J., Han, G., Guo, H., Shu, L., & Rodrigues, J. (2016). Geographic multipath routing based on geospatial division in duty-cycled underwater wireless sensor networks. *International Journal of Networking and Virtual Organisations, 59*(1), 4–13.

Kay, S. M. (1993). *Fundamentals of statistical signal processing*. Press: Prentice-Hall Inc.

Liang, Q., & Cheng, X. (2008) Underwater acoustic sensor networks: Target size detection and performance analysis. In *IEEE International Conference on Communications* (pp. 3151–3155).

Niu, R., & Varshney, P. (2005). Decision fusion in a wireless sensor network with a random number of sensors. In *IEEE International Conference on Acoustics, Speech, and Signal Processing*.

Shen, Z., Yin, H., Jing, L., Liang, Y., & Wang, J. (2022). A cooperative routing protocol based on q-learning for Underwater optical acoustic hybrid wireless sensor networks. *IEEE Systems Journal, 22*(1), 1041–1050.

Song, X., Willett, P., Glaz, J., & Zhou, S. (2012). Active detection with a barrier sensor network using a scan statistic. *IEEE Journal of Oceanic Engineering, 37*(1), 66–74.

Tian, Y., Liu, M., Zhang, S., & Zhou, T. (2022). Underwater multi target passive detection based on transient signals using adaptive empirical mode decomposition. *Applied Acoustics,190*(1), Article 108641.

Wan, Y., Namuduri, K., Akula, S., & Varanasi, M. (2013). The impact of multi-group network structure on the performance of distributed consensus building strategies. *International Journal of Robust and Nonlinear Control, 23*(6), 653–662.

Wan, Y., Yan, J., Lin, Z., Sheth, V., & Das, S. (2019). On the structural perspective of computational effectiveness for quantized consensus in layered UAV networks. *IEEE Transactions on Control of Network Systems, 6*(1), 276–288.

Xu, Y., Zheng, R., Liu, M., & Zhang, S. (2021). Robust inertial aided underwater localization and navigation based on imaging sonar keyframes. *IEEE Transactions on Instrumentation and Measurement, 71*(1), 7501812–7501814.

Yan, L., Ma, X., Li, X., & Lu, J. (May2021). Shot interference detection and mitigation for underwater acoustic communication systems. *IEEE Transactions on Communications,69*(5), 3274–3285.

Yan, J., Yang, X., Luo, X., & Chen, C. (2018). Energy efficient data collection over AUV-assisted underwater acoustic sensor network. *IEEE Systems Journal, 12*(4), 3519–3530.

Yan, J., Zhao, H., Pu, B., Luo, X., Chen, C., & Guan, X. (2020). Energy efficient target tracking with UASNs: A consensus based bayesian approach. *IEEE Transactions on Automation Sciences and Engineering, 17*(3), 1361–1375.

Yemeni, Z., Wang, H., Ismael, W., Hawbani, A., & Chen, Z. (2022). CFDDR: A centralized faulty data detection and recovery approach for WSN with recognizing the type of faults. *IEEE Systems Journal, 16*(2), 3001–3012.

Zhang, D., Gao, L., Teng, T., & Jia, Z. (2021). Underwater moving target detection using track-before-detect method with low power and high refresh rate signal. *Applied Acoustics,174*(1), Article 107750.

Chapter 3
Integrated Design of Detection and Communication

3.1 Introduction

Underwater communications and sensing serve as core enablers of modern ocean exploration and operations. However, existing systems typically adopt separate architectures for sensing and communication devices, thereby limiting spectral efficiency and system-level coordination. On the one hand, separate frequency allocation leads to spectral fragmentation and mutual interference that degrades operational robustness. On the other hand, operating two independent systems requires duplicated transducers, signal processing modules, and power units, increasing payload, volume, and energy consumption on AUV platforms, which in turn constrains endurance and maneuverability. Consequently, this decoupled "sensing-communication" architecture has become a key bottleneck hindering the development of efficient and intelligent deep-sea operations, underscoring the necessity and urgency of integrated physical-layer design for underwater detection and communication.

Inspired by integrated sensing and communication (ISAC) architectures in radar systems, waveform designs in this field are commonly categorized into four modes: time division (Rou et al. 2024), frequency division (Huang et al. 2024), beam division (Huang et al. 2025), and full sharing (Chen et al. 2023; Wang et al. 2023). However, due to the narrow band nature of underwater acoustic channels and the limited sonar aperture, ISAC solutions that are effective in electromagnetic radar do not directly transfer to the underwater domain. For example, the much lower propagation speed of sound yields excessive latency and potential detection blind zones in time-division strategies (Yan et al. 2024; Zheng et al. 2023). The extremely scarce acoustic spectrum makes frequency-division schemes (Huang et al. 2024) fragment bandwidth, thereby reducing spectral efficiency. Beam-division approaches (Liu et al. 2024, 2023) split transmit power across multiple spatial beams, diminishing the signal-to-noise ratio of each beam and degrading detection performance. In addition, Doppler spread and severe multipath in the ocean further complicate synchronization and equalization. By contrast, full sharing waveforms allow detection and communication signals to

J. Yan and X. Guan, *Maritime Unmanned Systems*, Cognitive Intelligence and Robotics 1, https://doi.org/10.1007/978-981-95-6907-6_3

coexist within the same physical resource, avoiding resource partitioning and cross-system coordination overhead. As a result, full sharing is widely regarded as the most promising pathway to achieve simultaneous sensing and communication in underwater environments.

The design approaches for shared waveforms can be classified into two categories: one is detection-centric, and the other is communication-centric. However, several challenges remain: (1) For long-range active sonar applications, low-frequency and long-duration broadband transmissions are commonly adopted to improve coverage and detection gain. Taking the linear frequency modulated (LFM) signal (Mahipathi et al. 2024) as an example, one of the most widely used detection waveforms, integrated designs include frequency-modulation-based schemes (Nowak et al. 2016) as well as intra-band phase-modulation schemes (Zhang et al. 2018). Such detection-centric constructions can effectively minimize degradation of detection performance; however, the limited degrees of freedom for information embedding inherently constrain achievable communication rates. (2) As underwater information systems evolve toward clustered and networked architectures, efficient multi-node cooperation requires high-throughput data exchange and minimal mutual interference among nodes. Orthogonal frequency-division multiplexing (OFDM) (Wang et al. 2025) provides a mature theoretical and practical foundation in communication. Nevertheless, the underwater channel exhibits severe energy attenuation, waveform distortion, Doppler spread, and time-varying multipath, which pose challenges to OFDM-such as Doppler-induced inter-carrier interference and increased complexity in synchronization and equalization. Therefore, it is imperative to design integrated waveforms with strong communication performance that are tailored to networked underwater scenarios. (3) No single waveform can simultaneously achieve sensing-optimal and communication-optimal performance across all underwater applications. Performance trade-offs differ across task requirements, channel conditions, and platform constraints, making it infeasible to rely on a single waveform in a one-size-fits-all manner. Considering constrained sensor resources, how to design scenario-driven, fully shared waveforms together with associated processing pipelines remains an open issue deserving to be solved.

In this chapter, two fully shared waveforms (i.e., a communication-based waveform and a detection-based waveform) are both investigated. Based on this, an integrated experimental platform is implemented for performance validation. The main contributions are given as follows.

- **Two complementary fully shared integrated waveforms are designed.** Two fully shared waveforms are proposed: a communication-based waveform and a detection-based waveform.
  Communication-based waveform: By jointly employing channel estimation and cyclic prefix insertion, channel fading and multipath interference are effectively mitigated. Compared with Lu et al. (2018), Yin et al. (2020), Yan et al. (2022), the introduced hybrid-frequency multicarrier structure further improves spectral efficiency under the same bandwidth constraint.

Detection-based waveform: Leveraging the narrow mainlobe and high sidelobe-suppression characteristics of the GSFM ambiguity function, this design preserves detection accuracy under Doppler perturbations. In comparison with fully shared waveforms in Cui et al. (2024), Niu et al. (2023), Wang et al. (2024), Cui et al. (2022), it offers superior detection resolution and interference robustness while sustaining reliable communication performance. Compared with traditional single-carrier schemes (Yan et al. 2024; Qin et al. 2024a, b; Sun et al. 2023), the proposed dual-track design accommodates heterogeneous MUS node requirements, alleviates spectrum contention, and enhances network-level cooperative efficiency.
- **A modular integrated platform is developed.** A modular hardware platform is implemented, together with the corresponding end-to-end signal-processing chain. Extensive simulation and experimental results on spectral efficiency, bit-error rate, and detection resolution substantiate the approach's engineering feasibility and practical applicability.

3.2 Integrated Wave Design

In fully shared integrated systems, the primary objective is to realize simultaneous, co-frequency target detection and data communication within a unified signaling framework. In this context, the joint design of integrated detection and communication waveforms is the foundational prerequisite for constructing integrated systems. Specifically, in complex underwater environments, the detection waveforms must account for target resolution, propagation loss, and scattering characteristics, whereas the communication waveforms should prioritize multipath propagation, background noise, and frequency-selective fading. Addressing these factors jointly ensures that a shared waveform can effectively satisfy the dual requirements of detection and communication. Moreover, an integrated coordination mechanism is critical to system robustness. Well-crafted integration strategies can leverage the complementarity between detection and communication signals, enhance adaptability to environmental dynamics, and maintain stable performance under underwater uncertainty. Guided by these principles, we consider two representative of integrated shared waveforms.

- Communication-based shared waveform extends the conventional communication signal to incorporate sensing functionality, thereby enabling simultaneous data transmission and target detection.
- Detection-based shared waveform embeds data modulation and coding into detection-oriented signals, achieving reliable communication while maintaining sensing performance.

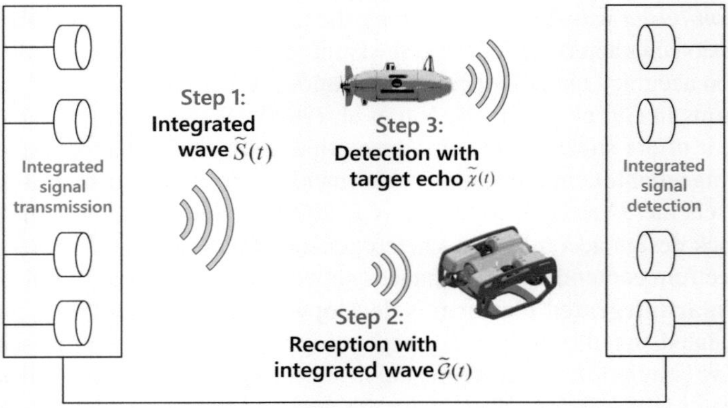

Fig. 3.1 The structure diagram of the integrated system

3.2.1 Communication-Based Shared Waveform

As illustrated in Fig. 3.1, the integrated protocol adopts a time-division modulation scheme in which time slots are allocated according to system requirements, task priorities, and scheduling constraints. This mechanism enables resource sharing between communication and detection while minimizing mutual interference. The detailed procedure is outlined below.

Step 1. Design of underwater integrated protocol

In underwater acoustic channels, sound waves undergo reflection, refraction, and scattering during propagation, producing multiple echoes with different arrival times, amplitudes, and phases at the receiver. These components superimpose in the time domain, creating interference known as the multipath effect, which disrupts subcarrier orthogonality and significantly degrades transmission quality. The OFDM is a multicarrier modulation technique that partitions the data stream into multiple subcarriers that are orthogonal in frequency, enabling parallel transmission within the same bandwidth. To mitigate inter-carrier interference and inter-symbol interference caused by multipath, the communication-based shared waveform prepends each OFDM symbol with a cyclic prefix of one-quarter of the symbol duration. This converts the channel's linear convolution into an approximate circular convolution, thereby improving temporal alignment, enhancing anti-interference capability, and restoring near-orthogonality under practical underwater conditions. To this end, let the pth and qth subcarriers be written as e^{j2f_pt} and $\mathrm{e}^{j2\pi f_qt}$, respectively, where f_p and f_q denote their subcarrier frequencies. Without loss of generality, assume all subcarriers share the same amplitude and phase reference. Based on this, the relationship between the above two subcarriers in the symbol period T_s can be represented as

$$\frac{1}{T_s}\int_0^{T_s} \mathrm{e}^{j2f_pt}\left[\mathrm{e}^{j2\pi f_qt}\right]^* \mathrm{d}t = \begin{cases} 1, & \text{if } p = q \\ 0, & \text{if } p \neq q \end{cases} \tag{3.1}$$

where $[\cdot]^*$denotes complex conjugation, its role is to ensure the linearity and symmetry properties of subcarrier inner products. A value of "1" denotes that each subcarrier is mutually independent when $p = q$, whereas a value of "0" indicates orthogonality between distinct subcarriers when $p \neq q$.

To improve symbol rate and energy efficiency, we employ a predefined mapping dictionary $\varpi = \Omega(\bar{m})$ for symbols $\bar{m} = 0, 1, \ldots, 2^M - 1$, which maps the communication bits onto OFDM subcarriers. Meanwhile, a linear frequency-modulated (LFM) waveform is used as the frame-synchronization preamble for the OFDM signal. Building on (3.1) and the mapping ϖ, the integrated shared waveforms can be constructed as $\widetilde{S}(t)$, i.e.

$$\widetilde{S}(t) = \sum_{n=0}^{N-1} \sum_{m=0}^{M-1} \Psi(n, m) \operatorname{rect}\left(\frac{t - nT_2}{T_s}\right) U_l(t), \tag{3.2}$$

where f_k denotes the frequency of the kth subcarrier and $\Psi(n, m)$ the modulated symbol assigned to the mth subcarrier of the nth OFDM symbol. The function rect $(\cdot)$ is the unit rectangular window that gates the waveform over the symbol interval T_s. The carrier term $U_l(t) = \mathrm{e}^{[j2\pi(f_k(t-nT_n)+\mu t^2/2)]}$ incorporates both the OFDM component and the LFM chirp.

Step 2. Signal reception on the communication unit:
At time t_0, the integrated waveform $\tilde{S}(t)$ is transmitted to the monitoring region, it is subsequently received by the communications unit at time t_c. On the basis of (3.2), the received waveform is denoted by

$$\tilde{\mathcal{G}}(t) = \tilde{S}(t) \otimes \delta(t) + n_o(t), \tag{3.3}$$

where $\otimes$ denotes the convolution operation, whose role is to describe the interaction between input signals and system response. $\delta(t)$ is the impulse response of the underwater acoustic multipath channel, and $n_o(t)$ is the additive white noise with zero mean and variance σ_r^2.

The communication waveform demodulator is adopted to sample, quantize and convert the received signal $\tilde{\delta}(t)$ into a discrete-time signal. The LFM synchronization pulse is marked the start of each data frame. Subsequently, the matched filtering is adopted to extract the OFDM communication signal. Following (3.1), the orthogonal demodulation and subsequent integration are performed to recover the mth symbol on the pth subcarrier. The mth symbol for the pth subcarrier can be represented as

$$\hat{d}_m = \frac{1}{T_s} \sum_{p=0}^{P-1} \sum_{m=0}^{M-1} \Psi(p, m) \int_{T_s} \mathrm{e}^{j2\pi \Delta f(t-pT_s)} \mathrm{d}t = d_m, \tag{3.4}$$

where the subcarrier spacing is chosen as $\Delta f = 1/T_s$, so that $f_k = f_0 + k\,\Delta f$ and $f_p - f_q = (p - q)\Delta f = (p - q)/T_s$, where T_s denotes the useful symbol duration (excluding the cyclic prefix). After cyclic prefix removal and an N-point FFT, OFDM demodulation recovers the per-subcarrier symbols $\{\hat{S}[k]\}$. A parallel-to-serial operation then yields the detected symbol stream $\hat{s}[n]$. The inverse constellation mapping $\Omega^{-1}(\cdot)$ demaps each symbol to its M-bit label, i.e., $\hat{\mathbf{b}}[n] = \Omega^{-1}(\hat{s}[n]) \in \{0, 1\}^M$ for a 2^M-ary constellation. Finally, channel decoding produces the estimate of the transmitted bitstream $\hat{\mathbf{m}}$.

Step 3. Target detection on the echo detector:
At time t_d, the detection unit receives the target echo at the echo-signal detector. Particularly, the echo signal can be expressed as

$$\widetilde{\chi}(t) = \widetilde{S}(t) \otimes \delta(t - \tau)e^{j2\pi f_d t} + n_o(t), \tag{3.5}$$

where τ, f_d is the propagation delay, $\delta(t - \tau)$ is the doppler frequency shift and the channel impulse response, respectively. Based on (3.5), the same operations as in Step 2 are performed to convert the $\widetilde{\chi}(t)$ into discrete-time signal. The echo detector then estimates the propagation delay τ by comparing t_0 from Step 1. Simultaneously, spectral analysis of the LFM component is performed to extract the Doppler shift f_d. Finally, the detection ability can be guaranteed to estimate the parameter information of target, i.e., distance d and velocity v.

3.2.2 Detection-Based Shared Waveform

The generalized sinusoidal frequency modulation (GSFM) signal is a nonlinearly frequency-modulated detection waveform characterized by flexible bandwidth configuration, strong anti-interference capability, high time resolution, and excellent approximate orthogonality. Owing to these advantages and its unique frequency modulation properties, GSFM has been widely applied in radar, sonar, and other signal processing fields. Building upon this, a detection-based shared waveform design is proposed, employing the GSFM signal. The core concept is to embed communication information into a detection waveform while ensuring high-resolution detection and low sidelobe levels, thereby achieving integrated sensing-communication functionality within a single signal. The design process consists of three main steps.

Step 1. Design of the detection waveform
The phase, instantaneous frequency, and time-domain expressions of the GSFM signal are given by

$$\begin{aligned}
\varphi_{GSFM}(t) &= \frac{\beta}{t^{(\rho-1)}} \sin\left(\frac{2\pi\alpha t^{\rho}}{\rho}\right) \\
f_{GSFM}(t) &= \frac{B}{2} \sin\left(2\pi\alpha t^{\rho}\right) + f_c \\
S_{GSFM}(t) &= \frac{\mathrm{rect}(t)}{\sqrt{T}} \exp\left(j\left(\varphi_{GSFM}(t) + 2\pi f_c t\right)\right)
\end{aligned} \tag{3.6}$$

where $\beta = \frac{B}{2\alpha}$ is the modulation index, B is the signal bandwidth, and f_c is the carrier frequency. The dimensionless shape parameter ρ controls the nonlinearity of the instantaneous frequency. Specifically, $\rho > 1$ corresponds to GSFM, while $\rho = 1$ yields SFM. When $\rho = 2$, the phase and frequency trajectories resemble those of an LFM chirp rather than the strictly periodic SFM form, resulting in lower autocorrelation sidelobes and improved range resolution. The frequency modulation factor α (units: $s^{-\rho}$) determines the number of frequency cycles $C = \alpha T^{\rho}$. From (3.6), it can be seen that the GSFM waveform's shape is primarily governed by the parameters ρ and α. To balance mainlobe width and sidelobe suppression, typical design ranges are $\rho \in [2.0, 2.5]$ and $\alpha \in [150, 400]$.

Step 2. Modulation for integrated waveform
To enable data transmission within the sensing-dominant framework, minimum shift keying (MSK) modulation is introduced. MSK is a form of continuous phase modulation and can be regarded as an improved version of FSK, achieving higher spectral efficiency by maintaining the minimum possible frequency separation between adjacent symbols while ensuring phase continuity. Its expression is given by

$$s_{MSK}(t) = \cos\left(2\pi f_c t + \frac{a_k \pi}{2T_B} t + \varphi_k\right), 0 \le t \le T \tag{3.7}$$

where f_c is the carrier frequency, $a_k = \{+1, -1\}$ represents th kth binary symbol, φ_k is the initial phase of the kth symbol, and T_B denotes the symbol duration. To ensure phase continuity at symbol boundaries, the phase condition must satisfy

$$\frac{a_{k-1}\pi}{2T_B} nT_B + \varphi_{k-1} = \frac{a_k \pi}{2T_B} nT_B + \varphi_k, \tag{3.8}$$

which simplifies to

$$\varphi_k = \varphi_{k-1} + \frac{n\pi}{2}(a_{k-1} - a_k) = \begin{cases} \varphi_{k-1}, a_{k-1} = a_k \\ \varphi_{k-1} \pm \pi, a_{k-1} \ne a_k \end{cases} \tag{3.9}$$

In general, the initial phase φ_0 is set to either 0 or π for analytical simplicity. Since GSFM signals with different parameter sets are approximately orthogonal, they can be used as composite carriers for MSK modulation. Under fixed parameter settings, multiple GSFM waveform types can be generated, as shown in Fig. 3.2. By calculating the maximum cross-correlation coefficients among candidate GSFM waveforms(see Table 3.1), a forward-time waveform and its time-folded version are selected as the I-channel and Q-channel carriers for MSK, respectively, to construct the GSFM-MSK integrated waveform. Based on this, the I/Q decomposition of the GSFM-MSK signal can be expressed as

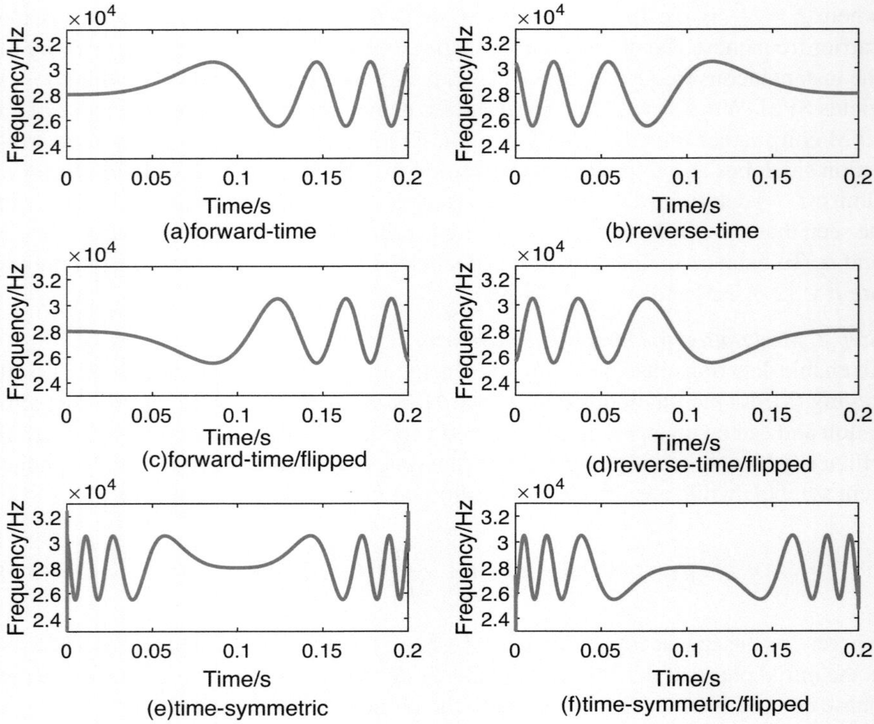

Fig. 3.2 Frequency characteristics of six GSFM waveforms

Table 3.1 Correlation coefficient of six GSFM signals

	Ga	Gb	Gc	Gd	Ge	Gf
Ga	1	0.1253	0.0783	0.1080	0.1071	0.1045
Gb	0.1253	1	0.1080	0.0783	0.1071	0.1046
Gc	0.0783	0.1080	1	0.1253	0.1044	0.1071
Gd	0.1080	0.0783	0.1253	1	0.1046	0.1071
Ge	0.1071	0.1071	0.1045	0.1046	1	0.0941
Gf	0.1045	0.1046	0.1072	0.1071	0.0941	1

$$\begin{aligned} s_{GM}(t) &= \cos(\varphi_k)\cos\left(\frac{a_k\pi}{2T_B}t\right)\exp\left(j2\pi\left(f_c+\frac{B}{2}\sin(2\pi\alpha t^{\rho})\right)t\right) \\ &\quad - \cos(\varphi_k)\sin\left(\frac{a_k\pi}{2T_B}t\right)\exp\left(j2\pi\left(f_c-\frac{B}{2}\sin(2\pi\alpha t^{\rho})\right)t\right) \end{aligned} \tag{3.10}$$

Let $p_k = \cos(\varphi_k) \in \{\pm 1\}$ and $q_k = a_k\cos(\varphi_k) \in \{\pm 1\}$. Then, for each symbol interval $kT_B \leq t \leq (k+1)T_B$, the I/Q decomposition is given by

$$\begin{aligned} e_{GMk}(t) &= p_k \cos\left(\frac{\pi}{2T_B}t\right)\exp\left(j2\pi\left(f_c + \frac{B}{2}\sin\left(2\pi\alpha t^{\rho}\right)\right)t\right) \\ &\quad - q_k \sin\left(\frac{a_k\pi}{2T_B}t\right)\exp\left(j2\pi\left(f_c - \frac{B}{2}\sin\left(2\pi\alpha t^{\rho}\right)\right)t\right) \\ kT_B &\le t \le (k+1)T_B \end{aligned} \tag{3.11}$$

Step 3. Demodulation for integrated waveform

After propagation through the underwater acoustic channel, the received signal at the receiver can be modeled as

$$r(t) = s_{GM}(t) \otimes h(t) + w(t) \tag{3.12}$$

where $h(t)$ denotes the channel impulse response between the transmitter and receiver, and $w(t)$ represents the underwater ambient noise. By demodulating the received signal with the carrier waveform, the in-phase (I) component is obtained as

$$\begin{aligned} r'_I(t) &= (s_{GM}(t) \otimes h(t)) \cdot \text{conj}\left[S_{\text{GSFM}}(t)\right] + w'(t) \\ &= \left(\left(a(t) \cdot \sum_{n=1}^{N} g_n(t)\right) \otimes h(t)\right) \cdot \sum_{n=1}^{N} g'_n(t) + w'(t) \\ &= \sum_i \sum_j a(t) \cdot g_i(t) \otimes h(t) \cdot g'_j(t) + w'(t) \end{aligned} \tag{3.13}$$

where $w'(t) = w(t) \cdot \text{con}\, j\left[S_{GSFM}(t)\right]$ denotes the demodulated noise component. Since the GSFM waveform is a pulse frequency-modulated signal, it can be expressed as a superposition of multiple single-frequency components, i.e.,

$$S_{GSFM}(t) = \sum_{n=1}^{N} A_n e^{j2\pi f_n t} = \sum_{n=1}^{N} g_n(t) \tag{3.14}$$

where A_n and f_n denote the amplitude coefficient and frequency of nth subcomponent, respectively. Substituting this expression into the demodulation equation yields

$$\begin{aligned} r'_I(t) &= (s_{LM}(t) \otimes h(t)) \cdot \text{conj}\left[S_{\text{GSFM}}(t)\right] + w'(t) \\ &= \left(\left(a(t) \cdot \sum_{n=1}^{N} g_n(t)\right) \otimes h(t)\right) \cdot \sum_{n=1}^{N} g'_n(t) + w'(t) \\ &= \sum_i \sum_j a(t) \cdot g_i(t) \otimes h(t) \cdot g'_j(t) + w'(t) \end{aligned} \tag{3.15}$$

where $g'_n(t)$ denotes the complex conjugate of $g_n(t)$. When $i = j$, the term $a(t) \cdot g_i(t) \otimes h(t) \cdot g'_i(t)$ corresponds to the conventional single-carrier demodulation process.

Subsequently, the signal $r_I'(t)$ is processed by a low-pass filter with frequency response given in (3.15). The filtered signal is expressed as

$$\begin{aligned} r_I''(t) &= \int_{-\infty}^{\infty} r_I'(t) \cdot h'(t-\tau)d\tau \\ &= M \cdot a(t) \otimes h(t) + \sum_i \sum_j a(t) \cdot g_i(t) \otimes h(t) \cdot g_j'(t) + w''(t) \\ &= b(t) + c_{i,j}(t) + w''(t) \end{aligned} \tag{3.16}$$

where $h'(t)$ denotes the time-domain response of the ideal low-pass filter, $b(t) = M \cdot a(t) \otimes h(t)$ represents the baseband component after carrier suppression, $w''(t)$ is the filtered noise term, and $c_{i,j}(t) = \sum_i \sum_j a(t) \cdot g_i(t) \otimes h(t) \cdot g_j'(t)$ denotes the inter-carrier interference term. In a frequency-selective fading channel, a larger carrier bandwidth B leads to stronger demodulation interference. Finally, after low-pass filtering, the signal undergoes integration-based detection, parallel-to-serial conversion, and differential decoding to recover the transmitted baseband data.

3.3 Implementation of the Integration System

The integrated sensing-communication system comprises four principal components: the integrated transmitter, the communication receiver, the echo-based target detection unit, and the target. The integrated transmitter performs data encoding, modulation of the fully shared integrated waveform, and transmit-power control. The communication receiver handles front-end filtering and synchronization, signal demodulation, symbol decoding and recovery, and data delivery. The echo detection unit focuses on echo processing, target-feature extraction, and state estimation. Following a modular design philosophy, the embedded system consists of six hardware modules and an integrated operational workflow, as summarized below.

3.3.1 System Hardware Architecture

To realize the above functions, a modular hardware architecture is developed, enabling efficient signal generation, transmission, reception, and processing under real-time constraints. The prototype integrates six hardware modules-processing, modulation, shared waveform generation, transducer/hydrophone interface, demodulation, and power supply-described in the following subsections.

- **Signal processing unit:** A STM32F407 microcontroller serves as the processing core, aided by the arm cortex microcontroller software interface standard—digital signal processing software library for accelerated fixed- and floating-point computation. This configuration fulfills real-time and energy-efficiency requirements on a Cortex-M platform without resorting to high-power processors.

- **Signal modulator:** The modulator consists of an external digital-to-analog converter (DAC), the STM32, and a clock circuit. A 16-bit DAC8550 performs D/A conversion with a programmable output sampling rate up to 200 kHz. The STM32 transmits the integrated signal to the DAC via serial peripheral interface and provides timing pulses through an internal timer. The microcontroller unit(MCU) runs at 168 MHz driven by a 25 MHz external crystal.
- **Shared waveform generator:** This analog front-end implements both communication and detection paths, providing impedance matching between the modulator and transducers. To enhance signal quality, the network includes a programmable attenuator, buffer amplifier, and Butterworth band-pass filter. Core components include the LM358, AD8310, and AD8227.
- **Signal transmission and detection sensors:** Underwater acoustic transducers act as transmit/receive sensors for acoustic signaling, while a hydrophone is employed for echo reception. Each transducer operates around a 35 kHz center frequency within 1-70 kHz. The hydrophone operates around 40 kHz within 1-80 kHz and features a voltage sensitivity of -204 ± 1.5 dB re 1 $V/\mu Pa$.
- **Signal demodulator:** To simplify hardware complexity, the on-chip 12-bit analog-to-digital converter of the STM32 is used with direct memory access to directly transfer sampled data to memory. Because the MCU cannot sample negative voltages, a direct-current offset circuit is introduced at the input stage. Following Step 2, the demodulator extracts the embedded information and forwards it to the receiver.
- **Power supply:** A high-capacity 18 Ah/266 Wh lithium battery serves as the power source. An LM2596S direct current buck module delivers regulated output voltages tailored to the system components.

3.3.2 Operational Procedure

As illustrated in Fig. 3.3, the system block diagram demonstrates the overall workflow of the integrated detection-communication process.

- **Integrated transmission terminal:**
  Step 1—Scenario deployment: In the detection-communication region, an integrated platform and a communication receiver are separately deployed. The platform performs communication transmission and target detection through its built-in sensing module.
  Step 2—Shared waveform transmission and propagation: The integrated platform broadcasts the shared waveform. After propagating through the underwater channel, the signal is decoded by the communication receiver, while the sensing module exploits the reflected echoes for target detection and localization.

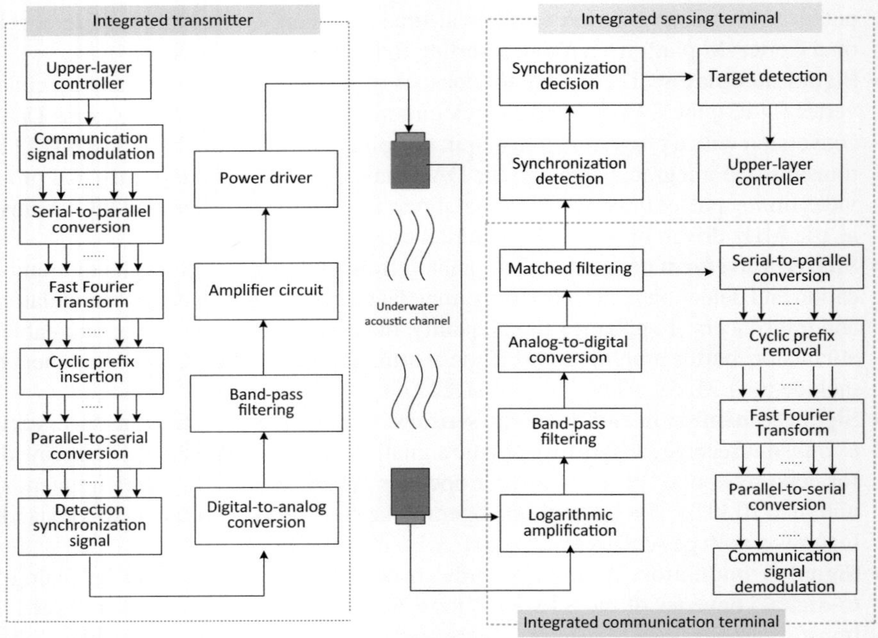

Fig. 3.3 The structure diagram of the integrated system

- **Communication receiving terminal:**
  Step 3—Front-end and matched filtering: The receiver applies logarithmic amplification and band-pass filtering to enhance SNR and suppress out-of-band noise. Matched filtering is then performed to emphasize MSK-like modulation features, enabling effective separation of the communication component.
  Step 4—Coherent demodulation and decoding: The separated signal is fed into a coherent demodulator for carrier and phase estimation, producing the baseband sequence, which is subsequently decoded to recover the original data.
- **Integrated detection terminal:**
  Step 5—Synchronous detection: The detection chain performs synchronization and decision-making over the target region, maintaining a unified time-frequency reference.
  Step 6—Echo verification and pulse recovery: When target echoes are detected, matched filtering reconstructs the transmitted pulse and other detection features; otherwise, the process returns to Step 5.
  Step 7—Fusion decision and parameter estimation: The fused detection and estimation results yield the echo's time delay and Doppler characteristics, from which the target parameters are obtained.

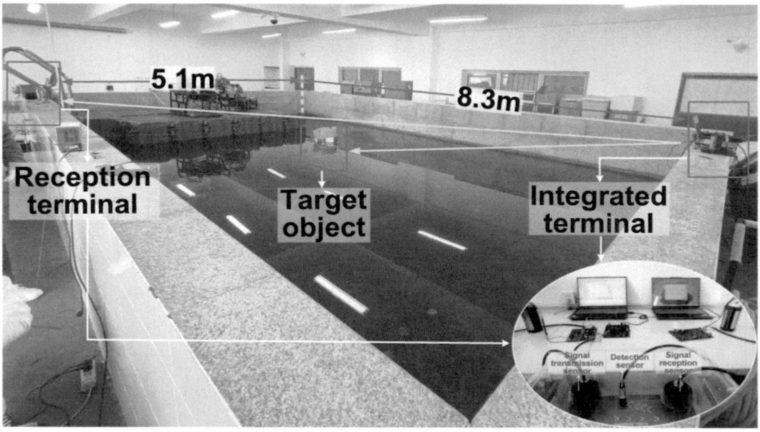

Fig. 3.4 Experimental setup of the embedded system in pool

3.4 Technology Practice

3.4.1 Case Study I: Integrated Design with Communication Signal

(1) Experimental validation in the pool
As illustrated in Fig. 3.4, the signal transmission and detection sensors are co-located on one side of the integrated terminal, while the signal reception sensor is placed at the remote communication terminal. A 16-ary symbol dictionary ϖ is employed within a 64-subcarrier OFDM framework, with each OFDM symbol having a duration of 10 ms. The estimated channel response is presented in Fig. 3.5a, where the average measured signal-to-noise ratio (SNR) of communication packets reaches approximately 15 dB. Figure 3.5b depicts the modulation of different data packets at the integrated terminal, corresponding to the dataset $1, 9, 9, 9, k, k, k, a, a, a, \ldots$. The reception results in Fig. 3.5c confirm that the communication terminal successfully demodulates the transmitted dataset, thereby verifying the communication capability of the system.

Subsequently, the detection functionality is evaluated. The frequency range used for detecting the synchronized LFM signal is set to 20 Hz-60 kHz. Two target positions are tested, i.e., $\mathbf{P}_1 = 4.24$ m, 2.45 m, 1 m and $\mathbf{P}_2 = 4.24$ m, 2.45 m, 1 m. The corresponding detection results are shown in Fig. 3.5d–e. The estimated propagation delays from the targets to the integrated terminal are $\tau = 6.6$ ms and $\tau = 7.3$ ms, respectively, with a Doppler frequency shift of $f_d = 0.01$ Hz. Based on these measurements, the estimated target positions are obtained as $\widetilde{\mathbf{P}}_1 = 4.33$ m, 2.51 m, 1 m and $\widetilde{\mathbf{P}}_2 = 4.84$ m, 2.82 m, 1 m, corresponding to localization errors of $e_1 = 0.11$ m and $e_2 = 0.10$ m. The above results demonstrate that the integrated communication-

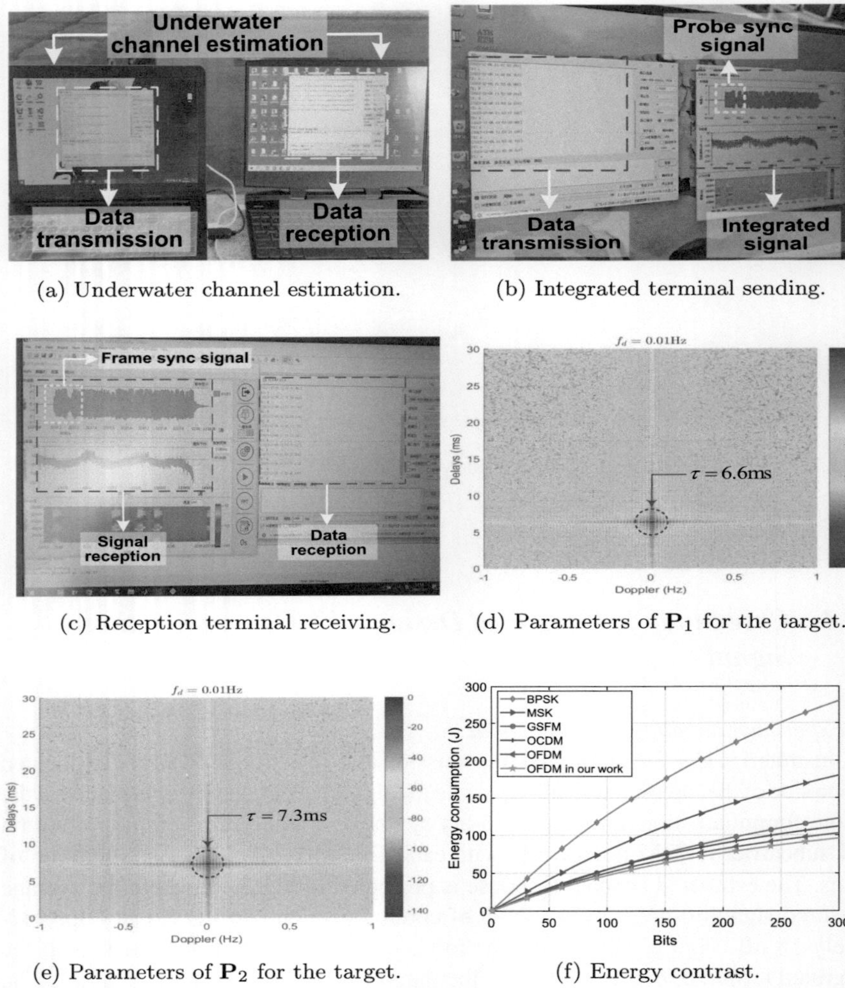

(a) Underwater channel estimation. (b) Integrated terminal sending. (c) Reception terminal receiving. (d) Parameters of $\mathbf{P}_1$ for the target. (e) Parameters of $\mathbf{P}_2$ for the target. (f) Energy contrast.

Fig. 3.5 Experiment results for the embedded integrated system

detection system can achieve accurate underwater detection in addition to the reliable communication.

In order to further assess performance, we compare our results with those in Lu et al. (2018), Yin et al. (2020), Yan et al. (2022). The symbol rate is defined as the number of transmitted symbols per unit time, and the spectral efficiency is defined as $\xi = Q/(TB)$, where B is the one-sided bandwidth and Q represents the number of bits transmitted over duration T. For normalization, the one-sided bandwidth of each signal is set 1 Hz. Under identical subcarrier and symbol-period conditions, Table 3.2 summarizes the performance comparison of different integrated waveform designs.

Table 3.2 Performance comparison of candidate waveforms

Waveforms	Symbol rate (sym/s)	Bit rate (bit/s)	Spectral eff. (bit/(s·Hz))	Energy (J/bit)
BPSK	1	1	1	1.21
MSK	1	2	2	0.73
GSFM in Lu et al. (2018)	1	4	4	0.56
OCDM in Yan et al. (2022)	2	4	4	0.47
OFDM in Yin et al. (2020)	2	4	4	0.43
This chapter	2	4	4	0.41

It is observed that OFDM- and OCDM-based integrated waveforms achieve the highest symbol rate, bit rate, and spectral efficiency. Notably, the proposed OFDM-based approach yields the lowest energy consumption per bit, as illustrated in Fig. 3.5f. Moreover, the predefined mapping dictionary ϖ further enhances the energy efficiency, albeit at the cost of potentially increased bit error rate due to the added decoding complexity. To mitigate this effect, the CRC-16-MODBUS checksum algorithm is employed, significantly improving decoding robustness. It is worth noting that the comparison in Lu et al. (2018), Yin et al. (2020), Yan et al. (2022) primarily focuses on communication performance, whereas the detection mechanisms are conceptually similar, thereby leading to comparable detection accuracy. Owing to space limitations, the detailed detection comparison is omitted here.

(2) Experimental validation in a real ocean

Field experiments were conducted in the Bohai Sea, China, to verify the system's real-world applicability. The experimental configuration is illustrated in Fig. 3.6a, where the signal transmission and detection sensors were deployed at 5.01 m, 0 m, -2 m, and the receiving sensor was positioned at -18.02 m, 0 m, -2 m. The sea-state conditions during the experiment included wind forces of 4–5 Beaufort scale, a water temperature of approximately 26 , wave heights around 0.5 m, and a measured communication SNR of 10 dB. Figure 3.6b displays the modulation of transmitted data packets $3, 6, 9, a, A, d, D, q, Q, 8$ at the integrated terminal. The received dataset at the communication terminal, as shown in Fig. 3.6c, confirms successful data decoding, thereby validating the robustness of the communication link in dynamic marine environments. For detection verification, the target was placed at $\mathbf{P}_3 = -1.99$ m, 0 m, -1 m, with detection results shown in Fig. 3.6d. The measured delay and Doppler shift were $\tau = 9.64$ ms and $f_d = 0.01$ Hz, respectively, yielding an estimated target position of $\widetilde{\mathbf{P}}_3 = -2.22$ m, 0 m, -1 m and an estimation error of $e_3 = 0.23$ m. These results confirm that the integrated design maintains its sensing and communication performance in open-sea conditions, validating the practical effectiveness and environmental adaptability of the proposed approach in this chapter.

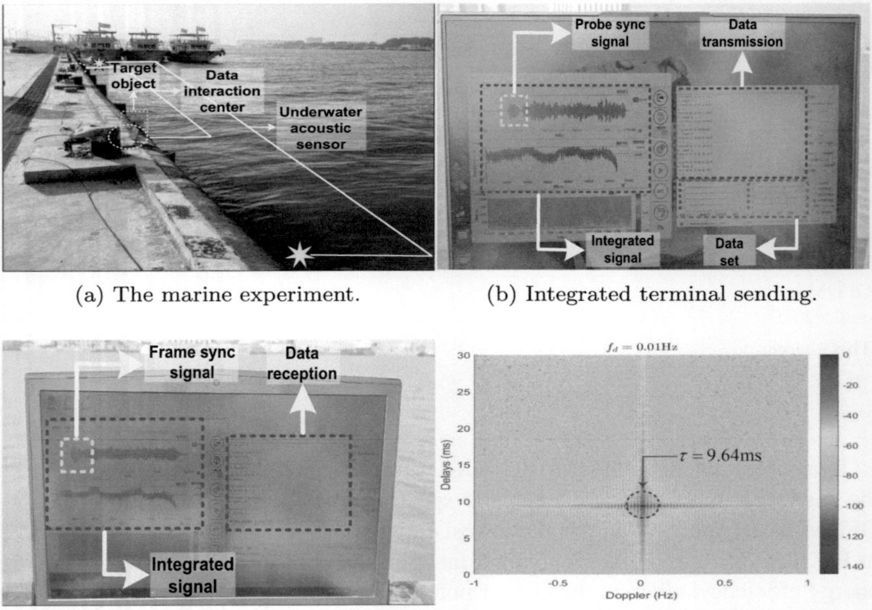

(a) The marine experiment. (b) Integrated terminal sending.

(c) Reception terminal receiving. (d) Parameters of $\mathbf{P}_3$ for the target.

Fig. 3.6 Experiment results for the embedded integrated system

3.4.2 Case Study II: Integrated Design with Detection Signal

(1) Detection performance analysis of integrated waveforms

For MUSs, the GSFM signal is configured with the following parameters: the bandwidth $B = 5$kHz, the carrier frequency $f_c = 28$kHz, the pulse duration $T_B = 0.2$s, and the shape parameters $\rho = 2.3$, $\alpha = 200$. A total of $M = 100$ symbols are are generated at random. With a transmit pulse width $T_r = 0.2$ s, the symbol period is given as $T_B = T_r/M = 2$ ms. The absolute symbols $a(n)$ are first mapped via XOR to bipolar symbols, which form the in-phase (I) and quadrature (Q) bit streams. The I/Q symbols are then modulated onto their respective GSFM carriers, producing two branch waveforms, as shown in Fig. 3.7. The magnitude spectrum and time-frequency representation of the resulting integrated waveform are shown in Fig. 3.8. The signal energy is primarily concentrated within 25–31 kHz, with spectral components decaying outside this band. This indicates that most of the energy is confined to a limited frequency span, which improves spectral efficiency an advantage that is particularly important in underwater communications where the available acoustic bandwidth is scarce.

Meanwhile, the ambiguity function characterizes two-dimensional correlation for a signal over time delay and Doppler shift, providing a direct measure of range and velocity resolution. It is a key indicator of detection performance and is defined as:

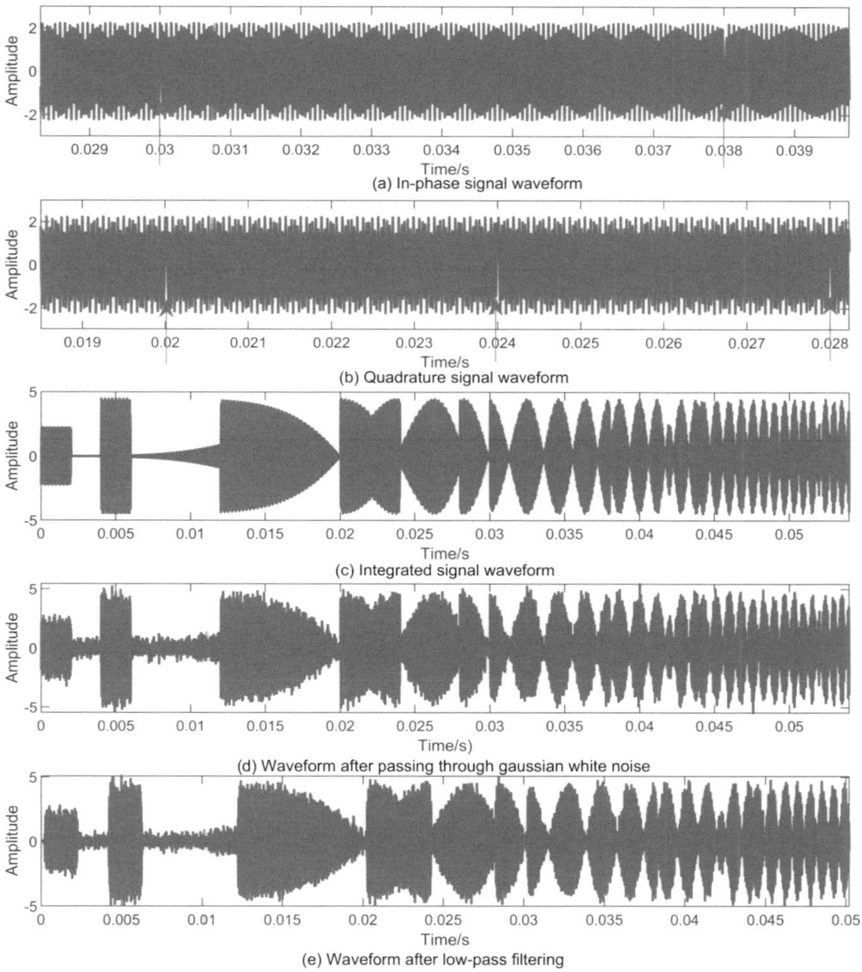

Fig. 3.7 Modulated integrated waveform

$$\left|\chi(\tau, f_d)\right|^2 = \left|\frac{1}{E}\int_{-\infty}^{\infty} u(t)\, u^*(t-\tau)\, e^{j2\pi f_d t}\, dt\right|^2, \tag{3.17}$$

where $u(t)$ denotes the transmitted signal envelope, $u^*(t-\tau)$ is its delayed complex conjugate corresponding to a target at delay τ, f_d is the Doppler shift, and E represents the total signal energy. The zero-Doppler cut ($f_d = 0$) reflects the range resolution, whereas the zero-delay cut ($\tau = 0$) reflects the velocity resolution. Analysis of Doppler slices of the ambiguity function shows that the integrated waveform inherits the excellent time-frequency concentration of GSFM and the constant-envelope prop-

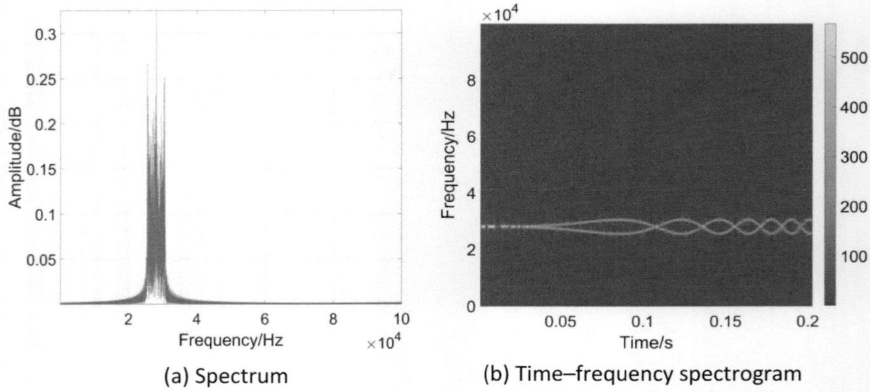

Fig. 3.8 Time frequency spectrogram of the integrated waveform

erty of MSK, thereby achieving strong temporal and Doppler resolution, as shown in Fig. 3.9. A comparative evaluation of the mainlobe width and sidelobe suppression ratio for GSFM, the integrated waveform, LFM, and the LFM-modulated integrated waveform yields the following observations, as shown in Fig. 3.10.

GSFM provides markedly better delay and bandwidth resolution than LFM, with a modest improvement in sidelobe suppression. The LFM integrated waveform exhibits a narrower mainlobe, offering superior discrimination of closely spaced targets; however, its sidelobe suppression degrades, attributable to imperfect orthogonality between its carrier components. The GSFM-MSK integrated waveform performs overall on par with GSFM across the considered metrics.

(2) Communication performance analysis of integrated waveforms

The bit error rate serves as a standard measure of system transmission reliability and is defined as

$$BER = \frac{M}{N}, \tag{3.18}$$

where M denotes the number of erroneous bits and N is the total number of received bits. Statistical averaging over multiple measurements provides a robust indicator of overall communication performance. The comparison of communication performance is primarily reflected in the bit-error rate. Figure 3.11 plots the SNR-BER characteristics of quadrature phase-shift keying (QPSK), differential phase-shift keying (DPSK), MSK, and GSFM-MSK. As observed from Fig. 3.11, both QPSK and DPSK exhibit rapidly decreasing BER in the high-SNR region, indicating superior performance; moreover, QPSK clearly outperforms DPSK at low SNR due to the detection gain offered by coherent demodulation, whereas differential encoding suffers from error propagation under noisy conditions. MSK demonstrates relatively high BER at low SNR but decreases sharply as SNR increases; in the high-SNR region, its performance approaches that of QPSK while benefiting from the power-amplifier efficiency and linearity associated with its constant-envelope property. GSFM-MSK achieves

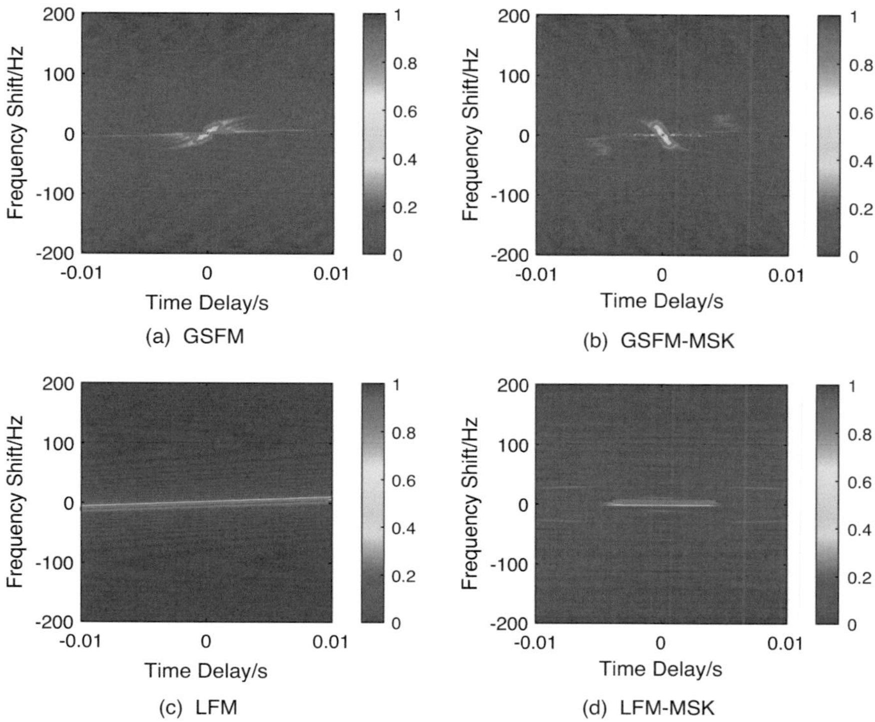

Fig. 3.9 Mainlobe of the ambiguity function

significantly lower BER than the other three schemes in the low-SNR region, showcasing enhanced robustness against interference and multipath distortion attributed to the wideband, time frequency diversity of GSFM combined with the constant-envelope characteristics of MSK. However, in the high-SNR region, its BER reduction saturates and eventually remains slightly above MSK, primarily because the underlying carrier/symbol set is not strictly orthogonal.

3.5 Summary

This chapter investigated the design of integrated sensing-communication waveforms under the constraints of underwater acoustic channels characterized by multipath propagation, Doppler shifts, and heterogeneous node requirements in MUSs. Two complementary fully shared waveform designs were developed: a communication-based shared waveform to enhance spectral efficiency and robustness against channel distortion, and a detection-based shared waveform that preserves detection resolu-

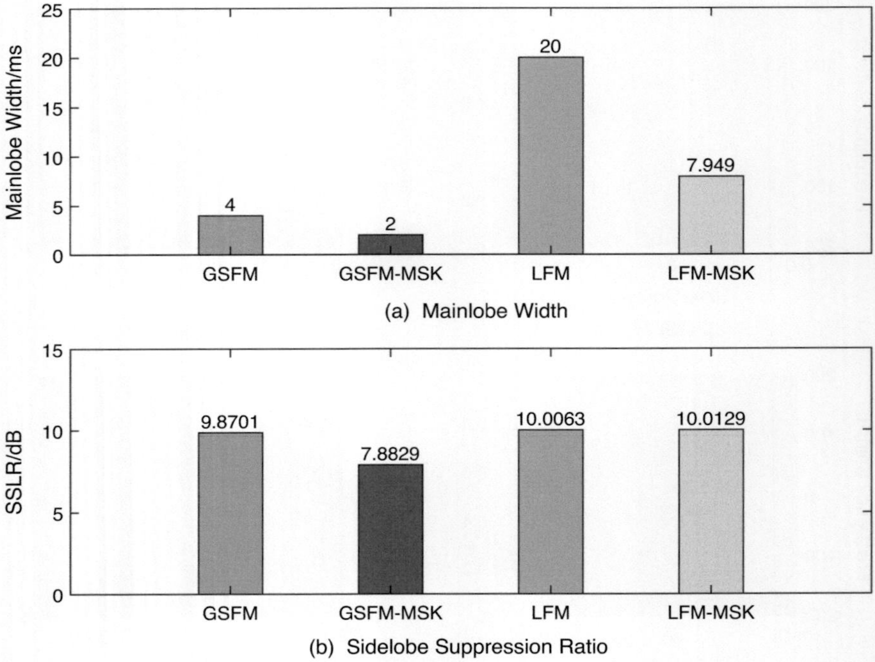

Fig. 3.10 Mainlobe width and sidelobe suppression ratio

Fig. 3.11 Bit error rate under different SNRs and signals

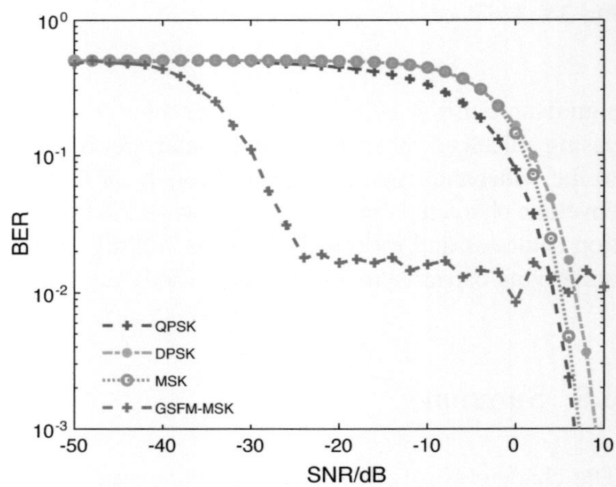

tion while supporting reliable data transmission. In addition, a modular hardware platform with an end-to-end processing chain was constructed to support practical deployment. Finally, comprehensive simulation and experimental results validated the effectiveness and engineering feasibility of the proposed waveforms.

References

Chen, F., Zhang, Y., Wang, H. et al. (2023) "Co-design of underwater detection and communication-part I: Detection-based shared waveform. In: *International Conference on Intelligent Autonomous Systems (ICoIAS), Qinhuangdao, China* (pp. 123–128).

Cui, Y., Jing, X., & Mu, J. (2022). Integrated sensing and communications via 5G NR waveform: Performance analysis. In: *Proceedings of IEEE International Conference on Acoustics, Speech and Signal Process. (ICASSP)*, Singapore (pp. 8747–8751).

Cui, X., Cai, Y., Li, J., et al. (2024). An integrated waveform design method for underwater acoustic detection and communication. *Internet Technology Letters, 7*(6), 1–6.

Huang, G., Zhang, S., Jin, S., Liu, Y., Ding, Y., Li, X., & Nallanathan, A. (2025). A novel multi-beam directional modulation OFDM waveform design for integrated sensing and communication. *IEEE Transactions on Vehicular Technology*, 1-6. https://doi.org/10.1109/TVT.2025.3576252

Huang, G., Zhang, K., Zhang, Y., et al. (2024). Orthogonal Frequency Division Multiplexing Directional Modulation Waveform Design for Integrated Sensing and Communication Systems. *IEEE Internet Things Journal, 11*(18), 29588–29599.

Huang, G., Zhang, K., Zhang, Y., et al. (2024). Orthogona frequency division multiplexing directional modulation waveform design for integrated sensing and communication systems. *IEEE Internet Things Journal, 11*(18), 29588–29599.

Liu, H., Li, Y., Cheng, W., et al. (2024). Sensing-efficient transmit beamforming for ISAC with MIMO radar and MU-MIMO communication. *Remote Sensors, 16*(16), 1–22.

Liu, Z., Li, X., Ji, H., et al. (2023). Toward STAR-RIS-empowered integrated sensing and communications: Joint active and passive beamforming design. *IEEE Transactions on Vehicular Technology, 72*(12), 15991–16005.

Lu, J., Zhang, Q., Zhang, L., & Shi, W. (2018). Detection performance of active sonar based on underwater acoustic communication signals. In: *IEEE International Conference on Computer, Communication and Signal Processing*, Qingdao, China (pp. 1–5).

Mahipathi, A. C., Pardhasaradhi, B., Lingadevaru, P., et al. (2024). A survey on waveform design for radar-communication convergence. *IEEE Access, 12*(1), 75442–75461.

Niu, Q., Zhang, Q., & Shi, W. (2023). An Waveform design and signal processing method for integrated underwater detection and communication system. *IET Radar Sonar Navigation, 17*(4), 617–627.

Nowak, M., Wicks, M., Zhang, Z., & Wu, Z. (2016). Co-designed radar-communication using linear frequency modulation waveform. *IEEE Aerospace and Electronic Systems Magazine, 31*(10), 28–35.

Qin, Y., Liu, H., Yin, R., et al. (2024). Robust multi-model mobile target localization scheme based on underwater acoustic sensor networks. *Ocean Eng., 291*(1), 1–15.

Qin, Y., Liu, H., Yin, R., et al. (2024). Robust multi-model mobile target localization scheme based on underwater acoustic sensor networks. *Ocean Eng., 291*(1), 1–15.

Rou, H. S., Zhang, H., Zhang, X., et al. (2024). From orthogonal time frequency space to affine frequency-division multiplexing: A comparative study of next-generation waveforms for integrated sensing and communications in doubly dispersive channels. *IEEE Signal Processing Magazine, 41*(5), 71–86.

Sun, D., Zhang, K., Mei, J., et al. (2023). Low frequency three-dimensional DOA estimation for underwater gliders using an arbitrary tetrahedral array. *Applied Acoustics, 214*(1), 1–14.

Wang, H., Luo, X., Chen, F. et al. (2023). Co-design of underwater detection and communication-Part II: Communication-based shared waveform. In: *International Conference on Intelligent Autonomous Systems(ICoIAS), Qinhuangdao, China* (pp. 129–134).

Wang, X., Shi, X., Tian, J., Wang, J., & Sun, Z. (2025). Orthogonal hyperbolic frequency division multiplexing modulation for underwater acoustic communications. *IEEE Transactions on Wireless Communications, 24*(7), 6180–6194.

Wang, Y., Tao, M., & Sun, S. (2024). Cramé-Rao bound analysis and beamforming design for integrated sensing and communication with extended targets. *IEEE Transactions on Wireless Communications, 23*(11), 15987–16000.

Yan, J., Cui, L., Yang, X., et al. (2024). Design of an embedded system for integrated underwater communication and detection. *IEEE Embedded Systems Letters, 17*(2), 83–86.

Yan, J., Xu, Z., Luo, X., Chen, C., & Guan, X. (2022). A joint sonar-communication system based on multicarrier waveforms. *IEEE Signal Processing Letters, 29*(1), 777–781.

Yan, J., Zhang, T., You, K., et al. (2024). Research on underwater target localization algorithm considering the uncertainty of anchor position. *Dianzi Yu Xinxi Xuebao, 46*(1), 67–73.

Yin, J., Men, W., Han, X., & Guo, L. (2020). Integrated waveform for continuous active sonar detection and communication. *IET Radar Sonar Navigation, 14*(9), 1382–1390.

Zhang, Z., Qu, Y., Wu, Z. et al. (2018). Rf steganography via lfm chirp radar signals. *IEEE Transactions on Vehicular Technology*, *54*(3), 1221–1236.

Zheng, J., Chu, P., Wang, X., et al. (2023). Inner-frame time division multiplexing waveform design of integrated sensing and communication in 5G NR system. *Sensors, 23*(15), 1–19.

Chapter 4
Integrated Design of Communication and Control

4.1 Introduction

MUSs have been revolutionizing maritime operations, from military missions to scientific research (Peng et al. 2021; Vasilijević et al. 2017; Jawhar et al. 2019; Su et al. 2019; Yan et al. 2025) and are now transforming commercial industries by enhancing efficiency and slashing operational costs. Two fundamental tasks need to be addressed for the integrated design of communication and control in MUSs. Task I: Achieve communication-efficient and collision-free motion planning for individual underwater vehicles in fading channels and obstacle environments. Task II: Realize coordinated formation control for multiple AUVs with unknown interaction communication topology. For that reason, this section is organized to review the existing solutions and their limitations related with the above tasks. Followed by this, the main contributions of this chapter are presented.

4.1.1 Task I: Communication-Efficient and Collision-Free Motion Planning

For MUSs, one of the most critical issues is to plan paths for underwater vehicles. For instance, an energy-efficient motion planning strategy was provided in Yan et al. (2018b) to balance the communication energy consumption and prolong the network lifetime. Yetkin et al. (2020) incorporated the environment information into the path planning of underwater vehicles, through which a decision-theoretic based subsea search algorithm was designed. In Khan et al. (2019), the end-to-end data freshness constraint was conducted to determine the paths of underwater vehicles, whose aim was to retrieve the collected data to control center as soon as possible. Followed by this, a heuristic algorithm was provided in Yordanova and Gips (2020) to optimize the paths of underwater vehicles, with respect to data quality and underwater coverage

J. Yan and X. Guan, *Maritime Unmanned Systems*, Cognitive Intelligence and Robotics 1, https://doi.org/10.1007/978-981-95-6907-6_4

efficiency. Wang et al. (2021) employed acoustic camera to capture the position and shape of unknown underwater pipelines. These schemes are well developed, however, they do not take the collision avoidance into consideration. As we have seen already, obstacles such as wrecks and plankton inevitably exist in water, while at the same time the collision between vehicles may occur when they work together. The above collision constraint has a strong impact on the motion safety and communication channel of underwater vehicles (Boban et al. 2011). Therefore, it is necessary to plan a collision-free path for each underwater vehicle.

In order to resolve the above problem, Song et al. (2017) developed a joint flocking and guidance scheme for underwater vehicles evolving in environments with obstacles. In Lin et al. (2020a), an artificial potential-based motion planning strategy was conducted, where the software-defined technology was employed to improve the scalability and controllability. A multilayered motion planning scheme was presented in Mahmoudzadeh et al. (2019) for underwater navigation, wherein a local motion planner was employed to avoid collision with obstacles. Note that the dynamics models of underwater vehicles in Song et al. (2017), Lin et al. (2020a) and Mahmoudzadeh et al. (2019) treat each vehicle as a point mass, such that a second-order nonlinear equation can be conducted to describe the kinematics of underwater vehicles. However, the kinematics model cannot capture the low-level interactions during the implementation of motion planning on actual vehicle. As such, Heshmati-Alamdari et al. (2021) jointly considered the kinematics and dynamics models of underwater vehicles, through which a model predictive controller was developed to steer each underwater vehicle to the desired trajectory with collision avoidance. In Shou et al. (2022), an adaptive motion controller was provided to achieve finite-time formation control and obstacle avoidance for underwater vehicles. Nevertheless, the above motion planning schemes rely on full or partial knowledge of underwater vehicle dynamic model. Due to the harsh ocean conditions, it is difficult if not impossible to acquire the accurate dynamics model of underwater vehicles. With regard to this, an artificial potential function based motion planning scheme was developed in Sahu and Subudhi (2018) to relax the dependence of model parameters for underwater vehicles. Also of relevance, the Astar algorithms were designed in Wang et al. (2021) and Yu et al. (2019) to steer underwater vehicles to the target points with collision avoidance. Although the artificial potential function and Astar algorithms are simple to implement, they are easy to get stuck at locally optimal value. More recently, Refs. Jiang et al. (2020) and Kontoudis and Vamvoudakis (2019) employed the distributed learning algorithms to reduce the dependency on model parameters and achieve global optimization, however they are not developed in the context of collision-free motion planning for underwater vehicles. Due to the complex dynamics of underwater vehicles, the issue of how to adopt the learning strategy to design a collision-free motion planning scheme without relying on models for underwater vehicles is largely unexplored.

Apart from that, most of existing motion planning schemes focus on the control techniques and they ignore the influence of underwater acoustic communication channel. To be specific, they assume that the channel quality is approximated by a deterministic disk model. The above assumption is reasonable for terrestrial vehicles, however it is not valid for underwater vehicles. As has been pointed out in Stefanov and Stojanovic (2011) and He et al. (2022), underwater acoustic communication suffers from stronger shadowing and multipath fading than the terrestrial radio wave communication. Ignoring the shadowing and multipath fading factors may result in the deterioration of communication quality during the motion planning process. Thereby, we need to incorporate the underwater communication quality into the motion planning procedure metric, such that a communication-efficient and collision-free motion planning scheme can be developed to improve the communication capacity via the control feedback of underwater vehicles. The above idea is similar to the co-design of estimation and communication for multi-agent systems, e.g., Cao et al. (2018) and Cao et al. (2013). To this end, we notice that some communication-efficient motion planning schemes have been developed for terrestrial vehicles. For instance, a gradient estimation based motion controller was developed in Dixon and Frew (2012) to optimize the communication chain. In Guo et al. (2021) and Wu et al. (2018), two co-design frameworks for aerial vehicle motion planning and communication efficiency were constructed. Yan et al. (2012) integrated the probabilistic signal-to-noise ratio (SNR) prediction approach into the router planning of robots. Followed by this, Ali et al. 2019 extended the first-order linear kinematic model of vehicles to the second order, through which a motion-communication co-optimization solution was designed. Note that the dynamics model of vehicles in Dixon and Frew (2012), Guo et al. (2021), Wu et al. (2018), Yan and Mostofi (2012) and Ali et al. (2019) is reduced to a first-order or second-order kinematic equation, however it cannot capture the actual dynamics model of underwater vehicles. In Licea et al. (2020), the kinematic and dynamics models of robots were incorporated into the communication-aware motion planning. Nonetheless, the least square estimators are developed in the above literatures to seek the SNR parameters, which are easy to trap in local optimum. To compensate these shortcomings, the distributed learning can offer us with a feasible solution, since it seeks a global optimum solution via online learning and iteration Chen et al. (2021). Our previous works Yan et al. (2022) and Yan et al. (2021) employed the learning algorithm to solve the underwater localization problem. Nevertheless, how to develop a learning-based solution that can jointly solve collision-free motion planning and global-optimum SNR estimation is still an unsolved issue.

The main contributions of Task I in this chapter are summarized as follows:

(1) Integrated optimization framework for the co-design of communication efficiency and motion planning. We develop an integrated optimization framework, including underwater vehicle dynamics, communication capacity, collision avoidance and position control. As far as we know, this is the first integrated optimization framework for the communication-efficient and collision-free motion planning of underwater vehicles that involves realistic communication channel and dynamic model. It connects the communication capacity of sensor nodes with the motion planning of underwater vehicles, which can improve the communication capacity.

(2) Model-based learning estimator for online SNR prediction. A model-based integral reinforcement learning (IRL) estimator is designed to predict the SNR of underwater vehicles. It can effectively predict the probabilistic SNR parameters with limited channel knowledge. Compared with least squares or maximum likelihood based estimators, e.g., Yan and Mostofi (2012), Ali et al. (2019) and Licea et al. (2020), the IRL estimator in this chapter can avoid local minimum. Meanwhile, the shadowing and multipath fading effects are considered in this chapter, which are ignored by terrestrial vehicles, e.g., Dixon and Frew (2012).

(3) Model-free learning algorithm for collision-free motion planning. With the predicted SNR information, a model-free IRL algorithm is developed to steer underwater vehicles to the desired position points while avoiding collision with obstacles and the other vehicles. Different from the motion planning controllers in Song et al. (2017), Lin et al. (2020a) and Mahmoudzadeh et al. (2019), the developed motion planning algorithm in this chapter not only considers the collision avoidance, but also takes into account the dynamics model of underwater vehicles. Compared with the solutions in Heshmati-Alamdari et al. (2021) and Shou et al. (2022), it relaxes the dependency on vehicle model parameters.

4.1.2 *Task II: Optimally Persistent Formation with Unknown Interaction*

Beyond individual motion planning, the formation control is another critical issue for multiple AUVs, due to its improved efficiency, enlarged sensing coverage and enhanced robustness as compared to the single-AUV system. As shown by Fig. 4.1, the general idea of formation control is to enable multiple AUVs to maintain a specific shape while steering the group to arrive at the destination position. For terrestrial vehicles, a lot of formation controllers have been developed. For instance, a finite-time formation controller was designed in Guo and Zhang (2023), and a prescribed-time formation controller was designed in Chen et al. (2023) for surface vehicles with external disturbances. Also of relevance, the work in Liu and Bucknall (2018) provided a survey on the formation control of unmanned vehicles within different environment domains. The above works have great significance for solving the formation control issue of terrestrial/surface vehicles. However, they cannot be directly used to AUVs due to the complex underwater environment and weak acoustic communication (Stojanovic and Preisig 2009).

Based on the above properties, we should resolve the following issues: (1) how to reveal the interaction topology of AUVs when the sensing capability of each AUV is limited? (2) how to guarantee the topology connectivity and improve the energy efficiency of AUVs? (3) how to keep the desired formation shape when the model parameters of AUVs are uncertain? For the first issue, a topology-efficient estimation algorithm was designed in Diamant et al. (2018) to discover underwater acoustic links. The work in Liu and Zhao (2018) proposed a conflict-free topology discovery protocol to reveal the interaction topology of sensor nodes. In Jiang et al. (2023), an RL based neighbor discovery scheme was developed to identify the topology of underwater sensor networks. The work in Liu et al. (2021) presented a survey on

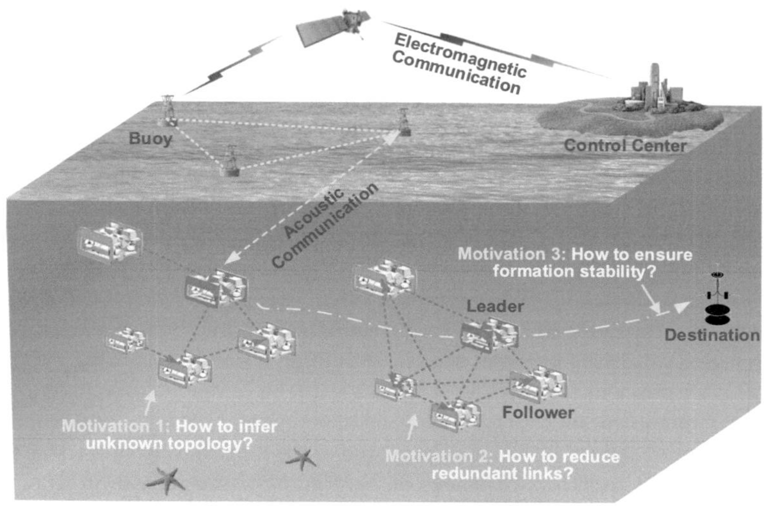

Fig. 4.1 Description of the formation control for multiple AUVs

the topology discovery protocols of MUSs. However, the above works mainly focus on static sensor networks, which cannot capture the mobility property of AUVs. If we directly apply the static network-suited protocols into dynamic network, a large amount of communication energy consumption will be required to tackle the time-variant topology among AUVs. For that reason, Vasquez et al. (2018) considered the mobility property of robotics, and then designed the topology inference algorithm from the closed-loop control perspective. Followed by this, a local topology inference algorithm for robotics was developed in Li et al. (2023), whose formation input was not required to be known in advance. Note that the first-order or second-order differential equations are conducted in Mendíez et al. (2018) and Li et al. (2023) to describe the motions of robotics, where each robot is treated as a point mass. The above description is not suitable for AUVs, since AUV typically has multiple degrees of freedom to maneuver various tasks in underwater environment. Besides that, the least squares-based estimator was adopted in Li et al. (2023) to seek topology parameters, which may easily converge to a local optimum. With consideration of the global optimization estimation, the design of a topology inference algorithm for multi-DOF AUVs is still lacking.

Once the interaction topology of AUVs is inferred, the next urgent issue is to optimize the network topology. With regard to this, many topology optimization algorithms have been proposed from different perspectives. For example, the work in Xu et al. (2023) utilized the angle of arrival measurements to construct an optimal topology for AUVs. In Mao and Yan (2018), the thermal-stress coupled topology optimization was conducted to alter the structure of battery rack and reduce the energy consumption of AUVs. In Chen et al. (2022), the topology of sensor nodes was optimized by an AUV-assisted routing mechanism. Note that the so-called neighboring rule is employed in Xu et al. (2023), Mao and Yan (2018) and Chen et al. (2022) to depict the connectivity of AUVs, i.e., each AUV requires to communicate with its neighbors within its sensing range. However, a good connectivity can result in the redundancy of network topology, which further leads to the reduction of network lifetime. Although the sparse deployment can improve the energy efficiency, it can lead to weak connectivity and even topology failure. So, there is a trade-off between topology connectivity and energy efficiency. To balance this trade-off, Xia et al. (2017) employed the graph theory to describe the topology relationship of underwater nodes, such that a global rigidity algorithm was designed to judge the localizability. Followed by this, a minimally rigid generation scheme was developed in Pang et al. (2023) for the adaptive formation reconfiguration of AUVs, and the node triangles were used in Wu et al. (2020) for high reliability-efficiency formation of AUVs. Our previous works (Yan et al. 2024, 2018b) employed the rigid topology to reduce communication energy consumption and guarantee topology connectivity for the formation control of AUVs. Compared to the neighboring rule-based formation strategies, the total link distances of rigid graph-based formation strategy can reach minimum on the basic of rigid structure. Despite all this, the solutions in Pang et al. (2023), Wu et al. (2020), Yan et al. (2024) and Yan et al. (2018b) are all based on the undirected rigid topologies, which cannot fully capture the complexity of a communication system. As far as we know, the direction of data flow is often crucial in understanding and optimizing the performance of communication networks. For that reason, it is necessary to construct a directed rigid graph to optimize the network topology of AUVs during the formation procedure.

After the above two issues are solved, the final and most important issue is to guarantee the formation stability. Till now, many formation controllers have been developed. For example, an active disturbance rejection-based formation controller for AUVs was presented in Wang et al. (2022). In Bhat et al. (2023), the model predictive control was employed to address the input constraint of AUVs during the formation procedure. Also of relevance, some other formation controllers can be found in Yuan et al. (2018) and Shou et al. (2021). However, the formation controllers in Wang et al. (2022), Bhat et al. (2023), Yuan et al. (2018) and Shou et al. (2021) require partial knowledge of AUV's dynamic parameters, which is difficult or impossible to accomplish due to the complex underwater environment. In view of this, the RL strategy has been widely employed to design model-free formation controllers for AUVs. Specifically, our previous work (Cao et al. 2023) considered the communication quality of AUVs, through which the RL strategy was conducted to achieve model-free formation control of AUVs. In Wu et al. (2019), a model-free RL framework for the depth control of AUVs in discrete time was studied. Besides that, Wang et al. (2023) considered the model uncertainty into the design of formation controller for AUVs. Note that the solutions in Cao et al. (2023), Wu et al. (2019) and Wang and Zhang (2023) require a pre-specified reward function. Nevertheless, in many underwater scenarios, it can be difficult or impossible to define a reward function, which means one should learn the reward function from demonstrations provided by an expert, without requiring explicit definition. In addition, the traditional RL strategy is sensitive to changes in the environment, as the optimal policy learned may no longer be effective if the environment changes. To make up these shortcomings, we find that the inverse reinforcement learning (IRL) can provide us an appropriate solution, due to its robustness to environment changes and the generalization to new environments. It should be emphasized that the IRL framework was first proposed by Ng et al. 2000 in 2000. Followed by this, the works in Xue et al. (2022) and Lian et al. (2023) applied the IRL into the cooperation control of multi-agent systems, where the nodes were considered as mass points. As mentioned above, AUVs have nonlinear property and multiple DOF, and hence, the solutions in Xue et al. (2022) and Lian et al. (2023) cannot be directly applied to the formation control of AUVs. To the best of our knowledge, how to design an IRL-based model-free formation controller that can both infer and optimize the interaction topology of AUVs is still largely unsolved.

The main contributions of Task II in this chapter are given as follows:

(1) IRL-based topology inference estimator. An IRL-based topology inference estimator is designed to acquire the interaction relationship of AUVs. Different from Diamant et al. (2018), Liu and Zhao (2018), Jiang et al. (2023) and Liu et al. (2021), the proposed estimator in this chapter can capture the mobility property of multi-DOF AUVs. Compared to the estimators in Mendíez et al. (2018) and Li et al. (2023), the proposed estimator can acquire the global optimization value by policy evaluation and improvement during the IRL procedure.

(2) Topology optimization via optimally persistent graph. To reduce redundant links, an optimally persistent topology is generated by graph direction and contraction. Different with the neighboring rule-based methods Xu et al. (2023), Mao and Yan (2018) and Chen et al. (2022), the proposed solution can balance topology connectivity and energy efficiency. Compared to the undirected rigid graph in Yan et al. (2024) and Yan et al. (2018b), it can further reduce the communication energy consumption by the persistent graph.

(3) IRL-based formation controller. With consideration of the unknown model parameters, an IRL-based formation controller is designed for AUVs. Different from the controllers in Wang et al. (2022), Bhat et al. (2023), Yuan et al. (2018) and Shou et al. (2021), the proposed controller does not need the model parameters of AUVs. Compared to the traditional RL solutions (Cao et al. 2023; Wu et al. 2019; Wang and Zhang 2023), it has better environment adaptability by learning from expert demonstrations and inferring the underlying reward function.

4.2 Communication-Efficient and Collision-Free Motion Planning

We would like to transmit the data information from source sensor node to destination sensor node, as shown in Fig. 4.2. To this end, the following four types of nodes are provided: (1) *Source sensor node*. The position of source sensor node is fixed, and it sends the collected data to the destination sensor node through the relays of

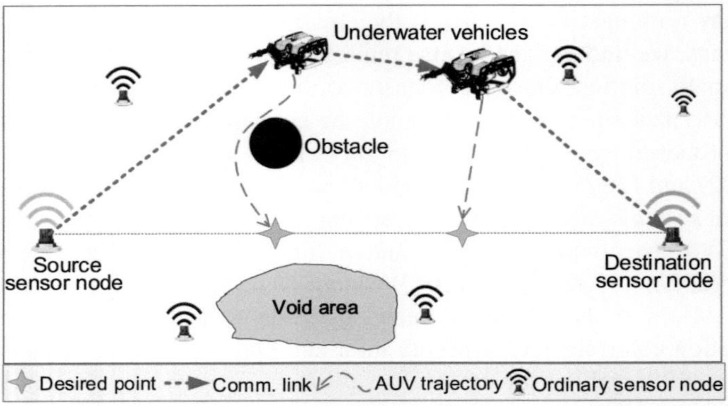

Fig. 4.2 Communication links from source sensor node to destination sensor node through the motion relays of underwater vehicles

underwater vehicles; (2) *Destination sensor node*. The position of destination sensor node is also fixed, whose objective is to indirectly receive the data from source sensor node; (3) *Underwater vehicle*. Underwater vehicles act as mobile communication relaying nodes; (4) *Ordinary sensor node*. Ordinary sensor nodes are provided to sense and collect channel measurements to underwater vehicles, which donot undertake the relay task.

On the basis of the above framework, we employ a team of ordinary sensor nodes to predict the stochastic SNR parameters. In view of this, let $\mathbf{p}_\mathrm{s} = \{\mathbf{p}_{\mathrm{s},1}, \ldots, \mathbf{p}_{\mathrm{s},N}\}$ be the team position set of ordinary sensor nodes and $\mathbf{p}_{\mathrm{s},i} = [x_{\mathrm{s},i}, y_{\mathrm{s},i}, z_{\mathrm{s},i}]^\mathrm{T}$ be the position vector of ordinary sensor node $i \in \mathcal{V}_\mathrm{s} = \{1, \ldots, N\}$, where $x_{\mathrm{s},i}$, $y_{\mathrm{s},i}$ and $z_{\mathrm{s},i}$ are the positions on X, Y and Z axis, respectively. Let $\mathcal{E} = \{R_0, R_1, \ldots, R_{M+1}\}$ be the end-to-end communication chain, where R_0 denotes the source sensor node, R_{M+1} denotes the destination sensor node, and the others are the underwater vehicles. Specifically, $\mathcal{V} = \{R_1, \ldots, R_M\}$ represents the set of underwater vehicles, where each underwater vehicle R_i relays data from its single source neighbor R_{i-1} to its single destination neighbor R_{i+1}. For underwater vehicle R_i, its position vector can be defined as $\mathbf{p}_i = [x_i, y_i, z_i]^\mathrm{T}$, while the position vectors of source senor node and destination senor node are defined as $\mathbf{p}_0 = [x_0, y_0, z_0]^\mathrm{T}$ and $\mathbf{p}_{M+1} = [x_{M+1}, y_{M+1}, z_{M+1}]^\mathrm{T}$, respectively. Besides that, N and M are the total numbers of ordinary sensor node and underwater vehicles.

The inertial reference frame (IRF) and body-fixed reference frame (BRF) are jointly utilized to depict the dynamic model of underwater vehicles. The position and orientation vector for underwater vehicle $R_i \in \mathcal{V}$ in IRF is defined as $\eta_i = [\mathbf{p}_i; \psi_i]$, where ψ_i is the angle on yaw. The linear and angle velocity vector in BRF is $\mathbf{v}_i = [u_i, v_i, w_i, r_i]^\mathrm{T}$, where u_i, v_i and w_i are the linear velocities on surge, sway and heave, respectively. In addition, r_i is the angle velocity on yaw. From Cui et al. (2017) and Cui et al. (2017), the dynamic model of underwater vehicle R_i can be described by

$$\begin{aligned}\dot{\eta}_i &= \mathbf{J}_i(\eta_i)\mathbf{v}_i,\\ \mathbf{M}_i\dot{\mathbf{v}}_i + \mathbf{C}_i(\mathbf{v}_i)\mathbf{v}_i &+ \mathbf{D}_i(\mathbf{v}_i)\mathbf{v}_i + \mathbf{g}_i(\eta_i) = \tau_i,\end{aligned} \tag{4.1}$$

where $\mathbf{M}_i \in \mathcal{R}^{4\times4}$, $\mathbf{C}(\mathbf{v}_i) \in \mathcal{R}^{4\times4}$ and $\mathbf{D}(\mathbf{v}_i) \in \mathcal{R}^{4\times4}$ are the inertia, Coriolis-centripetal and damping matrices, respectively. $\mathbf{J}_i(\eta_i) \in \mathcal{R}^{4\times4}$ is the rotation matrix, $\mathbf{g}_i(\eta_i) \in \mathcal{R}^4$ is the hydrostatic force, and $\tau_i = [\tau_{u_i}, \tau_{v_i}, \tau_{w_i}, \tau_{r_i}]^{\mathrm{T}}$ is the control input, where $\tau_{u_i}, \tau_{v_i}, \tau_{w_i}$ and τ_{r_i} are the control forces on surge, sway, heave and yaw, respectively.

Define $\mathbf{X}_i = [\eta_i; \mathbf{v}_i]$, and hence, model (4.1) is rearranged as

$$\dot{\mathbf{X}}_i = \mathbf{A}_i\mathbf{X}_i + \mathbf{B}_i\tau_i + \mathbf{G}_i, \tag{4.2}$$

with

$$\mathbf{A}_i = \begin{bmatrix} \mathbf{0} & \mathbf{J}_i(\eta_i) \\ \mathbf{0} & -\mathbf{M}_i^{-1}(\mathbf{C}_i(\mathbf{v}_i)+\mathbf{D}_i(\mathbf{v}_i)) \end{bmatrix}, \tag{4.3}$$

$$\mathbf{B}_i = \begin{bmatrix} \mathbf{0} \\ \mathbf{M}_i^{-1} \end{bmatrix}, \ \mathbf{G}_i = \begin{bmatrix} \mathbf{0} \\ -\mathbf{M}_i^{-1}\mathbf{g}_i(\eta_i) \end{bmatrix}. \tag{4.4}$$

In order to improve the communication capacity, the link capacity $\mathcal{C}_i$ of underwater vehicle R_i is introduced. Referring to Rappaport (2002), one knows the link capacity of an end-to-end communication chain is equal to the capacity of the worst link. Along with this, the link capacity $\mathcal{C}_i$ is defined as

$$\mathcal{C}_i = \min\{c_{i-1,i}, c_{i,i+1}\}, \tag{4.5}$$

where $c_{i-1,i}$ denotes the link capacity between source neighbor R_{i-1} and underwater vehicle R_i. Meanwhile, $c_{i,i+1}$ denotes the link capacity between underwater vehicle R_i and destination neighbor R_{i+1}.

Note that the link capacity is a function of bandwidth and communication quality SNR (Rappaport 2002). Hence, the SNR between source neighbor R_{i-1} and underwater vehicle R_i can be expressed as

$$\begin{aligned} &\mathrm{SNR}_{\mathrm{dB}}(\mathbf{p}_{i-1}, \mathbf{p}_i) \\ &= K_{\mathrm{dB}} - 10n_{\mathrm{PL}}\log_{10}(l_{i-1,i}) - 10l_{i-1,i}\log_{10}(\alpha(f)) \\ &\quad -N_{\mathrm{dB}}^0 + \sigma_{\mathrm{SH}}(\mathbf{p}_{i-1}, \mathbf{p}_i) + \mu_{\mathrm{MP}}(\mathbf{p}_{i-1}, \mathbf{p}_i), \end{aligned} \tag{4.6}$$

where $10\log_{10}\alpha(f) = \frac{0.11f^2}{1+f^2} + \frac{44f^2}{4100+f^2} + 2.75\times10^{-4}f^2 + 0.003$. In addition, K_{dB} denotes the average energy consumption of transmitting one bit data in dB, n_{PL} denotes the spreading coefficient, $l_{i-1,i} = \|\mathbf{p}_{i-1} - \mathbf{p}_i\|$ denotes the relative distance between source neighbor R_{i-1} and underwater vehicle R_i, N_{dB}^0 denotes the noise power spectral density in dB, and $\alpha(f)$ is the acoustic absorption with frequency f. Moreover, $\sigma_{\mathrm{SH}}(\mathbf{p}_{i-1}, \mathbf{p}_i)$ and $\mu_{\mathrm{MP}}(\mathbf{p}_{i-1}, \mathbf{p}_i)$ represent the location-related stochastic parameters, which reflect the effects of shadowing and multipath fading, respectively.

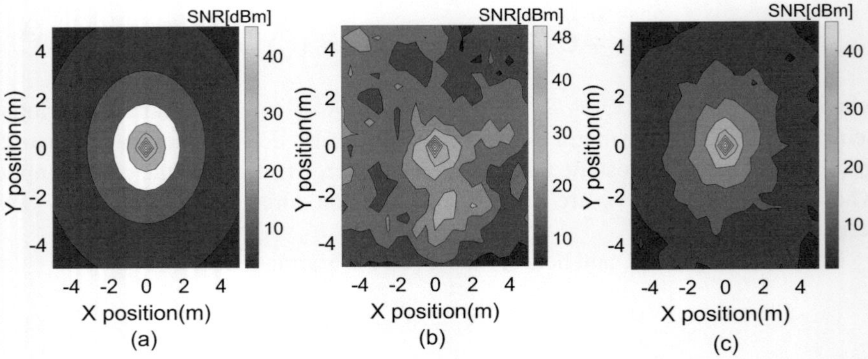

Fig. 4.3 **a** Only the path loss is considered; **b** Path loss and shadowing are considered; **c** Path loss, multipath fading are considered

Remark 4.1 Different from the simplified SNR models in Zhang et al. (2021) and Lucani et al. (2008), the shadow underwater fading parameter σ_{SH} and the multipath parameter μ_{MP} are both considered in this chapter, which can well capture the realistic underwater environment. To explain more clearly, an example of the shadowing and multipath fading is shown in Fig. 4.3.

From (4.6) and noting with the Shannon-Hartley Theorem (Dixon and Frew 2012), the link capacity $c_{i-1,i}$ which provides the theoretical upper bound is denoted by

$$c_{i-1,i} = B \log_2(1 + S_{i-1,i}), \tag{4.7}$$

with

$$S_{i-1,i} = 10^{\frac{\text{SNR}_{\text{dB}}(\mathbf{p}_{i-1}, \mathbf{p}_i)}{10}}, \tag{4.8}$$

where B denotes the communication bandwidth. Similarly, the detailed expression of $c_{i,i+1}$ can also be acquired.

Assumption 4.1 *The obstacles in underwater environment can be covered by convex cylinders. The mth*$(m = 1, 2, \ldots)$ *obstacle in the environment is denoted as* $B_m(O_m, \rho_m)$, *where* O_m *is its center and* $\rho_m(\rho_m > 0)$ *is its radius.*

Definition 4.1 Obstacle set: For underwater vehicle R_i at time t, the detected obstacle set can be defined as the subset $\Omega_i^t \subset \{B_1(O_1, \rho_1), \ldots, B_m(O_m, \rho_m), \ldots\}$ in the detected range of underwater vehicle $R_i \in \mathcal{V}$.

Assumption 4.2 *The obstacles are sparsely exist in the underwater environment, and hence, the impact of obstacles on communication channel is ignored. Of note, this assumption has been made in some existing works, e.g., Yan and Mostofi (2012) and Hurst et al. (2021).*

Assumption 4.3 *Without loss of generality, the offline map of the target area is to be known by the sonar installed on underwater vehicle. Meanwhile, underwater vehicle is equipped with camera, which is capable of online detecting obstacles within a certain range.*

Accordingly, the following two problems are formulated.

Problem 1 (SNR Prediction in Fading Channel). The shadow fading and multipath parameters in SNR model cannot be obtained previously. In view of this, we attempt to design a model-based IRL estimator to capture the unknown channel parameters. This problem can be reduced to the estimation of K_{dB}, n_{PL}, σ_{SH} and μ_{MP} with the limited channel measurements from ordinary sensor nodes $i \in \{1, \ldots, N\}$.

Problem 2: (Collision-Free Motion Planning). It is impossible to acquire the accurate dynamic model of underwater vehicles. Meanwhile, obstacles increase the difficulty of motion planning of underwater vehicles. In view of this, we aim to employ IRL to develop a model-free and collision-free motion planning algorithm. This problem is reduced to maximize $\mathcal{C}_i$ while guaranteeing $\|\mathbf{p}_i - O_m\| > \rho_m$ and $\|\mathbf{p}_i - \mathbf{p}_j\| > 0$.

4.2.1 SNR Prediction in Fading Channel

We first design an IRL-based estimator to capture the unknown shadowing and multipath parameters. Along with this, the IRL is adopted to develop a model-free and collision-free motion planning algorithm for underwater vehicles. Finally, the theoretical analysis for our solution are presented.

Initially, underwater vehicle $R_i \in \mathcal{V}$ at location $\mathbf{p}_i$ broadcasts an initiator message to its neighboring ordinary sensor nodes. Then, underwater vehicle R_i switches into the listening mode. For any ordinary sensor node $j \in \mathcal{N}_i$, it senses the SNR of underwater vehicle R_i, denoted by $\text{SNR}_{\text{dB}}(\mathbf{p}_i, \mathbf{p}_{\text{s},j})$ where $\mathcal{N}_i$ is the neighboring ordinary sensor set of underwater vehicle R_i. After that, ordinary sensor node $j \in \mathcal{N}_i$ replies its position and SNR measurement to underwater vehicle R_i. Repeating the above procedure, the collected messages on underwater vehicle $R_i \in \mathcal{V}$ can be expressed as

$$\left\{\mathbf{p}_{\text{s},j},\ \text{SNR}_{\text{dB}}(\mathbf{p}_i, \mathbf{p}_{\text{s},j})\right\}_{j\in\mathcal{N}_i}. \tag{4.9}$$

For clear of expression, the ordinary sensor nodes in set $\mathcal{N}_i$ are labeled as $1_i, 2_i, \ldots, |\mathcal{N}_i|_i$. We stack the above SNR measurements into a vector $\mathbf{Y}_{\text{dB}}^{R_i} = [\text{SNR}_{\text{dB}}(\mathbf{p}_i, \mathbf{p}_{\text{s},1_i}), \ldots, \text{SNR}_{\text{dB}}(\mathbf{p}_i, \mathbf{p}_{\text{s},|\mathcal{N}_i|_i})]^{\text{T}}$. Noting with (4.6), one can have

$$\mathbf{Y}_{\text{dB}}^{R_i} = \mathbf{H}_{R_i}\theta_{R_i} - \varepsilon_{R_i} + \sigma_{R_i} + \mu_{R_i}, \tag{4.10}$$

with

$$\mathbf{H}_{R_i} = \begin{bmatrix} 1 & -10\log_{10}(\|\mathbf{p}_{s,1_i} - \mathbf{p}_i\|) \\ \vdots & \vdots \\ 1 & -10\log_{10}(\|\mathbf{p}_{s,|\mathcal{N}_i|_i} - \mathbf{p}_i\|) \end{bmatrix},$$

$$\varepsilon_{R_i} = \begin{bmatrix} 1 & \|\mathbf{p}_{s,1_i} - \mathbf{p}_i\| \\ \vdots & \vdots \\ 1 & \|\mathbf{p}_{s,|\mathcal{N}_i|_i} - \mathbf{p}_i\| \end{bmatrix} \begin{bmatrix} N_{\text{dB}}^0 \\ 10\log_{10}(\alpha(f)) \end{bmatrix},$$

where $\theta_{R_i} = [K_{\text{dB},i}, n_{\text{PL},i}]^{\text{T}}$, $\sigma_{R_i} = [\sigma_{\text{SH}}(\mathbf{p}_i, \mathbf{p}_{s,1_i}), \ldots, \sigma_{\text{SH}}(\mathbf{p}_i, \mathbf{p}_{s,|\mathcal{N}_i|_i})]^{\text{T}}$ and $\mu_{R_i} = [\mu_{\text{MP}}(\mathbf{p}_i, \mathbf{p}_{s,1_i}), \ldots, \mu_{\text{MP}}(\mathbf{p}_i, \mathbf{p}_{s,|\mathcal{N}_i|_i})]^{\text{T}}$. It is worth mentioning that N_{dB}^0 and f can be acquired by the priori knowledge.

In (4.6), θ_{R_i} is a deterministic vector, representing the offset and slope of path loss, while σ_{R_i} and μ_{R_i} are location-related random shadowing and multipath fading parameter vectors, respectively. We can estimate θ_{R_i} by the available measurements, however one cannot estimate σ_{R_i} and μ_{R_i} due to their randomness. For that reason, we estimate the statistical characteristics of σ_{R_i} and μ_{R_i}, rather than the real-time values of σ_{R_i} and μ_{R_i}. It is assumed that σ_{R_i} is captured by a zero-mean Gaussian noise with an exponential spatial correlation. Similar to the assumption in Yan and Mostofi (2012) and Goldsmith (2005), μ_{R_i} is captured by lognormal distribution without the spatial correlation. Accordingly, we employ the spatial correlation to predict the SNR parameters. Then, the covariance matrices of σ_{R_i} and μ_{R_i} can be expressed as

$$\Upsilon_{R_i} = \xi_{R_i}^2 \begin{bmatrix} 1 & \cdots & \exp\{\frac{d_{1_i,|\mathcal{N}_i|_i}}{-\varphi_{R_i}}\} \\ \exp\{\frac{d_{2_i,1_i}}{-\varphi_{R_i}}\} & \cdots & \exp\{\frac{d_{2_i,|\mathcal{N}_i|_i}}{-\varphi_{R_i}}\} \\ \vdots & \ddots & \vdots \\ \exp\{\frac{d_{|\mathcal{N}_i|_i,1_i}}{-\varphi_{R_i}}\} & \cdots & 1 \end{bmatrix}$$

$$\Psi_{R_i} = \rho_{R_i}^2 \mathbf{I}_{|\mathcal{N}_i|_i}, \tag{4.11}$$

where $\xi_{R_i}^2$ is the power of shadowing, φ_{R_i} is the parameter controlling the spatial correlation, $\rho_{R_i}^2$ is the multipath fading power, and $d_{j_i,\mathbf{j}_i}$ is the distance between ordinary sensor nodes $j_i \in \{1_i, \ldots, |\mathcal{N}_i|_i\}$ and $\mathbf{j}_i \in \{1_i, \ldots, |\mathcal{N}_i|_i\}/\{j_i\}$.

Since σ_{R_i} and μ_{R_i} are independent, one can define the variance of $\mathbf{Y}_{\text{dB}}^{R_i}$ as $\Omega_{R_i} = \Upsilon_{\mathbf{R_i}} + \Psi_{R_i}$. Hence, the estimations of θ_{R_i} and Ω_{R_i} can be acquired by maximizing the following maximum likelihood function, i.e.,

$$\begin{aligned} & f(\mathbf{Y}_{\text{dB}}^{R_i} | \theta_{R_i}, \Omega_{R_i}) \\ & = \frac{1}{\sqrt{(2\pi)^{|\mathcal{N}_i|_i} |\Omega_{R_i}|}} \exp\left\{-\frac{1}{2}(\mathbf{Y}_{\text{dB}}^{R_i} - \mathbf{H}_{R_i}\theta_{R_i} \right. \\ & \left. + \varepsilon_{R_i})^{\text{T}} \Omega_{R_i}^{-1} (\mathbf{Y}_{\text{dB}}^{R_i} - \mathbf{H}_{R_i}\theta_{R_i} + \varepsilon_{R_i})\right\}, \end{aligned} \tag{4.12}$$

and hence, for $\Delta_{R_i} = \mathbf{Y}_{\mathrm{dB}}^{R_i} - \mathbf{H}_{R_i}\theta_{R_i} + \varepsilon_{R_i}$, it yields

$$\begin{aligned}&\ln f(\mathbf{Y}_{\mathrm{dB}}^{R_i}\left|\theta_{R_i}, \Omega_{R_i}\right.)\\&= -\tfrac{1}{2}\Delta_{R_i}^{\mathrm{T}}\Omega_{R_i}^{-1}\Delta_{R_i} - \tfrac{1}{2}\ln\left|\Omega_{R_i}\right| - \tfrac{|\mathcal{N}_i|_i}{2}\ln(2\pi).\end{aligned} \tag{4.13}$$

In our previous work (Cao et al. 2020), a separate design strategy was developed, where θ_{R_i} was estimated in Phase I, $\xi_{R_i}^2$, $\rho_{R_i}^2$ and φ_{R_i} were estimated in Phase II, and the iteration was conducted in Phase III. The above separate design has high computational complexity, while its estimation accuracy is sensitive to the measurement noise. To cover these deficiencies, this chapter jointly estimates θ_{R_i}, $\xi_{R_i}^2$, $\rho_{R_i}^2$ and φ_{R_i}. To this end, we differentiate (4.13) with respect to variable θ_{R_i} and variance Ω_{R_i}, and hence, one can further have

$$\frac{\partial \ln f(\mathbf{Y}_{\mathrm{dB}}^{R_i}\left|\theta_{R_i}, \Omega_{R_i}\right.)}{\partial\theta_{R_i}} = -\mathbf{H}_{R_i}^{\mathrm{T}}\Omega_{R_i}^{-1}\Delta_{R_i}, \tag{4.14}$$

$$\frac{\partial \ln f(\mathbf{Y}_{\mathrm{dB}}^{R_i}\left|\theta_{R_i}, \Omega_{R_i}\right.)}{\partial\Omega_{R_i}} = \frac{\Omega_{R_i}^{-1}\Delta_{R_i}\Delta_{R_i}^{\mathrm{T}}\Omega_{R_i}^{-1} - \Omega_{R_i}^{-1}}{2}. \tag{4.15}$$

From (4.14) and (4.15), the optimization of θ_{R_i} and Ω_{R_i} is acquired by solving $\Delta_{R_i} = \mathrm{argmin}_{\Delta_{R_i}}||\frac{\partial \ln f(\mathbf{Y}_{\mathrm{dB}}^{R_i}|\theta_{R_i}, \Omega_{R_i})}{\partial\theta_{R_i}}||$ and $\frac{\partial \ln f(\mathbf{Y}_{\mathrm{dB}}^{R_i}|\theta_{R_i}, \Omega_{R_i})}{\partial\Omega_{R_i}} = 0$, which are equal to $\Delta_{R_i} = \varkappa_{R_i}$ and $\Omega_{R_i} = \Delta_{R_i}\Delta_{R_i}^{\mathrm{T}}$. Of note, $\varkappa_{R_i}$ is the sum vector of multipath and shadow fading. Based on this and with the definition in (4.11), one rearranges Ω_{R_i} as

$$\begin{aligned}&\begin{bmatrix} \xi_{R_i}^2 + \rho_{R_i}^2 & \cdots & \xi_{R_i}^2\exp\{\frac{d_{1_i,|\mathcal{N}_i|_i}}{-\varphi_{R_i}}\} \\ \xi_{R_i}^2\exp\{\frac{d_{2_i,1_i}}{-\varphi_{R_i}}\} & \cdots & \xi_{R_i}^2\exp\{\frac{d_{2_i,|\mathcal{N}_i|_i}}{-\varphi_{R_i}}\} \\ \vdots & \ddots & \vdots \\ \xi_{R_i}^2\exp\{\frac{d_{|\mathcal{N}_i|_i,1_i}}{-\varphi_{R_i}}\} & \cdots & \xi_{R_i}^2 + \rho_{R_i}^2 \end{bmatrix}\\ &= \begin{bmatrix} ([\Delta_{R_i}]_{1_i})^2 & \cdots & [\Delta_{R_i}]_{1_i}[\Delta_{R_i}]_{|\mathcal{N}_i|_i} \\ [\Delta_{R_i}]_{2_i}[\Delta_{R_i}]_{1_i} & \cdots & [\Delta_{R_i}]_{2_i}[\Delta_{R_i}]_{|\mathcal{N}_i|_i} \\ \vdots & \ddots & \vdots \\ [\Delta_{R_i}]_{|\mathcal{N}_i|_i}[\Delta_{R_i}]_{1_i} & \cdots & ([\Delta_{R_i}]_{|\mathcal{N}_i|_i})^2 \end{bmatrix}\end{aligned} \tag{4.16}$$

where $[\Delta_{R_i}]_j$ is the j-th element of Δ_{R_i}.

Based on (4.14), (4.15) and (4.16), we can easily obtain

$$\mathbf{Y}_{\mathrm{dB}}^{R_i} - \mathbf{H}_{R_i}\theta_{R_i} + \varepsilon_{R_i} = \varkappa_{R_i}, \tag{4.17}$$

$$\xi_{R_i}^2 + \rho_{R_i}^2 = \frac{1}{|\mathcal{N}_i|_i} \sum_{j_i=1_i}^{|\mathcal{N}_i|_i} ([\Delta]_{j_i})^2, \tag{4.18}$$

$$\xi_{R_i}^2 \exp\left\{\frac{d_{j_i,\mathbf{j}_i}}{-\varphi_{R_i}}\right\} = [\Delta]_{j_i}[\Delta]_{\mathbf{j}_i}. \tag{4.19}$$

We define $\chi_{R_i} = [\theta_{R_i}; \ln \xi_{R_i}^2; 1/\varphi_{R_i}; \rho_{R_i}^2]$, while the estimations of θ_{R_i}, $\xi_{R_i}^2$, φ_{R_i} and $\rho_{R_i}^2$ are denoted by $\hat{\theta}_{R_i}$, $\hat{\xi}_{R_i}^2$, $\hat{\varphi}_{R_i}$ and $\hat{\rho}_{R_i}^2$, respectively. Based on the above results, the optimization of θ_{R_i}, $\xi_{R_i}^2$, φ_{R_i} and $\rho_{R_i}^2$ is conducted as

$$\begin{aligned}
&\hat{\chi}_{R_i}^* \\
&= \operatorname{argmin} \Bigg\{ \underbrace{\lambda_1 \left(\frac{1}{|\mathcal{N}_i|_i} \Delta_{R_i}^{\mathrm{T}} \Delta_{R_i} - \hat{\xi}_{R_i}^2 - \hat{\rho}_{R_i}^2\right)^2}_{\text{Part 1}} \\
&\quad \underbrace{+ \sum\nolimits_{(j_i,\mathbf{j}_i)\in\mathbf{G}} \lambda_2 \left[\ln \hat{\xi}_{R_i}^2 - \frac{d_{j_i,\mathbf{j}_i}}{\hat{\varphi}_{R_i}} - \ln([\Delta]_{j_i}[\Delta]_{\mathbf{j}_i})\right]^2}_{\text{Part 2}} \\
&\quad \underbrace{+ (\mathbf{Y}_{\mathrm{dB}}^{R_i} - \mathbf{H}_{R_i}\theta_{R_i} + \varepsilon_{R_i})^{\mathrm{T}} \mathbf{Q}_1 (\mathbf{Y}_{\mathrm{dB}}^{R_i} - \mathbf{H}_{R_i}\theta_{R_i} + \varepsilon_{R_i}) \Bigg\}}_{\text{Part 3}},
\end{aligned} \tag{4.20}$$

where $\mathbf{G} = \{(1_i, 2_i), \ldots, (j_i, \mathbf{j}_i), \ldots, (|\mathcal{N}_i|_i, |\mathcal{N}_i|_i - 1)\}$ denotes the set of ordinary sensor nodes pairs for j_i and $\mathbf{j}_i$. In addition, $\lambda_1 > 0$ and $\lambda_2 > 0$ are the tuning indexes of shadowing and multipath cost function terms, respectively. Moreover, $\mathbf{Q}_1$ is a positive definite matrix.

Let $\hat{\chi}_{R_i}$ be the estimation of χ_{R_i}, and $\mathbf{u} \in \mathcal{R}^4$ be the increment input vector of $\hat{\chi}_{R_i}$. Hence, the estimation procedure of χ_{R_i} can be described as

$$\dot{\hat{\chi}}_{R_i} = \mathbf{u}. \tag{4.21}$$

A model-based IRL estimator is developed to seek $\hat{\chi}_{R_i}^*$, whose basic idea is to minimize the integral temporal difference error (Liu et al. 2020; He et al. 2021). Then, the cost function is defined as

$$\begin{aligned}
&g_1(\hat{\chi}_{R_i}, \mathbf{u}) \\
&= \lambda_1 \left(\tfrac{1}{|\mathcal{N}_i|_i} \Delta_{R_i}^{\mathrm{T}} \Delta_{R_i} - \hat{\xi}_{R_i}^2 - \hat{\rho}_{R_i}^2\right)^2 \\
&\quad + \sum\nolimits_{(j_i,\mathbf{j}_i)\in\mathbf{G}} \lambda_2 [\ln \hat{\xi}_{R_i}^2 - \tfrac{d_{j_i,\mathbf{j}_i}}{\hat{\varphi}_{R_i}} - \ln([\Delta]_{j_i}[\Delta]_{\mathbf{j}_i})]^2 \\
&\quad + \Delta_{R_i}^{\mathrm{T}} \mathbf{Q}_1 \Delta_{R_i} + \mathbf{u}^{\mathrm{T}} \mathbf{R}_1 \mathbf{u},
\end{aligned} \tag{4.22}$$

where $\mathbf{R}_1$ is a positive definite matrix.

From (4.22), the value function for the estimation ofχ_{R_i} is

$$V_1(\hat{\chi}_{R_i}(t)) = \int_t^{t+T} g_1(\hat{\chi}_{R_i}, \mathbf{u})\mathrm{d}\tau + V_1(\hat{\chi}_{R_i}(t+T)), \tag{4.23}$$

and hence, the optimal value of χ_{R_i} is to select $\mathbf{u}$, such that an optimal update policy of $\mathbf{u}^*$ can be obtained, i.e.,

$$\mathbf{u}^* = \underset{\mathbf{u}}{\operatorname{argmin}} \left\{ \int_t^{t+T} g_1(\hat{\chi}_{R_i}, \mathbf{u})\mathrm{d}\tau + V_1(\hat{\chi}_{R_i}(t+T)) \right\}. \tag{4.24}$$

In the following, the IRL strategy includes two steps, i.e., policy evaluation and policy improvement. In policy evaluation, $V_1(\hat{\chi}_{R_i}(t))$ is evaluated by using (4.23), given the current update policy. In policy improvement, the optimal update policy is selected until the convergence are reached for the iteration procedure. The above steps are detailed as follows.

(1-a) *Initialization*. Initially, the policy and value function are set as $\mathbf{u}^{(0)}(0) = \mathbf{0}$ and $V^{(0)}(\hat{\chi}_{R_i}(0)) = 0$.

(1-b) *Policy Evaluation*. For each iteration s, one calculates the following value function

$$V_1^{(s)}(\hat{\chi}_{R_i}(t)) = \int_t^{t+T} g_1(\hat{\chi}_{R_i}, \mathbf{u}^{(s)})\mathrm{d}\tau + V_1^{(s)}(\hat{\chi}_{R_i}(t+T)). \tag{4.25}$$

(1-c) *Policy Improvement*. Find an updated control policy $\mathbf{u}_1^{(s+1)}$ through the following rule

$$\mathbf{u}^{(s+1)} = \underset{\mathbf{u}^{(s)}}{\operatorname{argmin}} \left\{ \int_t^{t+T} g_1(\hat{\chi}_{R_i}, \mathbf{u}^{(s)})\mathrm{d}\tau + V_1^{(s)}(\hat{\chi}_{R_i}(t+T)) \right\} \tag{4.26}$$

which yields $\mathbf{u}^{(s+1)} = -\frac{1}{2}\mathbf{R}_1^{-1}\frac{\partial V_1^{(s)}(\hat{\chi}_{R_i}(t))}{\partial \hat{\chi}_{R_i}}$.

In order to smoothly approximate the value function $V_1^{(s)}(\hat{\chi}_{R_i}(t))$, a critic network is introduced as

$$V_1^{(s)}(\hat{\chi}_{R_i}(t)) = \mathbf{W}_1^{(s)^{\mathrm{T}}}\phi_1(\hat{\chi}_{R_i}(t)), \tag{4.27}$$

where $\phi_1(\hat{\chi}_{R_i}(t))$ is basis function for weight vector $\mathbf{W}_1$.

Based on (4.27), one rearranges (4.25) as

$$\begin{aligned} &\mathbf{W}_1^{(s)^{\mathrm{T}}}\phi_1(\hat{\chi}_{R_i}(t)) \\ &= \int_t^{t+T} g_1(\hat{\chi}_{R_i}, \mathbf{u}^{(s)})\mathrm{d}\tau + \mathbf{W}_1^{(s)^{\mathrm{T}}}\phi_1(\hat{\chi}_{R_i}(t+T)), \end{aligned} \tag{4.28}$$

and its residual error can be expressed as

$$e_1(t) = \mathbf{W}_1^{(s)^{\mathrm{T}}} \left(\phi_1(\hat{\chi}_{R_i}(t)) - \phi_1(\hat{\chi}_{R_i}(t+T))\right) - \int_t^{t+T} g_1(\hat{\chi}_{R_i}, \mathbf{u}^{(s)})\mathrm{d}\tau. \quad (4.29)$$

The purpose of weight updating is to minimize the overall residual error, i.e., $\min \int_0^{\infty} e_1(\tau)\mathrm{d}\tau$. Hence, the recursive least square method is used to update the weight $\mathbf{W}_1$. Let $\varrho_1^{(s)} = \int_t^{t+T} g_1(\hat{\chi}_{R_i}, \mathbf{u}^{(s)})\mathrm{d}\tau$ denote the value of the cost function for a period of time under control policy $\mathbf{u}^{(s)}$. Along with this, the following law is adopted to update $\mathbf{W}_1^{(s)}$, i.e.,

$$\mathbf{W}_1^{(s)} = \mathbf{W}_1^{(s-1)} + \frac{\mathbf{P}_{s-1}\phi_1(t)(\varrho_1^{(s)} - \phi_1(t)^{\mathrm{T}}\mathbf{W}_1^{(s-1)})}{\kappa_1 + \phi_1(t)^{\mathrm{T}}\mathbf{P}_{s-1}\phi_1(t)}, \quad (4.30)$$

with variance matrix $\mathbf{P}_s$ to adjust the update speed, i.e.,

$$\mathbf{P}_s = \mathbf{P}_{s-1} - \frac{\mathbf{P}_{s-1}\phi_1(t)\phi_1(t)^{\mathrm{T}}\mathbf{P}_{s-1}}{\kappa_1 + \phi_1(t)^{\mathrm{T}}\mathbf{P}_{s-1}\phi_1(t)}, \quad (4.31)$$

where κ_1 is the forgetting factor in the weight update process. In addition, $\phi_1(t) = \phi_1(\hat{\chi}_{R_i}(t)) - \phi_1(\hat{\chi}_{R_i}(t+T))$ is the excitation function, whose role is to drive the weight to accomplish convergence. Moreover, $||\mathbf{W}_1^{(s)} - \mathbf{W}_1^{(s-1)}|| < \epsilon_1$ and $||\hat{\chi}_{R_i}(t + T) - \hat{\chi}_{R_i}(t)|| < \epsilon_2$ denote the termination conditions for the above two interaction functions, where ϵ_1 and ϵ_2 are small positive decimals.

When the above iteration procedure is ended, the optimal SNR parameters $\theta_{R_i}^*$, $\xi_{R_i}^{*2}$, $\varphi_{R_i}^*$ and $\rho_{R_i}^{*2}$ can be obtained. Accordingly, the predicted mean of SNR between source neighbor R_{i-1} and vehicle R_i is expressed as

$$\overrightarrow{\mathrm{SNR}}_{\mathrm{dB}}(\mathbf{p}_{i-1}, \mathbf{p}_i) = \mathbf{h}_{R_i}\theta_{R_i}^* - \varepsilon_{R_i} + \Xi_{R_i}^{\mathrm{T}}\Omega_{R_i}^{-1}(\mathbf{Y}_{\mathrm{dB}}^{R_i} - \mathbf{H}_{R_i}\theta_{R_i}^* + \varepsilon_{R_i}), \quad (4.32)$$

with

$$\begin{aligned}
\mathbf{h}_{R_i} &= [1, -10\log_{10}(\|\mathbf{p}_i - \mathbf{p}_{i-1}\|)], \\
\Xi_{R_i} &= \xi_{R_i}^{*2}[\mathrm{e}^{-\frac{\|\mathbf{p}_i - \mathbf{p}_{s,1_i}\|}{\varphi_{R_i}^*}}, \ldots, \mathrm{e}^{-\frac{\|\mathbf{p}_i - \mathbf{p}_{s,|\mathcal{N}_i|_i}\|}{\varphi_{R_i}^*}}]^{\mathrm{T}}, \\
\varepsilon_{R_i} &= N_{\mathrm{dB}}^0 + 10\log_{10}(\alpha(f)).
\end{aligned}$$

For SNR prediction, the optimal policy is given in (4.26), whose convergence is presented as follows.

Theorem 4.1 *Given an initial admissible policy* $\mathbf{u}^{(0)}(0)$, *the policy iteration (4.26) can make* $\mathbf{u}^{(s+1)}$ *converge to the optimal policy* $\mathbf{u}^*$, *such that* $\hat{\chi}_{R_i}^*$ *can also be obtained.*

Proof Given an initial admissible policy $\mathbf{u}^{(0)}(0)$ with the system trajectory of $\dot{\hat{\chi}}_{R_i} = \mathbf{u}^{(s+1)}$, the task of this proof is to prove $V_1^*(\hat{\chi}_{R_i}, \mathbf{u}^*) \leq V_1^{(s+1)}(\hat{\chi}_{R_i}, \mathbf{u}^{(s+1)}) \leq V_1^{(s)}(\hat{\chi}_{R_i}, \mathbf{u}^{(s)})$ where $V_1^*(\hat{\chi}_{R_i}, \mathbf{u}^*) = \min_{\mathbf{u}} \int_0^\infty g_1(\hat{\chi}_{R_i}, \mathbf{u}) \mathrm{d}\tau$.

To this end, we set $\mathbf{u}^{(s)}$ as an admissible policy. Based on this, we take the derivative of $V_1^{(s)}(\hat{\chi}_{R_i})$ along $\dot{\hat{\chi}}_{R_i} = \mathbf{u}^{(s+1)}$, through which one has

$$\dot{V}_1^{(s)}(\hat{\chi}_{R_i}, \mathbf{u}^{(s+1)}) = \left(\frac{\partial V_1^{(s)}(\hat{\chi}_{R_i})}{\partial \hat{\chi}_{R_i}}\right)^{\mathrm{T}} \mathbf{u}^{(s+1)}. \tag{4.33}$$

According to the definition of $V_1^*(\hat{\chi}_{R_i}, \mathbf{0})$, the Hamilton-Jacobi-Bellman (HJB) equation becomes

$$\left(\frac{\partial V_1^{(s)}(\hat{\chi}_{R_i})}{\partial \hat{\chi}_{R_i}}\right)^{\mathrm{T}} \mathbf{u}^{(s)} + g_1(\hat{\chi}_{R_i}, \mathbf{u}^{(s)}) = 0. \tag{4.34}$$

Combining (4.33) with (4.34), we can get

$$\begin{aligned}\dot{V}_1^{(s)}(\hat{\chi}_{R_i}, \mathbf{u}^{(s+1)}) = &\left(\frac{\partial V_1^{(s)}(\hat{\chi}_{R_i})}{\partial \hat{\chi}_{R_i}}\right)^{\mathrm{T}} \left(\mathbf{u}^{(s+1)} - \mathbf{u}^{(s)}\right) \\ &- g_1(\hat{\chi}_{R_i}, \mathbf{u}^{(s)}).\end{aligned} \tag{4.35}$$

With (4.26), we have $\left(\frac{\partial V_1^{(s)}(\hat{\chi}_{R_i})}{\partial \hat{\chi}_{R_i}}\right)^{\mathrm{T}} = -2(\mathbf{u}^{(s+1)})^{\mathrm{T}} \mathbf{R}_1$. Based on this, one can rearrange (4.35) as

$$\begin{aligned}&\dot{V}_1^{(s)}(\hat{\chi}_{R_i}, \mathbf{u}^{(s+1)}) \\ &= -2\mathbf{u}^{(s+1)\mathrm{T}} \mathbf{R}_1 \left(\mathbf{u}^{(s+1)} - \mathbf{u}^{(s)}\right) - g_1(\hat{\chi}_{R_i}, \mathbf{u}^{(s)}) \\ &= -\mathbf{u}^{(s+1)\mathrm{T}} \mathbf{R}_1 \mathbf{u}^{(s+1)} - \lambda_1 \left(\frac{1}{|\mathcal{N}_i|_i} \Delta_{R_i}^{\mathrm{T}} \Delta_{R_i} - \hat{\xi}_{R_i}^2 - \hat{\rho}_{R_i}^2\right)^2 \\ &\quad - \left(\mathbf{u}^{(s+1)} - \mathbf{u}^{(s)}\right)^{\mathrm{T}} \mathbf{R}_1 \left(\mathbf{u}^{(s+1)} - \mathbf{u}^{(s)}\right) \\ &\quad - \sum\nolimits_{(j_i, \mathbf{j}_i) \in \mathbf{G}} \lambda_2 \left[\ln \hat{\xi}_{R_i}^2 - \frac{d_{j_i, \mathbf{j}_i}}{\hat{\varphi}_{R_i}} - \ln([\Delta]_{j_i} [\Delta]_{\mathbf{j}_i})\right]^2 \\ &\quad - \Delta_{R_i}^{\mathrm{T}} \mathbf{Q}_1 \Delta_{R_i} \\ &= -\left(\mathbf{u}^{(s+1)} - \mathbf{u}^{(s)}\right)^{\mathrm{T}} \mathbf{R}_1 \left(\mathbf{u}^{(s+1)} - \mathbf{u}^{(s)}\right) - g_1(\hat{\chi}_{R_i}, \mathbf{u}^{(s+1)}).\end{aligned} \tag{4.36}$$

Clearly, $\dot{V}_1^{(s)}(\hat{\chi}_{R_i}, \mathbf{u}^{(s+1)}) \leq 0$ is always satisfied. Since $V_1^{(s)}(\hat{\chi}_{R_i}, \mathbf{u}^{(s+1)})$ is positive definite, continuous and differentiable, $V_1^{(s)}(\hat{\chi}_{R_i}, \mathbf{u}^{(s+1)})$ can be regarded as the Lyapunov function of $\mathbf{u}^{(s+1)}$, through which one knows $\mathbf{u}^{(s)}$ and $\mathbf{u}^{(s+1)}$ are both the admissible policies.

Based on the above conclusion, we require to prove $V_1^{(s+1)}(\hat{\chi}_{R_i}, \mathbf{u}^{(s+1)}) - V_1^{(s)}(\hat{\chi}_{R_i}, \mathbf{u}^{(s)}) \leq 0$ in the trajectory along $\dot{\hat{\chi}}_{R_i} = \mathbf{u}^{(s+1)}$. Combining (4.33) with (4.34), we have

$$
\begin{aligned}
& V_1^{(s)}(\hat{\chi}_{R_i}, \mathbf{u}^{(s)}) - V_1^{(s+1)}(\hat{\chi}_{R_i}, \mathbf{u}^{(s+1)}) \\
&= \int_0^{\infty} g_1(\hat{\chi}_{R_i}, \mathbf{u}^{(s)}) \mathrm{d}\tau - \int_0^{\infty} g_1(\hat{\chi}_{R_i}, \mathbf{u}^{(s+1)}) \mathrm{d}\tau \\
&= \int_0^{\infty} \left(\frac{\partial V_1^{(s+1)}(\hat{\chi}_{R_i})}{\partial \hat{\chi}_{R_i}} \right)^{\mathrm{T}} \mathbf{u}^{(s+1)} \mathrm{d}\tau - \int_0^{\infty} \left(\frac{\partial V_1^{(s)}(\hat{\chi}_{R_i})}{\partial \hat{\chi}_{R_i}} \right)^{\mathrm{T}} \mathbf{u}^{(s+1)} \mathrm{d}\tau \\
&= \int_0^{\infty} \left(\dot{V}_1^{(s+1)}(\hat{\chi}_{R_i}, \mathbf{u}^{(s+1)}) - \dot{V}_1^{(s)}(\hat{\chi}_{R_i}, \mathbf{u}^{(s+1)}) \right) \mathrm{d}\tau \\
&= \int_0^{\infty} \left(-g_1(\hat{\chi}_{R_i}, \mathbf{u}^{(s+1)}) - \dot{V}_1^{(s)}(\hat{\chi}_{R_i}, \mathbf{u}^{(s+1)}) \right) \mathrm{d}\tau \\
&= \int_0^{\infty} \left(\mathbf{u}^{(s+1)} - \mathbf{u}^{(s)} \right)^{\mathrm{T}} \mathbf{R}_1 \left(\mathbf{u}^{(s+1)} - \mathbf{u}^{(s)} \right) \mathrm{d}\tau.
\end{aligned}
\tag{4.37}
$$

Therefore, one can easily obtain $V_1^{(s+1)}(\hat{\chi}_{R_i}, \mathbf{u}^{(s+1)}) \leq V_1^{(s)}(\hat{\chi}_{R_i}, \mathbf{u}^{(s)})$. Noting with the definition of $V_1^*(\hat{\chi}_{R_i}, \mathbf{u}^*)$, one can further have $V_1^*(\hat{\chi}_{R_i}, \mathbf{u}^*) \leq V_1^{(s+1)}$ $(\hat{\chi}_{R_i}, \mathbf{u}^{(s+1)}) \leq V_1^{(s)}(\hat{\chi}_{R_i}, \mathbf{u}^{(s)})$, which means the optimal value of $\hat{\chi}_{R_i}$ (i.e., $\hat{\chi}_{R_i}^*$) can also be obtained. That completes the proof. □

4.2.2 *Model-Free and Collision-Free Motion Planning Algorithm*

The following cost function is defined for underwater vehicle R_i to maximize its channel capacity C_i, i.e.,

$$
E_{i,\mathrm{L}}(\mathbf{p}_i) = \frac{1}{C_i^2(\mathbf{p}_i)}. \tag{4.38}
$$

Then, the obstacle avoidance function for underwater vehicle R_i can be defined as

$$
E_{i,\mathrm{O}}(\mathbf{p}_i) = \sum\nolimits_{m \in \Omega_i^t} f(l_{i,m}) \left(\frac{1}{l_{i,m}} - \frac{1}{\rho_{om}} \right)^2, \tag{4.39}
$$

with

$$
f(l_{i,m}) = \exp\left(-\frac{1}{2} \left(\frac{l_{i,m}^2}{\nu^2} \right)^c \right), \tag{4.40}
$$

where $l_{i,m} = \rho_m + \rho_{i,m}$, $\rho_{i,m}$ is the minimum distance from underwater vehicle R_i to obstacle m, ρ_{om} is the radius of the m-th obstacle's influence boundary cylinder, c is the steepness of repulsive function, and ν is the repulsive range.

Remark 4.2 In (4.40), a large c causes a steep shape for obstacle avoidance, while a large ν causes a wide repulsive range. In view of this, a point $P_{\mathrm{A}} = [l_{\mathrm{left}}, \kappa_1]$ that tends to ρ_{om} from left is selected to capture the steepness, and a point $P_{\mathrm{B}} = [l_{\mathrm{right}}, \kappa_2]$

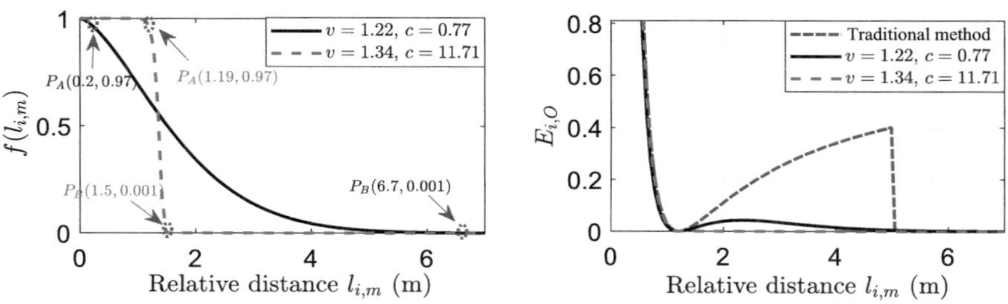

(a) The steepness of $f(l_{i,m})$ (b) The value of $E_{i,O}$

Fig. 4.4 Example of obstacle avoidance function where $\rho_{om} = 1.2\text{m}$

that stay off ρ_{om} from right is selected to capture the repulsive range. Of note, $0 < \kappa_2 < \kappa_1 < 1$. Then, c and ν are selected as $c = \frac{\ln(\log_{\kappa_1} \kappa_2)}{2 \ln l_{\text{right}} - \ln l_{\text{left}}}$ and $\nu = \frac{l_{\text{left}}}{\exp\left(\frac{\ln(-2 \ln \kappa_1)}{2c}\right)}$. An example of the above selection result is shown in Fig. 4.4a.

Most of the existing works (e.g., Zhang et al. (2021) and Lin et al. (2020a)) set $f(l_{i,m})$ as 1 if $l_{i,m} \leq \nu$, and $f(l_{i,m}) = 0$ otherwise. The above design leads to the discontinuity of repulsive potential field, which causes the system dithering. To avoid this shortage, a smooth coefficient $f(l_{i,m})$ is introduced to eliminate the impact of avoidance function $E_{i,\text{O}}(\mathbf{p}_i)$ beyond the safety distance. Clearly, $E_{i,\text{O}}(\mathbf{p}_i) = 0$ is equivalent to $l_{i,m} = \rho_{om}$, which means underwater vehicle R_i moves on the influence boundaries of the obstacles. If $l_{i,m} > \nu$, one regards that underwater vehicle R_i has already escaped the influences of obstacles, and hence, the value of $f(l_{i,m})$ is very small. If $l_{i,m} \leq \nu$, one regards that underwater vehicle R_i enters into the influences of obstacles, and hence, the value of $f(l_{i,m})$ is increased with the decreasing of $l_{i,m}$. The effect of $f(l_{i,m})$ on the obstacle avoidance function is depicted by Fig. 4.4b.

Underwater vehicle R_i also requires to avoid collision with its neighboring vehicle $j \in \mathcal{N}_i^*$, where $\mathcal{N}_i^*$ is its neighboring set. Similar to (4.39), the internal collision avoidance function for underwater vehicle R_i can be denoted by

$$E_{i,j}(\mathbf{p}_i) = \sum_{j \in \mathcal{N}_i^*} \exp\left(-\frac{1}{2}\left(\frac{l_{i,j}^2}{\nu_1^2}\right)^{c_1}\right)\frac{1}{l_{i,j}^2}, \tag{4.41}$$

where $l_{i,j} = \left\|\mathbf{p}_i - \mathbf{p}_j\right\|$ is the relative distance between underwater vehicle R_i and neighboring vehicle $j \in \mathcal{N}_i^*$. In addition, ν_1 and c_1 are the repulsive range and steepness for the internal collision avoidance, respectively.

With (4.2), (4.38), (4.39) and (4.41), the total cost function for the motion planning of underwater vehicle R_i is

$$
\begin{aligned}
&\bar{g}_i(\hat{\mathbf{p}}_i, \tau_i) \\
&= \beta_1 E_{i,\mathrm{L}}(\hat{\mathbf{p}}_i) + \beta_2 E_{i,\mathrm{O}}(\hat{\mathbf{p}}_i) + \beta_3 E_{i,j}(\hat{\mathbf{p}}_i) + \tau_i^{\mathrm{T}} \bar{\mathbf{R}}_i \tau_i, \\
&= \frac{\beta_1}{\mathcal{C}_i^2(\hat{\mathbf{p}}_i)} + \beta_2 \sum_{m \in \Omega_i^t} f(\hat{l}_{i,m}) \left(\frac{1}{\hat{l}_{i,m}} - \frac{1}{\rho_{om}} \right)^2 \\
&\quad + \beta_3 \sum_{j \in \mathcal{N}_i^*} \exp\left(-\frac{1}{2} \left(\frac{l_{i,j}^2}{\nu_1^2} \right)^{c_1} \right) \frac{1}{l_{i,j}^2} + \tau_i^{\mathrm{T}} \bar{\mathbf{R}}_i \tau_i,
\end{aligned} \tag{4.42}
$$

where $\hat{\mathbf{p}}_i$ and $\hat{l}_{i,m}$ denote the estimations of $\mathbf{p}_i$ and $l_{i,m}$, respectively. $\bar{\mathbf{R}}_i$ is a positive definite matrix. Beside that, β_1, β_2 and β_3 are positive constants, whose role is to balance the communication efficiency and collision avoidance.

Based on (4.42), the value function for control input τ_i is

$$
\bar{V}_i(\hat{\mathbf{X}}_i(t)) = \int_t^{t+T} \bar{g}_i(\hat{\mathbf{p}}_i, \tau_i) \mathrm{d}\tau + \bar{V}_i(\hat{\mathbf{X}}_i(t+T)), \tag{4.43}
$$

where $\hat{\mathbf{X}}_i$ denotes the estimated value of $\mathbf{X}_i$.

Hence, one can construct the following optimal problem

$$
\tau_i^*(t) = \underset{\tau_i}{\operatorname{argmin}} \int_t^{t+T} \bar{g}_i(\hat{\mathbf{p}}_i, \tau_i) \mathrm{d}\tau + \bar{V}_i(\hat{\mathbf{X}}_i(t+T)). \tag{4.44}
$$

Meanwhile, dynamic model (4.2) can be rearranged as

$$
\dot{\mathbf{X}}_i = \mathbf{A}_i \mathbf{X}_i + \mathbf{B}_i \tau_i^{(s)} + \mathbf{B}_i (\tau_i - \tau_i^{(s)}) + \mathbf{G}_i, \tag{4.45}
$$

where $\tau_i^{(s)}$ is the updated policy in the s-th iteration and τ_i is an admissible policy for the learning procedure.

Combining (4.43) with (4.45), the derivative of the value function in the s-th iteration can be calculated as

$$
\begin{aligned}
&\dot{\bar{V}}_i^{(s)}(\hat{\mathbf{X}}_i) \\
&= \nabla \bar{V}_{\mathbf{X}_i}^{(s)\mathrm{T}} (\mathbf{A}_i \hat{\mathbf{X}}_i + \mathbf{B}_i \tau_i^{(s)} + \mathbf{G}_i) \\
&\quad -2(-\frac{1}{2} \bar{\mathbf{R}}_i^{-1} \mathbf{B}_i^{\mathrm{T}} \nabla \bar{V}_{\mathbf{X}_i}^{(s)})^{\mathrm{T}} \bar{\mathbf{R}}_i (\tau_i - \tau_i^{(s)}) \\
&= -\bar{g}_i(\hat{\mathbf{p}}_i, \tau_i^{(s)}) - 2\tau_i^{(s+1)\mathrm{T}} \bar{\mathbf{R}}_i (\tau_i - \tau_i^{(s)}),
\end{aligned} \tag{4.46}
$$

where $\bar{\mathbf{R}}_i = \mathrm{diag}(r_{i,1}, r_{i,2}, r_{i,3}, r_{i,4})$ is a positive definite matrix, and $\nabla \bar{V}_{\mathbf{X}_i}^{(s)} = \partial \bar{V}^{(s)}(\mathbf{X}_i)/\partial \mathbf{X}_i$. The model parameter $\mathbf{B}_i$ can be eliminated through the model-based control policy $-\frac{1}{2}\bar{\mathbf{R}}_i^{-1}\mathbf{B}_i^{\mathrm{T}}\nabla \bar{V}_{\mathbf{X}_i}^{(s)}$ which is similar to the one in Sect. 4.2.1. Hence, the desired policy $\tau_i^{(s)}$ can be obtained by solving (4.46).

In the following, a model-free policy iteration algorithm is employed to seek the optimal update policy τ_i^*.

(2-a) *Initialization.* Initially, the policy and value function are set as $\tau_i^{(0)}(0) = \mathbf{0}$ and $\bar{V}_i^{(0)}(\hat{\mathbf{X}}_i(0)) = 0$, respectively.

(2-b) *Policy Evaluation.* For each iteration s, calculate the following value function obtained by integrating (4.46)

$$\begin{aligned}&\bar{V}_i^{(s)}(\hat{\mathbf{X}}_i(t+T)) - \bar{V}_i^{(s)}(\hat{\mathbf{X}}_i(t)) \\ &= -\int_t^{t+T} 2\tau_i^{(s+1)\mathrm{T}}\bar{\mathbf{R}}_i(\tau_i - \tau_i^{(s)})\mathrm{d}\tau - \int_t^{t+T} \bar{g}_i(\hat{\mathbf{p}}_i, \tau_i^{(s)})\mathrm{d}\tau. \end{aligned} \tag{4.47}$$

(2-c) *Policy Improvement.* Find an updated control policy $\tau_i^{(s+1)}$ through the following rule

$$\tau_i^{(s+1)} = \mathrm{argmin}_{\tau_i^{(s)}} \int_t^{t+T} e_2(\tau_i^{(s)})\mathrm{d}\tau, \tag{4.48}$$

with

$$\begin{aligned}e_2(\tau_i^{(s)}) = &\bar{V}_i^{(s)}(\hat{\mathbf{X}}_i(t+T)) - \bar{V}_i^{(s)}(\hat{\mathbf{X}}_i(t)) \\ &+ \int_t^{t+T} 2\tau_i^{(s+1)\mathrm{T}}\bar{\mathbf{R}}_i(\tau_i - \tau_i^{(s)})\mathrm{d}\tau \\ &+ \int_t^{t+T} \bar{g}_i(\hat{\mathbf{p}}_i, \tau_i^{(s)})\mathrm{d}\tau. \end{aligned} \tag{4.49}$$

In above process, the analytical forms of $\bar{V}_i^{(s)}$ and $\tau_i^{(s)}$ are unknown previously. In order to smoothly approximate the value function $\bar{V}_i^{(s)}$ and the desired policy $\tau_i^{(s)}$, the critic network and actor network are introduced as

$$\bar{V}_i^{(s)}(\hat{\mathbf{X}}_i) = \bar{\mathbf{W}}_i^{(s)\mathrm{T}}\phi_2^i(\hat{\mathbf{X}}_i), \tag{4.50}$$

$$\tau_i^{(s+1)} = [\tau_{i,1}^{(s+1)}; \tau_{i,2}^{(s+1)}; \tau_{i,3}^{(s+1)}; \tau_{i,4}^{(s+1)}], \tag{4.51}$$

where ϕ_2^i is the basis function vector for the weight vector $\bar{\mathbf{W}}_i$. In addition, $\tau_{i,\mathbf{i}}^{(s+1)} = \tilde{\mathbf{W}}_{i,\mathbf{i}}^{(s+1)^{\mathrm{T}}}\phi_{3,\mathbf{i}}^i(\hat{\mathbf{X}}_i)$, where $\phi_{3,\mathbf{i}}^i$ is the basis function vector for the weight vector $\tilde{\mathbf{W}}_{i,\mathbf{i}}$. Of note, $\tilde{\mathbf{W}}_{i,\mathbf{i}}$ is the $\mathbf{i}$-th policy weight for $\mathbf{i} \in \{1, \ldots, 4\}$.

Noting with (4.50) and (4.51), one can deduce the following result from (4.49), i.e.,

$$
\begin{aligned}
& e_2(\tau_i^{(s)}) \\
&= \bar{\mathbf{W}}_i^{(s)^{\mathrm{T}}}(\phi_2^i(\hat{\mathbf{X}}_i(t+T)) - \phi_2^i(\hat{\mathbf{X}}_i(t)) \\
&\quad +2\sum_{\mathbf{i}=1}^{4} r_{i,\mathbf{i}} \int_t^{t+T} \tilde{\mathbf{W}}_{i,\mathbf{i}}^{(s+1)^{\mathrm{T}}} \phi_{3,\mathbf{i}}^i(\hat{\mathbf{X}}_i(\tau))\bar{\tau}_{i,\mathbf{i}}(t)\mathrm{d}\tau \\
&\quad + \int_t^{t+T} \bar{g}_i(\hat{\mathbf{p}}_i, \tau_i^{(s)})\mathrm{d}\tau,
\end{aligned}
\tag{4.52}
$$

where $\bar{\tau}_{i,\mathbf{i}}(t) = \tau_{i,\mathbf{i}} - \tau_{i,\mathbf{i}}^{(s)}$. Specifically, $\tau_{i,\mathbf{i}}$ denotes the $\mathbf{i}$-th element of τ_i, and $\tau_{i,\mathbf{i}}^{(s)}$ denotes the $\mathbf{i}$-th element of $\tau_i^{(s)}$.

Let $\bar{\varrho}_i^{(s)} = \int_t^{t+T} \bar{g}_i(\hat{\mathbf{p}}_i, \tau_i^{(s)})\mathrm{d}\tau$ denote the value of the cost function for a period of time T under the given control policy $\tau_i^{(s)}$. From (4.52), the update weight vector can be updated by the following iteration procedure, i.e.,

$$
\bar{\mathbf{W}}_i^{(s)} = \bar{\mathbf{W}}_i^{(s-1)} + \frac{\bar{\mathbf{P}}_{i,s-1}\phi(t)(\bar{\varrho}_i^{(s)} - \phi(t)^{\mathrm{T}}\bar{\mathbf{W}}_i^{(s-1)})}{\bar{\kappa} + \phi(t)^{\mathrm{T}}\bar{\mathbf{P}}_{i,s-1}\phi(t)},
\tag{4.53}
$$

with variance matrix $\bar{\mathbf{P}}_i$ to adjust the update speed, i.e.,

$$
\bar{\mathbf{P}}_{i,s} = \bar{\mathbf{P}}_{i,s-1} - \frac{\bar{\mathbf{P}}_{i,s-1}\phi(t)\phi(t)^{\mathrm{T}}\bar{\mathbf{P}}_{i,s-1}}{\bar{\kappa} + \phi(t)\bar{\mathbf{P}}_{i,s-1}\phi(t)},
\tag{4.54}
$$

where $\bar{\kappa}$ is the forgetting factor. In addition, $\phi(t) = [\phi_2^i(\hat{\mathbf{X}}_i(t)) - \phi_2^i(\hat{\mathbf{X}}_i(t+T));$ $2r_{i,1}\int_t^{t+T} \phi_{3,1}^i(\hat{\mathbf{X}}_i)\bar{\tau}_{i,1}\mathrm{d}\tau;$ $2r_{i,2}\int_t^{t+T} \phi_{3,2}^i(\hat{\mathbf{X}}_i)\bar{\tau}_{i,2}\mathrm{d}\tau;$ $2r_{i,3}\int_t^{t+T} \phi_{3,3}^i(\hat{\mathbf{X}}_i)\times\ \bar{\tau}_{i,3}\mathrm{d}\tau;$ $2r_{i,4}\int_t^{t+T} \phi_{3,4}^i(\hat{\mathbf{X}}_i)\bar{\tau}_{i,4}\mathrm{d}\tau]$, which continuously stimulates the weight to converge to the appropriate value. Besides that, $||\bar{\mathbf{W}}_i^{(s)} - \bar{\mathbf{W}}_i^{(s-1)}|| < \epsilon_3$ and $||\frac{1}{C_i(t+T)} - \frac{1}{C_i(t)}|| < \epsilon_4$ are end conditions, where ϵ_3 and ϵ_4 are positive decimals.

Based on the above iteration procedure, the optimization control input τ_i^* can be obtained, as depicted by Algorithm 3, where $T_{\max,1}$ and $T_{\max,2}$ represent the maximum times for SNR prediction and motion planning, respectively.

Remark 4.3 In Sect. 4.2.1, a model-based IRL estimator is adopted, since the kinematic model (4.21) that is employed can be accurately known by underwater vehicles. By contrast, the dynamic model (4.2) is adopted in Sect. 4.2.2. Due to the harsh ocean environment, it is difficult to acquire the accurate dynamics model of underwater vehicles, e.g., $\mathbf{A}_i$, $\mathbf{B}_i$ and $\mathbf{G}_i$. In view of this, a model-based IRL estimator is developed in Sect. 4.2.2, even if it is complicated in implementation.

For the motion planning control of underwater vehicle, the model-free IRL is adopted to find the optimal control strategy τ_i^*, as provided by (4.44). With regard to this, the following convergence analysis is presented for the optimal control strategy τ_i^*.

Algorithm 3: IRL-based motion planning controller

Input: $\chi_{R_i}(0)$, $\mathbf{X}_i(0)$, $\tau_i(0)$
Output: The optimal control policy τ_i^*

while $||\frac{1}{\mathcal{C}_i(t+T)} - \frac{1}{\mathcal{C}_i(t)}|| < \epsilon_4$ **do**
  Collect SNR measurement data $\mathbf{Y}_{\text{dB}}^{R_i}$ and $\mathbf{p}_{\text{s},j}$
  $s \leftarrow 0$;
  for $t = 0 : T_{max,1}$ **do**
    Calculate (4.22), (4.26) and $\varrho_1^{(s)}$ in turn
    Collect data $\hat{\chi}_{R_i}(t)$, $\hat{\chi}_{R_i}(t+T)$,$\mathbf{u}_1^{(s)}$ for (4.30)
    if *The termination conditions are satisfied* **then**
      break;
    $s \leftarrow s + 1$;
  $s \leftarrow 0$;
  Obtain $\hat{\chi}_{R_i}^*$, and construct the link capacity $\mathcal{C}_i$
  for $t = 0 : T_{max,2}$ **do**
    Determine (4.42), (4.52) and $\bar{\varrho}_i^{(s)}$ in trun
    Collect data $\hat{\mathbf{X}}_i(t)$, $\hat{\mathbf{X}}_i(t+T)$, $\tau_i^{(s)}$ for (4.53)
    if $||\bar{\mathbf{W}}_i^{(s)} - \bar{\mathbf{W}}_i^{(s-1)}|| < \epsilon_3$ **then**
      break;
    $s \leftarrow s + 1$;

Theorem 4.2 *Given an initial admissible policy $\tau_i^{(0)}(0)$, the policy iterations (4.47) and (4.48) can make $\tau_i^{(s+1)}$ converge to the optimal policy τ_i^*.*

Proof Given an initial admissible policy $\tau_i^{(0)}(0)$ with system trajectory of $\dot{\mathbf{X}}_i = \mathbf{A}_i\mathbf{X}_i + \mathbf{B}_i\tau_i^{(s)} + \mathbf{G}_i$, the task of this proof is to prove the solution to the model-free Bellman equation is the same as the model-based Bellman equation.

Firstly, the form of model-based Bellman equation is established. Differentiating the value function (4.43), we have the following Bellman equation

$$\begin{aligned} &\bar{H}(\bar{V}_i^{(s)}, \tau_i^{(s)}) \\ &= \nabla \bar{V}_{\mathbf{X}_i}^{(s)\mathrm{T}}(\mathbf{A}_i\hat{\mathbf{X}}_i + \mathbf{B}_i\tau_i^{(s)} + \mathbf{G}_i) + \bar{g}_i(\hat{\mathbf{p}}_i, \tau_i^{(s)}) \\ &= 0. \end{aligned} \tag{4.55}$$

By the stationarity condition $\partial H(\bar{V}_i^{(s)}, \tau_i^{(s)})/\partial\tau_i^{(s)} = 0$, we can have the following optimal control, i.e.,

$$\tau_i^{(s+1)} = -\frac{1}{2}\bar{\mathbf{R}}_i^{-1}\mathbf{B}_i^{\mathrm{T}}\nabla\bar{V}_{\mathbf{X}_i}^{(s)}. \tag{4.56}$$

From (4.55) and (4.56), one obtains

$$\begin{aligned}
&\bar{H}(\bar{V}_i^{(s)}, \tau_i^{(s)}) \\
&= \nabla \bar{V}_{\mathbf{X}_i}^{(s)\mathrm{T}}(\mathbf{A}_i\hat{\mathbf{X}}_i + \mathbf{B}_i\tau_i^{(s)} + \mathbf{G}_i) + \tau_i^{(s)\mathrm{T}}\bar{\mathbf{R}}_i\tau_i^{(s)} \\
&\quad + \frac{\beta_1}{\mathcal{C}_i^2(\hat{\mathbf{p}}_i)} + \beta_2 \sum_{m\in\Omega_i^t} f(\hat{l}_{i,m}) \left(\tfrac{1}{\hat{l}_{i,m}} - \tfrac{1}{\rho_{om}}\right)^2 \\
&\quad + \beta_3 \sum_{j\in\mathcal{N}_i^*} \exp\left(-\frac{1}{2}\left(\frac{l_{i,j}^2}{\nu_1^2}\right)^{c_1}\right)\frac{1}{l_{i,j}^2} \\
&= 0.
\end{aligned} \tag{4.57}$$

On the other hand, the form of model-free Bellman equation can also be established. Form (4.46), we have

$$\begin{aligned}
&H(\bar{V}_i, \tau_i^{(s)}) \\
&= \nabla \bar{V}_{\mathbf{X}_i}^{(s)\mathrm{T}}(\mathbf{A}_i\hat{\mathbf{X}}_i + \mathbf{B}_i\tau_i^{(s)} + \mathbf{B}_i(\tau_i - \tau_i^{(s)}) + \mathbf{G}_i) \\
&\quad + \bar{g}_i(\hat{\mathbf{p}}_i, \tau_i^{(s)}) + 2\tau_i^{(s+1)\mathrm{T}}\bar{\mathbf{R}}_i(\tau_i - \tau_i^{(s)}) \\
&= 0.
\end{aligned} \tag{4.58}$$

Substituting (4.56) into (4.58), we can have the same Bellman equation as (4.55). Therefore, the optimal solution τ_i^* to the Bellman function in model-free equation is the same as that of the Bellman function in model-based equation. Note that the convergence of the solution to model-based Bellman equation (i.e., (4.56)) has been proved in Theorem 4.1. Based on this, one knows the policy iterations (4.47) and (4.48) can make $\tau_i^{(s+1)}$ converge to the optimal policy τ_i^*. That completes the proof. □

The underwater acoustic communication in this chapter is divided into the following two parts: (1) data collection of SNR measurements (see Sect. 4.2.1); (2) collision avoidance between different vehicles (see Sect. 4.2.2). Then, the communication complexity is studied by counting the transmitted and received scalars for each node, as similar to Yan et al. (2021) and Shi and Wu (2018).

Step 1 (Complexity in Part 1). Recall that underwater vehicle $R_i \in \mathcal{V}$ broadcasts an initiator message to its neighboring ordinary sensor nodes, through which underwater vehicle $R_i \in \mathcal{V}$ receives the replies from ordinary sensor node $j \in \mathcal{N}_i$, i.e., $\{\mathbf{p}_{s,j},\ \mathrm{SNR}_{\mathrm{dB}}(\mathbf{p}_i, \mathbf{p}_{s,j})\}_{j\in\mathcal{N}_i}$. Based on this, underwater vehicle $R_i \in \mathcal{V}$ transmits 1 scalars and receives $|\mathcal{N}_i|_i$ scalars during the data collection procedure. Along with this, any ordinary sensor node $j \in \mathcal{N}_i$ receives the initiator message from underwater vehicle R_i, and then it replies its position and SNR measurement to underwater vehicle R_i. Correspondingly, ordinary sensor node $j \in \mathcal{N}_i$ transmits 4 scalars and receives 1 scalars during the data collection procedure.

Step 2 (Complexity in Part 2). During the internal collision avoidance procedure, underwater vehicle $R_i \in \mathcal{V}$ transmits its position information to its neighboring underwater vehicles, and meanwhile it receives the position information from neighboring underwater vehicles. Thus, the transmitted and received scalars for underwater vehicle $R_i \in \mathcal{V}$ are given as 4 and 4, respectively. In addition, the ordinary sensor nodes donot implement communication task in this part, and hence, the transmitted and received scalars for ordinary sensor node $j \in \mathcal{N}_i$ are all zeros.

The collision avoidance analysis is presented as follows.

Corollary 4.1 *Given cost function (4.42) and value function (4.43), underwater vehicle R_i never collides with obstacles or neighbor R_j, if the following condition is satisfied, i.e.,*

$$\begin{aligned}\bar{g}_i(\hat{\mathbf{p}}_i, \tau_i^{(0)})|_{t=0} &= \beta_1 E_{i,L}(\hat{\mathbf{p}}_i)|_{t=0} + \tau_i^{(0)^T}\bar{\mathbf{R}}_i\tau_i^{(0)}|_{t=0} \\ &< \min\{\beta_2 E_{i,O}^{\max}, \beta_3 E_{i,j}^{\max}\},\end{aligned} \tag{4.59}$$

where $\beta_2 E_{i,O}^{\max} = \sum_{m\in\Omega_i^t}(\frac{\sqrt{\beta_2}}{\hat{l}_{i,m}} - \frac{\sqrt{\beta_2}}{\rho_{om}})^2$ *and* $\beta_3 E_{i,j}^{\max} = \sum_{j\in\mathcal{N}_i^*}\frac{\beta_3}{l_{i,j}^2}$.

Proof The proof includes two parts. First, we prove that $\bar{V}_i^{(s)}(\hat{\mathbf{X}}_i(t), \tau_i^{(s)})$ is decreasing function with policy $\tau_i^{(s+1)} = \frac{-1}{2}\bar{\mathbf{R}}_i^{-1}\mathbf{B}_i^{\mathrm{T}}\nabla\bar{V}_{\hat{\mathbf{X}}_i}^{(s)}$. Taking the derivative of $\bar{V}_i(\hat{\mathbf{X}}_i(t), \tau_i^{(s+1)})$ along $\dot{\hat{\mathbf{X}}}_i = \mathbf{A}_i\hat{\mathbf{X}}_i + \mathbf{B}_i\tau_i^{(s+1)} + \mathbf{G}_i$, we have

$$\dot{\bar{V}}_i^{(s)}(\hat{\mathbf{X}}_i(t), \tau_i^{(s+1)}) = \nabla\bar{V}_{\hat{\mathbf{X}}_i}^{(s)}(\mathbf{A}_i\hat{\mathbf{X}}_i + \mathbf{B}_i\tau_i^{(s+1)} + \mathbf{G}_i). \tag{4.60}$$

According to (4.45), we have

$$\nabla\bar{V}_{\mathbf{X}_i}^{(s)\mathrm{T}}\mathbf{G}_i = -\nabla\bar{V}_{\mathbf{X}_i}^{(s)\mathrm{T}}(\mathbf{A}_i\hat{\mathbf{X}}_i + \mathbf{B}_i\tau_i^{(s)}) - \bar{g}_i(\hat{\mathbf{p}}_i, \tau_i^{(s)}). \tag{4.61}$$

Combining $\tau_i^{(s+1)} = -\frac{1}{2}\bar{\mathbf{R}}_i^{-1}\mathbf{B}_i^{\mathrm{T}}\nabla\bar{V}_{\mathbf{X}_i}^{(s)}$ with (4.60) and (4.61), one can further have

$$\begin{aligned}&\dot{\bar{V}}_i^{(s)}(\hat{\mathbf{X}}_i(t), \tau_i^{(s+1)}) \\ &= -2\tau_i^{(s+1)^{\mathrm{T}}}\bar{\mathbf{R}}_i(\tau_i^{(s+1)} - \tau_i^{(s)}) - \bar{g}_i(\hat{\mathbf{p}}_i, \tau_i^{(s)}). \\ &= -\tau_i^{(s+1)^{\mathrm{T}}}\bar{\mathbf{R}}_i\tau_i^{(s+1)} - (\tau_i^{(s+1)} - \tau_i^{(s)})^{\mathrm{T}}\bar{\mathbf{R}}_i(\tau_i^{(s+1)} - \tau_i^{(s)}) \\ &\quad -\beta_2 E_{i,\mathrm{O}}(\hat{\mathbf{p}}_i) - \beta_3 E_{i,j}(\hat{\mathbf{p}}_i) \le 0,\end{aligned} \tag{4.62}$$

which means $\bar{V}_i^{(s)}(\hat{\mathbf{X}}_i(t), \tau_i^{(s+1)}) \le \bar{V}_i^{(s)}(\hat{\mathbf{X}}_i(t), \tau_i^{(s)})$. Similarly, the equivalence between the solution of IRL and the optimal policy solution $\tau_i^{(s+1)}$ can be proved by Theorem 4.2.

Next, we prove that the collision never occurs between underwater vehicle R_i and obstacles or neighbor R_j if (4.59) are satisfied. Assume that at $t = t^*$, underwater vehicle R_i collides with obstacles or neighbor R_j, then cost function becomes

$$\bar{g}_i(\hat{\mathbf{p}}_i, \tau_i^{(s)})|_{t=t^*}$$
$$= \beta_1 E_{i,\mathrm{L}}(\hat{\mathbf{p}}_i)|_{t=t^*} + \tau_i^{(s)^{\mathrm{T}}} \bar{\mathbf{R}}_i \tau_i^{(s)}|_{t=t^*} + \max\{\beta_2 E_{i,\mathrm{O}}^{\max}, \beta_3 E_{i,j}^{\max}\}. \quad (4.63)$$

Then we can conclude that

$$\bar{g}_i(\hat{\mathbf{p}}_i, \tau_i^{(s)})|_{t=t^*} > \min\{\beta_2 E_{i,\mathrm{O}}^{\max}, \beta_3 E_{i,j}^{\max}\}. \quad (4.64)$$

From (4.62), (4.63) and (4.64), we have

$$\begin{aligned}\bar{g}_i(\hat{\mathbf{p}}_i, \tau_i^{(s+1)})|_{t=t^*} &< \bar{g}_i(\hat{\mathbf{p}}_i, \tau_i^{(s)})|_{t=t^*} < \bar{g}_i(\hat{\mathbf{p}}_i, \tau_i^{(0)})|_{t=0} \\ &< \min\{\beta_2 E_{i,\mathrm{O}}^{\max}, \beta_3 E_{i,j}^{\max}\},\end{aligned} \quad (4.65)$$

which means (4.65) has contradiction with (4.64). Therefore, underwater vehicle R_i never collides with obstacles or neighbor R_j if (4.59) is satisfied. That completes the proof. □

4.3 Optimally Persistent Formation with Unknown Interaction Topology

As shown in Fig. 4.1, we would like to steer AUVs to arrive at the destination in a specific formation shape. Consider the formation control with one leader AUV and $N-1$ follower AUVs, where $N \geq 3$. Let x_i, y_i and z_i be the positions of AUV $i \in \{1, 2, \ldots, N\}$ on X, Y and Z axes, respectively. Meanwhile, ψ_i denotes the angle on yaw for AUV i. Define $\eta_i = [\mathbf{P}_i; \psi_i]$ as the position vector of AUV i in inertial reference frame, where $\mathbf{P}_i = [x_i, y_i, z_i]^{\mathrm{T}}$. Let $\mathbf{v}_i = [u_i, v_i, w_i, r_i]^{\mathrm{T}}$ be the linear and angle velocity vector of AUV i in body-fixed reference frame, where u_i, v_i, w_i and r_i are the velocities on surge, sway, heave and yaw, respectively. Referring to Yan et al. (2024) and Cui et al. (2017), the kinematic and dynamic model of AUV $i \in \{1, 2, \ldots, N\}$ can be formulated as

$$\begin{aligned}\dot{\eta}_i &= \mathbf{J}_i(\psi_i)\mathbf{v}_i, \\ \mathbf{M}_i\mathbf{v}_i + \mathbf{C}_i(\mathbf{v}_i)\mathbf{v}_i + \mathbf{D}_i(\mathbf{v}_i)\mathbf{v}_i + \mathbf{g}_i(\eta_i) &= \tau_i,\end{aligned} \quad (4.66)$$

where $\mathbf{M}_i = \tilde{\mathbf{M}}_i + \Delta\mathbf{M}_i$, $\mathbf{C}_i(\mathbf{v}_i) = \tilde{\mathbf{C}}_i(\mathbf{v}_i) + \Delta\mathbf{C}_i(\mathbf{v}_i)$, $\mathbf{D}_i(\mathbf{v}_i) = \tilde{\mathbf{D}}_i(\mathbf{v}_i) + \Delta\mathbf{D}_i(\mathbf{v}_i)$ and $\mathbf{g}_i(\eta_i) = \tilde{\mathbf{g}}_i(\eta_i) + \Delta\mathbf{g}_i(\eta_i)$. Specially, $\tilde{\mathbf{M}}_i \in \mathcal{R}^{4\times4}$ and $\Delta\mathbf{M} \in \mathcal{R}^{4\times4}$ represent the nominal and uncertainty inertia matrices, respectively. $\tilde{\mathbf{C}}_i(\mathbf{v}_i) \in \mathcal{R}^{4\times4}$ and $\Delta\mathbf{C}_i(\mathbf{v}_i) \in \mathcal{R}^{4\times4}$ are the nominal and uncertainty Coriolis-centripetal matrices, respectively. $\tilde{\mathbf{D}}_i(\mathbf{v}_i) \in \mathcal{R}^{4\times4}$ and $\Delta\mathbf{D}_i(\mathbf{v}_i) \in \mathcal{R}^{4\times4}$ donate the nominal and uncertainty hydrodynamic damping matrices, respectively. $\tilde{\mathbf{g}}_i(\eta_i) \in \mathcal{R}^4$ and $\Delta\mathbf{g}_i(\eta_i) \in \mathcal{R}^4$ are the nominal and uncertainty vectors for gravitational and buoyancy effects,

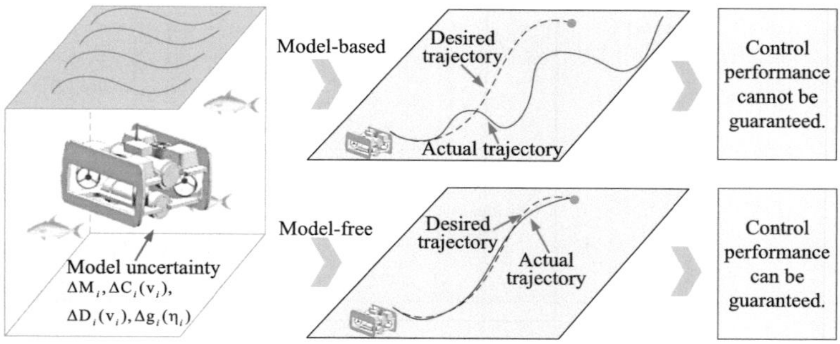

Fig. 4.5 Description of the model uncertainty and model-free terms

respectively. In addition, $\tau_i = \left[\tau_{u_i}, \tau_{v_i}, \tau_{w_i}, \tau_{r_i}\right]^{\mathrm{T}}$ is the control input vector. τ_{u_i}, τ_{v_i}, τ_{w_i} and τ_{r_i} are the control inputs on surge, sway, heave and yaw, respectively.

Remark 4.4 It is worth emphasizing that $\Delta\mathbf{M}_i$, $\Delta\mathbf{C}_i(\mathbf{v}_i)$, $\Delta\mathbf{D}_i(\mathbf{v}_i)$ and $\Delta\mathbf{g}_i(\eta_i)$ cannot be accurately obtained due to the complex underwater environment. The above phenomenon can be termed as "model uncertainty". Although some model-based controllers can achieve guaranteed performance under certain constraints on uncertainty (e.g. adaptive control (Li and Tong 2023, 2024; Li and Li 2024), model predictive control (Huang et al. 2024), and sliding mode control (Wang et al. 2025), these controllers still face some challenges. For instance, the adaptive controllers rely on real-time parameter estimation, however the accurate parameter identification is difficult due to noise, or insufficient excitation. The model predictive controllers require a precise mathematical model to predict future behavior, while the sliding mode controllers suffer from chattering oscillations around the sliding surface. Based on the above consideration, the model-free controller is selected as our solution to resolve the model uncertainty issue (Fig. 4.5).

The state of AUV $i \in \{1, 2, \ldots, N\}$ can be stacked into $\mathbf{X}_i = [\eta_i; \mathbf{v}_i]$, through which the model (4.66) can be rewritten as

$$\dot{\mathbf{X}}_i = \mathbf{A}_i\mathbf{X}_i + \mathbf{B}_i\tau_i + \mathbf{G}_i, \tag{4.67}$$

with

$$\mathbf{A}_i = \begin{bmatrix} \mathbf{0} & \mathbf{J}_i(\psi_i) \\ \mathbf{0} & -\mathbf{M}_i^{-1}\left(\mathbf{C}_i(\mathbf{v}_i) + \mathbf{D}_i(\mathbf{v}_i)\right) \end{bmatrix}, \tag{4.68}$$

$$\mathbf{B}_i = \begin{bmatrix} 0 \\ \mathbf{M}_i^{-1} \end{bmatrix}, \mathbf{G}_i = \begin{bmatrix} 0 \\ -\mathbf{M}_i^{-1}\mathbf{g}_i(\eta_i) \end{bmatrix}. \tag{4.69}$$

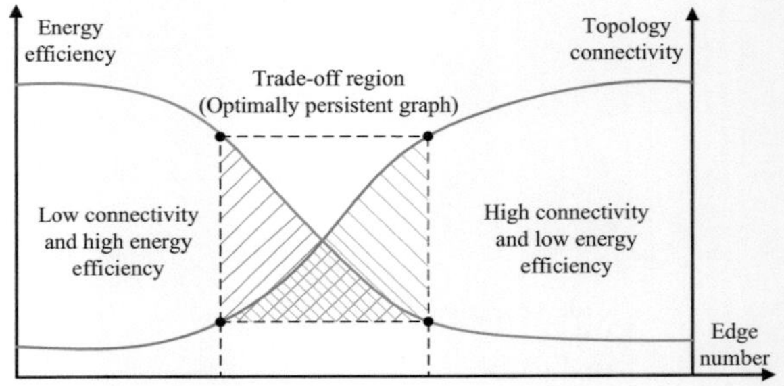

Fig. 4.6 Tradeoff between topology connectivity and energy efficiency

Different from the previous works (Yan et al. 2024, 2018b), the initial interaction relationship of AUVs in this chapter is assumed to be unknown. For that reason, a *directed graph* can be employed to describe the initial topology relationship of AUVs, where AUVs and their interaction links are considered as vertices and edges respectively. It should be pointed out that the edges of an undirected graph have no direction, while those of a directed graph have clear directions. Particularly, the directed graph can be denoted by $\mathbf{G} = (\mathcal{V},\ \mathbf{E})$, where $\mathcal{V} = \{1, 2, \ldots, N\}$ and $\mathbf{E} = \{\overrightarrow{(i, j)} : i, j \in \mathcal{V}, i \neq j\}$. The neighbor set of vertex i is defined by $\mathcal{N}_i = \{j \in \mathcal{V} : \overrightarrow{(i, j)} \in \mathbf{E}\}$. To describe the interaction relationship, the adjacency matrix is defined by $A = \left[a_{ij}\right] \in \mathcal{R}^{|\mathcal{V}|\times|\mathcal{V}|}$, where $a_{ij} = 1$ if $\overrightarrow{(i, j)} \in \mathbf{E}$, and $a_{ij} = 0$ otherwise. Accordingly, $L = D - A$ is the Laplacian matrix, where $D = \text{diag}\{d_i\}$ and $d_i = \sum_{j=1}^{|\mathcal{N}_i|} a_{ij}$.

With the inferred topology relationship, we attempt to use the optimally persistent graph to balance the topology connectivity and energy efficiency, whose tradeoff can be shown in Fig. 4.6. Accordingly, the following knowledge on rigid graph and rigidity matrix is introduced.

Definition 4.2 (*Rigid graph* Hendrickx et al. (2007)) A graph is *rigid* if it remains uniquely unchanged relative to a flexible graph.

Definition 4.3 (*Minimally rigid graph*Hendrickx et al. (2007)) A graph is *minimally rigid* if it has the least edges to maintain rigid compared with other graphs which have the same vertices. The minimally rigid graph $\mathcal{G} = (\mathcal{V}, \mathcal{E})$ in 3-D space satisfies $|\mathcal{E}| = 3N - 6$, where $N \geq 3$ is the number of vertices.

Definition 4.4 (*Optimally rigid graph*Ren et al. (2010)) A graph is *optimally rigid* if it satisfies the conditions: 1) the graph is minimally rigid; 2) the weighted sum of all edge lengths in the graph is minimal among all graphs with the same vertices.

Definition 4.5 (*Rigidity matrix* Hendrickx et al. (2007)) Let $\mathbf{P}_i, \mathbf{P}_j \in \mathcal{R}^{3\times1}$ be the position vector of vertices i and j, respectively. With the rows and columns corresponding to the edges and vertices in $\mathcal{G}$, the *rigidity matrix* $M(\mathcal{G}) \in \mathcal{R}^{|\mathcal{E}|\times3|\mathcal{V}|}$ is defined by

$$\Phi(\mathcal{G}) \tag{4.70}$$
$$= \begin{bmatrix} \cdots & \ddots & \vdots & \vdots & & \vdots & \vdots & \vdots & \ddots & \vdots \\ 0 & \ldots & 0 & \mathbf{P}_i^{\mathrm{T}} - \mathbf{P}_j^{\mathrm{T}} & \cdots & \mathbf{P}_j^{\mathrm{T}} - \mathbf{P}_i^{\mathrm{T}} & 0 & \ldots & 0 \\ \cdots & \ddots & \vdots & \vdots & & \vdots & \vdots & \vdots & \ddots & \vdots \end{bmatrix}.$$

Remark 4.5 In order to show more clearly, the rigid matrix of Fig. 4.7c can be represented as

$$\hat{\Phi}(\mathcal{G})$$
$$= \begin{bmatrix} \mathbf{r}_{1,1} & \mathbf{r}_{1,2} & 0 & 0\,0\,0 & \ldots & 0 & 0 \\ \mathbf{r}_{2,1} & 0 & \mathbf{r}_{2,3} & 0\,0\,0 & \ldots & 0 & 0 \\ \vdots & \vdots & \vdots & \vdots\,\vdots\,\vdots & \ddots & \vdots & \vdots \\ 0 & 0 & 0 & 0\,0\,0 & 0 & \mathbf{r}_{22,7} & \mathbf{r}_{22,8} \end{bmatrix},$$

where $\mathbf{r}_{1,1} = (\mathbf{P}_1 - \mathbf{P}_2)^{\mathrm{T}}$, $\mathbf{r}_{1,2} = (\mathbf{P}_2 - \mathbf{P}_1)^{\mathrm{T}}$, $\mathbf{r}_{2,1} = (\mathbf{P}_1 - \mathbf{P}_3)^{\mathrm{T}}$, $\mathbf{r}_{2,3} = (\mathbf{P}_3 - \mathbf{P}_1)^{\mathrm{T}}$, $\mathbf{r}_{22,7} = (\mathbf{P}_7 - \mathbf{P}_8)^{\mathrm{T}}$, and $\mathbf{r}_{22,8} = (\mathbf{P}_8 - \mathbf{P}_7)^{\mathrm{T}}$. The other elements in $\hat{\Phi}(\mathcal{G})$ can be analogized. We can see $\mathrm{rank}(\hat{\Phi}(\mathcal{G})) = 18$ and the graph is a rigid graph.

Remark 4.6 With the above definitions, the differences between rigid graphs and optimally rigid graphs are given as: (1) the optimally rigid graphs maintain rigidity with the minimal number of edges; (2) the optimally rigid graphs have the shortest total edge length compared to the rigid graphs with the same set of vertices. Examples of different topologies are shown in Fig. 4.7. We can find that the number of edges in Fig. 4.7c is 22. However the number of edges in Fig. 4.7e is 18.

It is worth mentioning that the optimally rigid graphs are oriented to undirected formation (e.g., Yan et al. (2024) and Yan et al. (2018b)), and they cannot handle the directionality of the topology relationship. For that reason, we attempt to adopt the optimally persistent graph to balance the topology connectivity and energy efficiency, whose definition can be presented as follows.

Definition 4.6 (*Optimally persistent graph*Luo et al. (2009)) A directed graph is *optimally persistent* if it satisfies the conditions: 1) its corresponding undirected graph is optimally rigid; 2) the vertex $i \in \{1, 2, \ldots, N\}$ satisfies $d_i^+ \leq 3$ in three-dimensional space, where d_i^+ is the out-degree of vertex i.

Without loss of generality, the following assumption is provided to depict the network topology relationship.

Assumption 4.4 *Assume that the network topology between AUVs remains connected during formation control.*

Let r be the relative distance from a node to its next-hop node. Referring to Jiang et al. (2016), the energy consumption of transmitting b bits over distance r can be denoted by

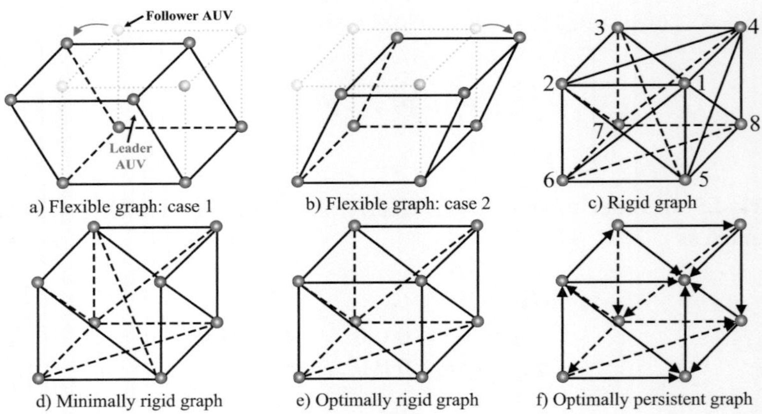

Fig. 4.7 Examples of the different topologies

$$E_{\text{tx}}(b, r) = bE_{\text{elec}} + 4.2 * bT_b He^{\alpha_f r} \cdot 10^{-9.5}, \quad (4.71)$$

where T_b is the bit duration, E_{elec} is the unit energy consumption of processing one bit message, H is the depth of the underwater node and α_f is the absorption coefficient.

Receiver's energy consumption by receiving b bits data is

$$E_{\text{rx}}(b, r) = bE_{\text{elec}}, \quad (4.72)$$

and hence, its communication energy consumption is

$$E_{\text{con}} = E_{\text{tx}} + E_{\text{rx}}. \quad (4.73)$$

Problem 1: Topology inference. As mentioned above, the sensing capability of follower AUVs is limited, which makes it hard to acquire the topology relationship in advance. For that reason, we attempt to estimate the Laplacian matrix L for the unknown topology relationship of AUVs.

Problem 2: Topology optimization. Due to the weak acoustic communication, it is hard to balance the topology connectivity and energy efficiency. To this end, we attempt to acquire the optimally persistent graph $\mathbf{G}^*$ with the adjacency matrix A, such that the tradeoff between topology connectivity and energy efficiency is balanced, as depicted by Fig. 4.6.

Problem 3: Formation controller. With the optimized network topology, an IRL-based formation controller is attempted to design for AUVs to guarantee the formation stability, i.e.,

$$\eta_1 \to \eta_{\text{tar}} \text{ and } (\eta_i - \eta_j) \to \eta_{i,j} \quad (4.74)$$

where η_{tar} is the state vector of destination, and $\eta_{i,j}$ is the desired relative state vector between AUVs i and j.

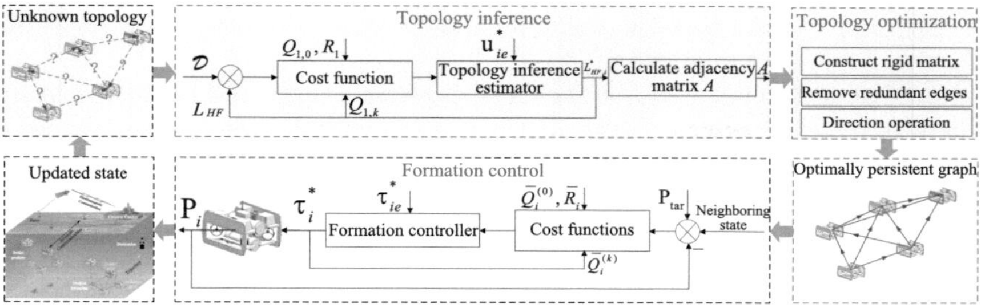

Fig. 4.8 Relationship of the topology inference, topology optimization and formation controller in this chapter

As shown by Fig. 4.8, we first design a local topology inference estimator to acquire the initial topology relationship. Then, an optimally persistent topology generation algorithm is presented to optimize the initial topology. After that, a model-free formation controller is designed for AUVs.

4.3.1 Topology Inference and Optimization

As shown in Fig. 4.9, the working principle of the local topology estimator is given as follows: (1) *Initialization.* Describe the motion of formation system and collect their coupling state information; (2) *Transform.* Construct the relationship between topology structure and the coupling state information, such that a topology optimization problem can be established; (3) *Estimation.* An IRL-based estimator is designed to solve the optimization problem and infer the topology structure for AUVs. Particularly, the above principle are detailed as the following steps.

Step 1. Patrol and collect the surrounding information

Before performing the formation task, the leader AUV 1 patrols the entire region, and then it collects the following state information from the follower AUVs, i.e.,

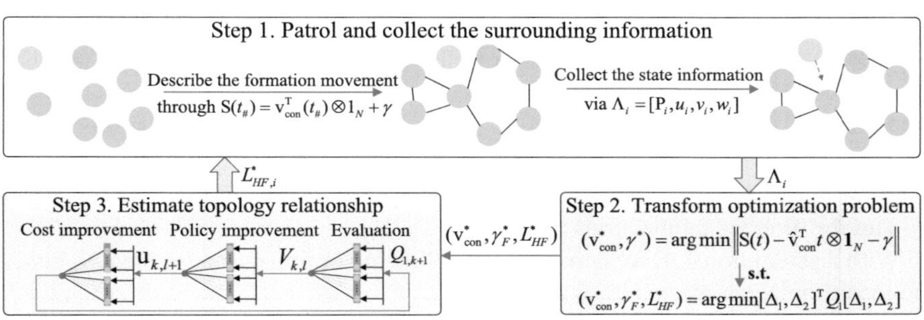

Fig. 4.9 Main structure of the topology inference estimator

$$\Lambda_i = [\mathbf{P}_i, u_i, v_i, w_i], \text{ where } i \in \{2, \ldots, N\}. \tag{4.75}$$

Note that the formation system should keep a fixed shape with a constant velocity. In view of this, we denote $t_\#$ as the time instant when the desired formation shape is achieved before reaching the destination. Meanwhile, the constant velocity vector is denoted by $\mathbf{v}_{\text{con}} \in \mathcal{R}^3$. Based on this, we introduce a desired matrix for the formation system, i.e.,

$$\begin{aligned}\mathbf{S}(t_\#) &= [\mathbf{P}_1(t_\#), \mathbf{P}_2(t_\#), \ldots, \mathbf{P}_N(t_\#)]^{\text{T}} \\ &= \mathbf{v}_{\text{con}}^{\text{T}} t_\# \otimes \mathbf{1}_N + \gamma,\end{aligned} \tag{4.76}$$

where γ is the offset vector, and it is unknown to AUVs.

From (4.76), one requires to achieve the following task, i.e.,

$$\lim_{t\to\infty} \left\| \mathbf{S}(t) - \mathbf{v}_{\text{con}}^{\text{T}} t \otimes \mathbf{1}_N - \gamma \right\| = 0. \tag{4.77}$$

With regard to (4.77), we denote $\hat{\mathbf{v}}_{\text{con}}$ and $\hat{\gamma}$ as the estimations of $\mathbf{v}_{\text{con}}$ and γ, respectively. Along with this, $\hat{\mathbf{v}}_{\text{con}}$ and $\hat{\gamma}$ can be acquired by solving the following optimization problem

$$\begin{aligned}&(\mathbf{v}_{\text{con}}^*, \gamma^*) \\ &= \operatorname{argmin}_{(\mathbf{v}_{\text{con}}, \gamma)} \left\| \mathbf{S}(t) - \hat{\mathbf{v}}_{\text{con}}^{\text{T}} t \otimes \mathbf{1}_N - \hat{\gamma} \right\|^2.\end{aligned} \tag{4.78}$$

Step 2. Transform the optimization problem

Note that the optimization problem in (4.78) does not explicitly reveal the topology relationship of AUVs. For that reason, it is necessary to incorporate the topology relationship into the optimization problem. To this end, we define R_{f} as the observation radius of leader AUV 1, and the set of follower AUVs within the observation radius is denoted by $\mathcal{V}_{\text{F}}$. Correspondingly, the subset $\mathcal{V}_{\text{H}} \subseteq \mathcal{V}_{\text{F}}$ is introduced to narrow down the inference range from $\mathcal{V}_{\text{F}}$ to $\mathcal{V}_{\text{H}}$. The purpose of doing this is to avoid the influence caused by AUVs outside the observation range. Then, we define the observation set of the leader AUV as $\mathcal{D} = \{\Lambda_i, i \in \mathcal{V}_{\text{F}}\}$, which is linearly modeled. Referring to Li et al. (2023), the observations in $\mathcal{D}$ satisfy

$$E(\mathbf{S}'_H - \mathbf{v}_{\text{con}}^{\text{T}} \otimes \mathbf{I}_H) = L_{HF} E(-\mathbf{S}_F + \hat{\gamma}_F - \hat{\gamma}_F^{[j,:]} \otimes \mathbf{1}_F), \tag{4.79}$$

where $E(\cdot)$ is the expectation. $\mathbf{S}'_H = [\ldots; u_i, v_i, w_i; \ldots]_{i\in\mathcal{V}_{\text{H}}}$ is the velocity matrix, $\mathbf{I}_H = [1, 0, \ldots, 0]^{\text{T}} \in \mathcal{R}^{|\mathcal{V}_{\text{H}}|}$ is the unit vector, and $\gamma_F^{[j,:]}$ is the j-th row of γ_F. In addition, $L_{HF} = [l_{ij}, i \in \mathcal{V}_H, \ j \in \mathcal{V}_F] \in \mathcal{R}^{|\mathcal{V}_{\text{H}}|\times|\mathcal{V}_{\text{F}}|}$ covers all connections within $\mathcal{V}_{\text{H}}$ and the directed connections from $\mathcal{V}_{\text{H}}$ to $\mathcal{V}_{\text{F}}/\mathcal{V}_{\text{H}}$.

With Assumption 4.4, we construct the following optimization problem to reveal the local Laplacian matrix L_{HF}, i.e.,

$$L^*_{HF} = \text{argmin}_{L_{HF}} \Big\| \mathbf{S}'_H - \hat{\mathbf{v}}^{\text{T}}_{\text{con}} \otimes \mathbf{I}_H + \hat{L}_{HF}(\mathbf{S}_F - \hat{\gamma}_F + \hat{\gamma}_F^{[j,:]} \otimes \mathbf{1}_F) \Big\|^2. \tag{4.80}$$

From (4.78) and (4.80), the local topology inference problem for AUVs can be formulated as

$$\begin{aligned}
&(\mathbf{v}^*_{\text{con}}, \gamma^*_F, L^*_{HF}) \\
&= \text{argmin}_{(\mathbf{v}_{\text{con}}, \gamma_F, L_{HF})} \Big(\beta_1 \left\|\mathbf{S}_F(t) - \hat{\mathbf{v}}^{\text{T}}_{\text{con}} t \otimes \mathbf{1}_F - \hat{\gamma}_F\right\|^2 \\
&\quad + \beta_2 \Big\|\mathbf{S}'_H - \hat{\mathbf{v}}^{\text{T}}_{\text{con}} \otimes \mathbf{I}_H + \hat{L}_{HF}(\mathbf{S}_F - \hat{\gamma}_F + \hat{\gamma}_F^{[j,:]} \otimes \mathbf{1}_F)\Big\|^2\Big) \\
&= \text{argmin}_{(\mathbf{v}_{\text{con}}, \gamma_F, L_{HF})} [\Delta_1, \Delta_2]^{\text{T}} Q_1 [\Delta_1, \Delta_2],
\end{aligned} \tag{4.81}$$

where $\Delta_2 = \Big\|\mathbf{S}'_H - \hat{\mathbf{v}}^{\text{T}}_{\text{con}} \otimes \mathbf{I}_H + \hat{L}_{HF}(\mathbf{S}_F - \hat{\gamma}_F + \hat{\gamma}_F^{[j,:]} \otimes \mathbf{1}_F)\Big\|$, $Q_1 = \text{diag}\{\beta_1, \beta_2\}$, and $\Delta_1 = \left\|\mathbf{S}_F(t) - \hat{\mathbf{v}}^{\text{T}}_{\text{con}} t \otimes \mathbf{1}_F - \hat{\gamma}_F\right\|$. Of note, β_1 and β_2 are positive coherence scalars.

Step 3. Estimate the topology relationship

We stack the desired formation velocity, offset vector and local topology into an augmented vector, i.e., $\mathcal{X} = [\mathbf{v}_{\text{con}}; \text{vec}(\gamma_F); \text{vec}(L_{HF})]$ where $\text{vec}(\cdot)$ is the vectorization operation on a matrix. Let $\hat{\mathcal{X}}$ be the estimation of $\mathcal{X}$, which is determined by an incremental input $\mathbf{u}$. Then the update process of $\mathcal{X}$ can be formulated as

$$\dot{\hat{\mathcal{X}}} = \mathbf{u}. \tag{4.82}$$

Based on (4.81), the cost function for local topology estimation of AUVs can be defined by

$$g(\hat{\mathcal{X}}, \mathbf{u}) = [\Delta_1, \Delta_2]^{\text{T}} Q_1 [\Delta_1, \Delta_2] + \mathbf{u}^{\text{T}} R_1 \mathbf{u}, \tag{4.83}$$

where R_1 is a positive definite matrix.

Based on (4.83), the value function is described by

$$V(\hat{\mathcal{X}}(t)) = \int_t^{t+T} g(\hat{\mathcal{X}}, \mathbf{u}) \mathrm{d}\tau + V(\hat{\mathcal{X}}(t+T)). \tag{4.84}$$

From (4.82) and (4.84), the optimal policy is selected by

$$\mathbf{u}^* = \text{argmin}_{\mathbf{u}} \left(\int_t^{t+T} g(\hat{\mathcal{X}}, \mathbf{u}) \mathrm{d}\tau + V(\hat{\mathcal{X}}(t+T))\right). \tag{4.85}$$

Next, we attempt to employ IRL to acquire the desired matrix Q_1. To this end, one rewrites (4.82) as

$$\dot{\hat{\mathcal{X}}} = \mathbf{u}^*_e + (\mathbf{u} - \mathbf{u}^*_e), \tag{4.86}$$

where $\mathbf{u}_e^*$ is the expert input policy. The expert knowledge is obtained by the following operations: (1) before performing the IRL algorithm, we deploy AUVs to interact with environment, through which the policies are updated by using reward functions; (2) we evaluate the strategies and record the best ones, which can be regarded as expert polices; (3) we apply the expert policies into AUVs, and then the expert demonstrations are generated to be the inputs of IRL algorithm.

Substituting (4.86) into the derivative of $V(\hat{\mathcal{X}}(t))$, one has

$$\begin{aligned}\dot{V}(\hat{\mathcal{X}}(t)) &= \nabla V(\hat{\mathcal{X}}(t))^{\mathrm{T}}\mathbf{u}_e^* \\ &\quad -(-R_1^{-1}\partial V_{k,l}(\hat{\mathcal{X}}(t))/\partial\hat{\mathcal{X}}(t))^{\mathrm{T}}R_1(\mathbf{u}-\mathbf{u}_e^*), \\ &= -g(\hat{\mathcal{X}},\mathbf{u}_e^*) - 2\mathbf{u}_e^{*\mathrm{T}}R_1(\mathbf{u}-\mathbf{u}_e^*).\end{aligned} \tag{4.87}$$

By integrating both sides of (4.87), one obtains

$$\begin{aligned}&V(\hat{\mathcal{X}}(t+T)) - V(\hat{\mathcal{X}}(t)) \\ &= -\int_t^{t+T} g(\hat{\mathcal{X}},\mathbf{u}_e^*)\mathrm{d}\tau - \int_t^{t+T} 2\mathbf{u}_e^{*\mathrm{T}}R_1(\mathbf{u}-\mathbf{u}_e^*)\mathrm{d}\tau.\end{aligned} \tag{4.88}$$

Substituting (4.83) into (4.88), then Q_1 can be solved by

$$\begin{aligned}&\int_t^{t+T}[\Delta_1,\Delta_2]^{\mathrm{T}}Q_1[\Delta_1,\Delta_2]\mathrm{d}\tau \\ &= \int_t^{t+T}(\mathbf{u}_e^{*\mathrm{T}}R_1\mathbf{u}_e^* - 2\mathbf{u}_e^{*\mathrm{T}}R_1\mathbf{u})\mathrm{d}\tau \\ &\quad -V(\hat{\mathcal{X}}(t+T)) + V(\hat{\mathcal{X}}(t)).\end{aligned} \tag{4.89}$$

Based on the above analysis, we present an IRL-based iterative algorithm to acquire the optimal value $\mathbf{u}^*$.

1. *Initialization*. Set $Q_{1,0} > 0$, $k = 0$, $l = 0$, $V_{k,l} = 0$ and $u_{k,l} = 0$. At the l-th iteration for update procedure k, the value function and policy are $V_{k,l}$ and $\mathbf{u}_{k,l}$, respectively.
2. *Policy Evaluation*. For each iteration $l = 0, 1, 2, \ldots$, one calculates the following value function:

$$V_{k,l}(\hat{\mathcal{X}}(t)) = \int_t^{t+T} g(\hat{\mathcal{X}},\mathbf{u}_{k,l})\mathrm{d}\tau + V_{k,l}(\hat{\mathcal{X}}(t+T)). \tag{4.90}$$

3. *Policy Improvement*. Find an updated policy $\mathbf{u}_{k,l+1}$ through the following rule:

$$\begin{aligned}&\mathbf{u}_{k,l+1} \\ &= \mathrm{argmin}_{\mathbf{u}_{k,l}}\left(\int_t^{t+T} g(\hat{\mathcal{X}},\mathbf{u}_{k,l})\mathrm{d}\tau + V_{k,l}(\hat{\mathcal{X}}(t+T))\right).\end{aligned} \tag{4.91}$$

which yields $\mathbf{u}_{k,l+1} = -\frac{1}{2}R_1^{-1}\partial V_{k,l}(\hat{\mathcal{X}}(t))/\partial\hat{\mathcal{X}}(t)$.

4. *Cost Function Improvement*. Substituting $\mathbf{u}_k = \mathbf{u}_{k,l+1}$ and $V_k = V_{k,l}$ into (4.89), one has

---

Algorithm 4: Local topology estimator

Input: The state information Λ_i;

Initialize $Q_{1,0} > 0$, $k = 0$, $l = 0$, $V_{k,l} = 0$, $u_{k,l} = 0$, small thresholds ϱ_1 and ϱ_2;

repeat

    $k \leftarrow k + 1$;

    repeat

        Solve the value function (4.90);

        Update $\mathbf{u}_{k,l+1}$ through (4.91);

    until $\|\mathbf{u}_{k,l+1} - \mathbf{u}_{k,l}\| \leq \varrho_1$

    Substitute $\mathbf{u}_k = \mathbf{u}_{k,l+1}$ and $V_k = V_{k,l}$ into (4.89);

    Update Q_k through (4.92);

    Set $\mathbf{u}_{k,l} = \mathbf{u}_k$;

until $\|Q_{1,k+1} - Q_{1,k}\| \leq \varrho_2$

Output: The optimal value $\mathbf{u}^*$.

$$
\begin{aligned}
&\int_t^{t+T} [\Delta_1, \Delta_2]^{\mathrm{T}} Q_{1,k+1} [\Delta_1, \Delta_2] \mathrm{d}\tau \\
&= \int_t^{t+T} (\mathbf{u}_e^{*T} R_1 \mathbf{u}_e^* - 2\mathbf{u}_e^{*T} R_1 \mathbf{u}_k) \mathrm{d}\tau \\
&\quad - V_k(\hat{\mathcal{X}}(t+T)) + V_k(\hat{\mathcal{X}}(t)),
\end{aligned}
\tag{4.92}
$$

then, one can obtain the matrix $Q_{1,k+1}$. Followed by this, one substitutes $Q_{1,k+1}$ and $\mathbf{u}_{k,l} = \mathbf{u}_k$ into the Step 2) for conducting the next update procedure k.

The pseudo code of local topology estimator is presented in Algorithm 4. Of note, the (4.90) and (4.92) can be solved by the BLSs, through which the optimal $\mathcal{X}^*$ can be obtained. That means the local topology L_{HF}^* can be permuted. Applying this strategy to the entire formation, one has

$$
L^* = L_{HF,1}^* \cup L_{HF,2}^* \cup \cdots \cup L_{HF,n}^*, \tag{4.93}
$$

where $L_{HF,i\in\{1,2,\ldots,n\}}^*$ is the local Laplacian matrix of the i-th observation, and n represents the number of observations.

Since $L = D - A$, and hence, the topology relationship of AUVs can be described by the adjacency matrix, i.e.,

$$
A = \mathrm{sign}(\mathrm{diag}\{l_{11}, l_{22}, \ldots, l_{NN}\} - L^*), \tag{4.94}
$$

where l_{ii} is the diagonal element of L^* for $i \in \{1, 2, \ldots, N\}$. Without loss of generality, the communication weight coefficient is set to 1 within the communication range, and 0 otherwise. This assumption is widely used in other works, such as Xia et al. (2017), Pang et al. (2023), Wu et al. (2020), Yan et al. (2024), Yan et al. (2018b), Cao et al. (2023), Wu et al. (2019) and Wang and Zhang (2023).

For Algorithm 4, the convergence of optimal policy $\mathbf{u}^*$ and matrix $Q_1^{(k+1)}$ can be presented as follows.

Theorem 4.3 *Given an initial admissible policy* $\mathbf{u}_{k,0}$ *at procedure* k *and* $Q_{1,k} > 0$. *The policy iteration in (4.91) can make* $\mathbf{u}_{k,l+1}$ *converge to the optimal policy* $\mathbf{u}_k^*$. *In addition, the matrix* $Q_{1,k+1}$ *and optimal policy* $\mathbf{u}^*$ *can converge to the desired matrix* $\tilde{Q}_1$ *and expert policy* u_e^*, *respectively.*

Proof We set $\mathbf{u}_{k,l}$ as an admissible policy at procedure k, and take the derivative of $V_{k,l}(\hat{\mathcal{X}})$ along $\dot{\hat{\mathcal{X}}} = \mathbf{u}_{k,l+1}$, i.e.,

$$\dot{V}_{k,l}(\hat{\mathcal{X}}, \mathbf{u}_{k,l+1}) = (\partial V_{k,l}(\hat{\mathcal{X}})/\partial \hat{\mathcal{X}})^{\mathrm{T}} \mathbf{u}_{k,l+1}. \tag{4.95}$$

According to $V_{k,l}(\hat{\mathcal{X}}(0), \mathbf{u}_{k,l}) = \int_0^\infty g(\hat{\mathcal{X}}, \mathbf{u}_{k,l})\mathrm{d}\tau$, the Hamilton–Jacobi–Bellman (HJB) equation is

$$(\partial V_{k,l}(\hat{\mathcal{X}})/\partial \hat{\mathcal{X}})^{\mathrm{T}} \mathbf{u}_{k,l} + g(\hat{\mathcal{X}}, \mathbf{u}_{k,l}) = 0. \tag{4.96}$$

Combining (4.95) and (4.96), one has

$$\begin{aligned} &\dot{V}_{k,l}(\hat{\mathcal{X}}, \mathbf{u}_{k,l+1}) \\ &= (\partial V_{k,l}(\hat{\mathcal{X}})/\partial \hat{\mathcal{X}})^{\mathrm{T}} (\mathbf{u}_{k,l+1} - \mathbf{u}_{k,l}) - g(\hat{\mathcal{X}}, \mathbf{u}_{k,l}). \end{aligned} \tag{4.97}$$

From (4.91), one can have $\partial V_{k,l}(\hat{\mathcal{X}}, \mathbf{u}_{k,l+1})/\partial \hat{\mathcal{X}} = -2\mathbf{u}_{k,l+1}^{\mathrm{T}} R_i$. Then, one rearranges (4.97) as

$$\begin{aligned} &\dot{V}_{k,l}(\hat{\mathcal{X}}, \mathbf{u}_{k,l+1}) \\ = \quad & -2\mathbf{u}_{k,l+1}^{\mathrm{T}} R_i (\mathbf{u}_{k,l+1} - \mathbf{u}_{k,l}) - g(\hat{\mathcal{X}}, \mathbf{u}_{k,l}) \\ = \quad & -2\mathbf{u}_{k,l+1}^{\mathrm{T}} R_i (\mathbf{u}_{k,l+1} - \mathbf{u}_{k,l}) \\ & -[\Delta_1, \Delta_2]^{\mathrm{T}} Q_1 [\Delta_1, \Delta_2] - \mathbf{u}_{k,l}^{\mathrm{T}} R_1 \mathbf{u}_{k,l} \\ = & -(\mathbf{u}_{k,l+1} - \mathbf{u}_{k,l})^{\mathrm{T}} R_1 (\mathbf{u}_{k,l+1} - \mathbf{u}_{k,l}) - g(\hat{\mathcal{X}}, \mathbf{u}_{k,l+1}). \end{aligned} \tag{4.98}$$

Apparently, $\dot{V}_{k,l}(\hat{\mathcal{X}}, \mathbf{u}_{k,l+1}) \leq 0$ is always satisfied, since $V_{k,l}(\hat{\mathcal{X}}, \mathbf{u}_{k,l+1}) \geq 0$ can be regarded as the Lyapunov function of $\mathbf{u}_{k,l+1}$, and hence (4.82) is asymptotically stable.

With the above analysis, we obtain $V_{k,l+1}(\hat{\mathcal{X}}, \mathbf{u}_{k,l+1}) \leq V_{k,l}(\hat{\mathcal{X}}, \mathbf{u}_{k,l})$, i.e.,

$$\begin{aligned} &V_{k,l}(\hat{\mathcal{X}}, \mathbf{u}_{k,l}) - V_{k,l+1}(\hat{\mathcal{X}}, \mathbf{u}_{k,l+1}) \\ &= \int_0^\infty g(\hat{\mathcal{X}}, \mathbf{u}_{k,l})\mathrm{d}\tau - \int_0^\infty g(\hat{\mathcal{X}}, \mathbf{u}_{k,l+1})\mathrm{d}\tau \\ &= \int_0^\infty (\dot{V}_{k,l+1}(\hat{\mathcal{X}}, \mathbf{u}_{k,l+1}) - \dot{V}_{k,l}(\hat{\mathcal{X}}, \mathbf{u}_{k,l}))\mathrm{d}\tau \\ &= \int_0^\infty (-g(\hat{\mathcal{X}}, \mathbf{u}_{k,l+1}) - \dot{V}_{k,l}(\hat{\mathcal{X}}, \mathbf{u}_{k,l+1}))\mathrm{d}\tau \\ &= \int_0^\infty (\mathbf{u}_{k,l+1} - \mathbf{u}_{k,l})^{\mathrm{T}} R_1 (\mathbf{u}_{k,l+1} - \mathbf{u}_{k,l})\mathrm{d}\tau. \end{aligned} \tag{4.99}$$

According to $V_k^*(\hat{\mathcal{X}}, \mathbf{u}_k^*) = \min_{\mathbf{u}_k} \int_0^\infty g(\hat{\mathcal{X}}, \mathbf{u}_k)\mathrm{d}\tau$, one can easily obtain $V_k^*(\hat{\mathcal{X}}, \mathbf{u}_k^*) \leq V_{k,l+1}(\hat{\mathcal{X}}, \mathbf{u}_{k,l+1}) \leq V_{k,l}(\hat{\mathcal{X}}, \mathbf{u}_{k,l})$, which means $\hat{\mathcal{X}}^*$ can also be acquired.

Next, we prove the convergence of $Q_{1,k}$, where $Q_{1,0} > 0$. Based on (4.83) and (4.84), one can obtain

$$\begin{aligned} &V_k(\hat{\mathcal{X}}(t+T)) - V_k(\hat{\mathcal{X}}(t)) \\ =& -\int_t^{t+T} ([\Delta_1, \Delta_2]^{\mathrm{T}} Q_{1,k}[\Delta_1, \Delta_2] + \mathbf{u}_k^{\mathrm{T}} R_1 \mathbf{u}_k)\mathrm{d}\tau. \end{aligned} \tag{4.100}$$

Combining (4.92) and (4.100), one yields

$$\begin{aligned} &\int_t^{t+T} [\Delta_1, \Delta_2]^{\mathrm{T}} Q_{1,k+1}[\Delta_1, \Delta_2]\mathrm{d}\tau \\ =& \int_t^{t+T} (\mathbf{u}_e^{*\mathrm{T}} R_1 \mathbf{u}_e^* - 2\mathbf{u}_e^{*\mathrm{T}} R_1 \mathbf{u}_k)\mathrm{d}\tau \\ &+ \int_t^{t+T} ([\Delta_1, \Delta_2]^{\mathrm{T}} Q_{1,k}[\Delta_1, \Delta_2] + \mathbf{u}_k^{\mathrm{T}} R_1 \mathbf{u}_k)\mathrm{d}\tau \\ =& \int_t^{t+T} (\mathbf{u}_e^* - \mathbf{u}_k)^{\mathrm{T}} R_i (\mathbf{u}_e^* - \mathbf{u}_k)\mathrm{d}\tau \\ &+ \int_t^{t+T} ([\Delta_1, \Delta_2]^{\mathrm{T}} Q_{1,k}[\Delta_1, \Delta_2])\mathrm{d}\tau. \end{aligned} \tag{4.101}$$

According to (4.101) and the positivity of $Q_{1,k}$, one has

$$0 < \left\| Q_{1,0} \right\| \leq \cdots \leq \left\| Q_{1,k} \right\| \leq \left\| Q_{1,k+1} \right\|. \tag{4.102}$$

Defining $\Upsilon_k = (\mathbf{u}_e^* - \mathbf{u}_k)^{\mathrm{T}} R_i (\mathbf{u}_e^* - \mathbf{u}_k)$, one yields

$$\begin{aligned} &\int_t^{t+T} [\Delta_1, \Delta_2]^{\mathrm{T}} Q_{1,k+1}[\Delta_1, \Delta_2]\mathrm{d}\tau \\ =& \int_t^{t+T} (\Upsilon_{k-1} + \cdots + \Upsilon_0 + [\Delta_1, \Delta_2]^{\mathrm{T}} Q_{1,0}[\Delta_1, \Delta_2])\mathrm{d}\tau. \end{aligned} \tag{4.103}$$

According to (4.103), one can know that $Q_{1,k+1}$ is increased to the vicinity of $\tilde{Q}_1$. That means, there exists a small threshold α holding $\left\| \tilde{Q}_i - Q_{1,k+1} \right\| \leq \alpha$. In view of this, the uniform approximate solution of u_{ie}^* can also be acquired, i.e., $\left\| \mathbf{u}_k - \mathbf{u}_e^* \right\| \leq \mu_i$, where the small threshold μ_i is used to avoid affecting the positive definiteness of $Q_{1,k}$. We make the algorithm converge at procedure k, and thus one has

$$\begin{aligned} &\left\| \int_t^{t+T} [\Delta_1, \Delta_2]^{\mathrm{T}} Q_{1,k+1}[\Delta_1, \Delta_2]\mathrm{d}\tau \right\| \\ =& \left\| \int_t^{t+T} (\Upsilon_{k-1} + \cdots + \Upsilon_0 + [\Delta_1, \Delta_2]^{\mathrm{T}} Q_{1,0}[\Delta_1, \Delta_2])\mathrm{d}\tau \right\| \\ \geq& \left\| \int_t^{t+T} (k\mu_i^2 R_i + [\Delta_1, \Delta_2]^{\mathrm{T}} Q_{1,0}[\Delta_1, \Delta_2])\mathrm{d}\tau \right\| \\ \geq& k\mu_i^2 T \lambda_{\min}(R_i) - \left\| \int_t^{t+T} [\Delta_1, \Delta_2]^{\mathrm{T}} Q_{1,0}[\Delta_1, \Delta_2])\mathrm{d}\tau \right\|, \end{aligned} \tag{4.104}$$

which implies

$$k \leq \left(\left\|\int_t^{t+T}[\Delta_1, \Delta_2]^{\mathrm{T}} Q_{1,0}[\Delta_1, \Delta_2])\mathrm{d}\tau\right\| + \left\|\int_t^{t+T} e_i^{\mathrm{T}} Q_i^0 e_i d\tau\right\|\right)/\mu_i^2 T\lambda_{\min}(R_i) \equiv k_0. \tag{4.105}$$

The iterative algorithm converges after a maximum of procedure k_0. Then the uniformly approximate solutions of $\tilde{Q}_1$ and $\mathbf{u}_e^*$ can be obtained. That completes the proof. □

Corollary 4.2 *Consider the system (4.82) with cost function (4.84). By using Algorithm 4, the system (4.82) is globally stable with* $\mathbf{u}_{k,l+1}$ *in (4.91) and* $\mathbf{u}_k$ *in (4.92) at each l- and k-loop, where* $k = 1, 2, \ldots,$ *and* $l = 1, 2, \ldots$.

Proof We first prove that system (4.82) is globally stable at the l-th loop with Algorithm 4. Then, we select (4.84) as the Lyapunov candidate for system (4.82). According to (4.90), one has

$$\dot{V}_{k,l} = -[\Delta_1, \Delta_2]^{\mathrm{T}} Q_{1,k}[\Delta_1, \Delta_2] - \mathbf{u}_{k,l}^{\mathrm{T}} R_1 \mathbf{u}_{k,l}, \tag{4.106}$$

Based on (4.106), one has $\dot{V}_{k,l} \leq 0$, $\forall l = 0, 1, 2, \ldots$. At the same time, one can deduce that $V_{k,l} \geq 0$. Note that $\dot{V}_{k,l} = V_{k,l} = 0$ holds only when $\mathcal{X} = \mathbf{0}$. Based on the above analysis, we can prove that the system (4.82) is globally asymptotically stable at l-loop, since $\mathcal{X} \in \mathcal{R}^{\mathcal{P}}$ for $\mathcal{P} \geq 1$.

Next, we prove system (4.82) is globally stable at the k-th loop with Algorithm 4. With regard to (4.100), the derivative of Lyapunov candidate for (4.82) can be given as

$$\dot{V}_k = -[\Delta_1, \Delta_2]^{\mathrm{T}} Q_{1,k}[\Delta_1, \Delta_2] - \mathbf{u}_k^{\mathrm{T}} R_1 \mathbf{u}_k. \tag{4.107}$$

From (4.102) and (4.107), $\dot{V}_k \leq 0$ and $V_k \geq 0$. The $\dot{V}_k = V_k = 0$ only holds when $\mathcal{X} = \mathbf{0}$. Thus, for $\mathcal{X} \in \mathcal{R}^{\mathcal{P}}$ and $\mathcal{P} \geq 1$, the system (4.82) is globally asymptotically stable at the k-th loop.

In summary, the system (4.82) is globally stable at both l-th and k-th loop with Algorithm 4. That completes the proof. ■

With the acquired adjacency matrix A, each AUV requires to communicate with neighbors in its sensing range. However, the above operation can lead to the redundancy of network topology. In view of this, we employ the graph contraction and direction to generate an optimal persistent topology for AUVs, as shown in Fig. 4.10. Particularly, the vertex addition and edge separation operations are adopted to realize the graph contraction and direction, which are defined as follows.

Definition 4.7 (*Vertex addition operation* Smith et al. (2009)) For a minimally rigid graph $\mathcal{G} = (\mathcal{V}, \mathcal{E})$, the vertex addition operation refers to adding the vertex i (i.e., AUV i), the edges (i, j), (i, k) and (i, l) into $\mathcal{G}$, as shown in Fig. 4.11a.

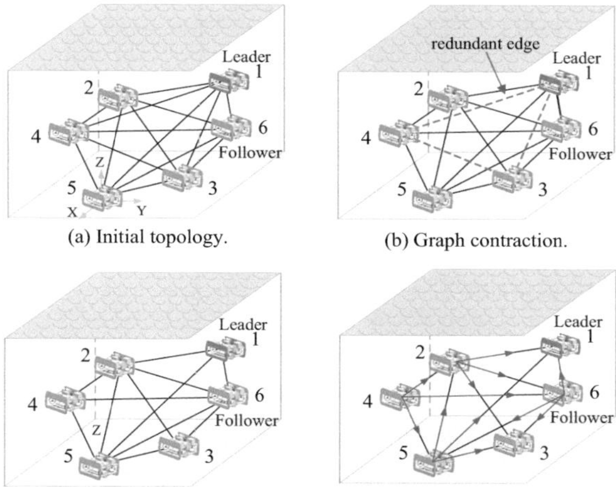

(a) Initial topology. (b) Graph contraction.
(c) Delete redundant edges. (d) Graph direction.

Fig. 4.10 Description of graph contraction and direction

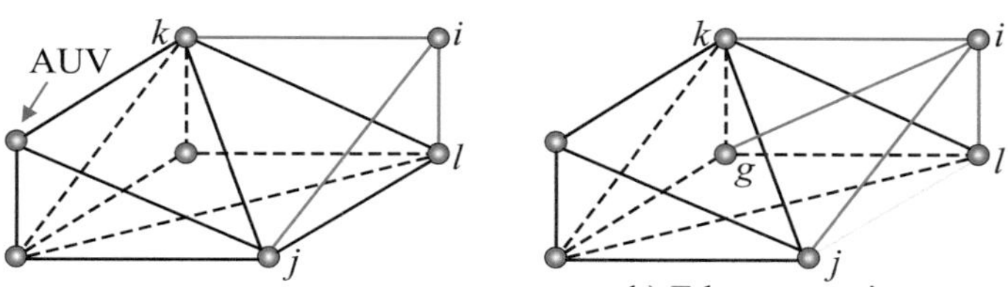

a) Vertex addition b) Edge separation

Fig. 4.11 Description of vertex addition and edge separation operations

Definition 4.8 (*Edge separation operation*Smith et al. (2009)) For a minimally rigid graph $\mathcal{G} = (\mathcal{V}, \mathcal{E})$, the edge separation operation refers to adding vertex i (i.e., AUV i), edges (i, j), (i, k), (i, l), (i, g), and removing edge (l, j). Of note, the vertex g is not in the same plane with j, k and l, as shown in Fig. 4.11b.

Based on the above definitions, the topology optimization algorithm is designed by the following stages.

Stage 1. Generate an optimally rigid graph

We first construct an optimally rigid graph by using the inferred topology relationship of AUVs. The main idea is to build a local rigid matrix with matrix A, such that the redundant edges are deleted to generate a global optimally rigid graph. To this end, the Definition 4.5 is used to build a local rigid matrix for AUV $i \in \{1, \ldots, N\}$, i.e.,

$$\Phi_i = [\mathbf{m}_{i,1};\ \mathbf{m}_{i,2};\ \ldots;\ \mathbf{m}_{i,|\mathcal{N}_i|}], \tag{4.108}$$

where $\mathbf{m}_{i,j} = [0, \ldots, \mathbf{P}_i^{\mathrm{T}} - \mathbf{P}_j^{\mathrm{T}}, \ldots, 0, \ldots, \mathbf{P}_j^{\mathrm{T}} - \mathbf{P}_i^{\mathrm{T}}, \ldots, 0]$ denotes the j-th row of Φ_i for $j \in \mathcal{N}_i$.

With regard to (4.108), we implement the following operations: (1) according to the edge lengths, the row sequence of Φ_i is arranged in the ascending order, and hence, a new local rigid matrix Φ_i^* is obtained; (2) if Φ_i^* is not full-rank, one deletes rows in a bottom-up order; (3) if Φ_i^* is full-rank, one records the edges corresponding to the deleted rows; 4) remove the above recorded edges, and then a local optimally rigid graph $\mathcal{G}_i^*$ is obtained; 5) one applies the above steps into other AUVs, then, a global optimally rigid graph $\mathcal{G}^*$ can be obtained.

Stage 2. Generate an optimally persistent graph

Since the optimally rigid graph $\mathcal{G}^*$ is undirected, it is necessary to transform $\mathcal{G}^*$ into an optimally persistent graph. From Definition 4.3 and Definition 4.4, one knows the optimally rigid graph $\mathcal{G}^*$ can remain rigidity after performing vertex addition and edge separation operations. In view of this, the Hannibal sequence of an optimally rigid graph is constructed, through which the direction and path negation operations are conducted to generate the optimally rigid graph.

Firstly, the Hannibal sequence for $\mathcal{G}^*$ is constructed as

$$\mathcal{G}^*(3) \rightarrow \cdots \rightarrow \mathcal{G}^*(a-1) \rightarrow \mathcal{G}^*(a) \rightarrow \cdots \rightarrow \mathcal{G}^*(N), \tag{4.109}$$

where $\mathcal{G}^*(a)$ denotes a graph with $a \in \{3, \ldots, N\}$ vertices.

Next, we perform a reverse operation on (4.109) for the convenience of graph direction. Followed by this, the reverse Hannibal sequence is described as

$$\mathcal{G}^*(3) \leftarrow \mathcal{G}^*(4) \leftarrow \cdots \leftarrow \mathcal{G}^*(N-1) \leftarrow \mathcal{G}^*(N). \tag{4.110}$$

Based on the order of graph contraction in (4.110), we direct $\mathcal{G}^*(N)$ through direction operation. As shown in Fig. 4.12, the edges corresponding to vertices with a connectivity of 2, 3, and 4 can be directed by using 2-degree, 3-degree, and 4-degree operations, respectively. Of note, the vertex g (i.e., AUV g) is not in the same plane with j, k and l. Apparently, an undirected edge remains in $\mathcal{G}^*(3)$ after the direction operation. By specifying a direction of the last undirected edge, the direction of $\mathcal{G}^*(N)$ can be accomplished.

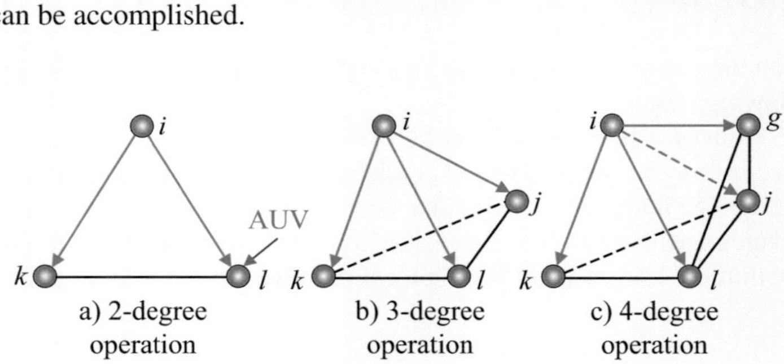

Fig. 4.12 Description of the direction operation

Algorithm 5: Topology optimization algorithm

Input: $\mathcal{G}_i = (\mathcal{V}_i, \mathcal{E}_i)$ for $i \in \{1, 2, \cdots, N\}$;

for $i = 1 : N$ **do**
    Construct Φ_i with $\mathcal{N}_i$, and initialize Φ_i as Φ_i^*;
    for $i = 1 : |\mathcal{E}_i|$ **do**
        while $rank(\Phi_i^*) < |\mathcal{E}_i|$ **do**
            Delete the last row of Φ_i^* to obtain $\Phi_i^{*'}$;
            if $rank(\Phi_i^{*'}) = 3|\mathcal{V}_i| - 6$ **then**
                Record the deleted rows and $\Phi_i^* = \Phi_i^{*'}$;
        Delete edges in $\mathcal{G}_i$ to obtain $\mathcal{G}_i^*$;
Integrate $\mathcal{G}_i^*$ into $\mathcal{G}^*$;
for $i = 1 : N$ **do**
    for $j = 1 : N$ **do**
        if $d_j^+ = 2,\ d_j^+ = 3\ or\ d_j^+ = 4$ **then**
            Perform direction operations to AUV i, and delete directed edges in $\mathcal{G}^*$;
    Construct the directed edges as a subgraph $\mathbf{G}_i^*$;
Integrate $\mathbf{G}_i^*$ into $\mathbf{G}^*$;
for $i = 1 : N$ **do**
    for $j = 1 : N$ **do**
        if $d_i^+ < 3\ and\ d_j^+ = 4$ **then**
            Negate the path between AUVs i and j;

Output: The optimally persistent graph $\mathbf{G}^*$.

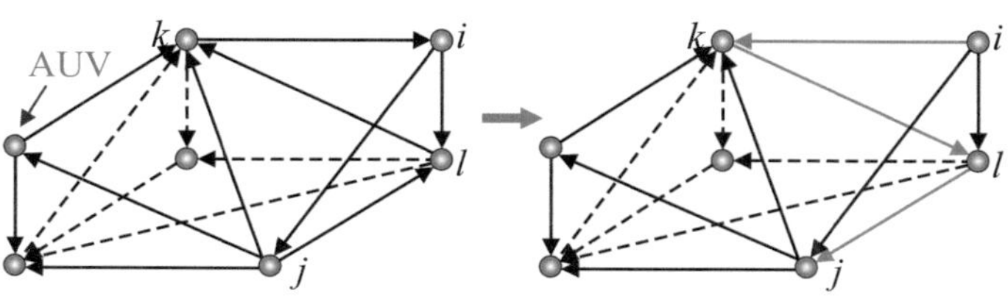

Fig. 4.13 Description of the path negation operation

Of note, the direction operation cannot satisfy the persistent conditions, since the out-degree of each vertex should meet $d^+ \leq 3$ (see Definition 4.6). For that reason, a path negation operation is implemented on the directed graph. Specifically, a path between AUVs i and j is searched to negate, which results in a decrease of out-degree from $d_j^+ = 4$ to $d_j^+ = 3$ for AUV j and an increase of out-degree from $d_i^+ = 2$ to $d_i^+ = 3$ for AUV i. As shown in Fig. 4.13, only the out-degree of the endpoints in a path is exchanged, while the overall out-degree remains unchanged. Finally, one applies the above direction method to all AUVs, through which a global optimally persistent graph $\mathbf{G}^*$ can be generated. To this end, the detailed procedure in Stages 1–2 is depicted by Algorithm 5.

With the above topology optimization scheme (see Algorithm 5), we have the following corollaries.

Corollary 4.3 *(a) Compared with the neighbor-rule based topology, the communication links via optimally persistent graph* $\mathbf{G}^*$ *in this chapter are minimum on the condition of rigidity maintaining, i.e., the sum of the energy consumption for AUVs is also minimum; (b) compared with the undirected optimally rigid graph* $\mathcal{G}^*$ *based optimization algorithm, the optimally persistent graph* $\mathbf{G}^*$ *in this chapter can further reduce communication consumption; (c) the graph contraction/direction operations in Algorithm 5 can reduce communication redundancy and ensure topology connectivity.*

Proof The proof is divided into three parts.

Part 1. Proof of Corollary 4.3*(a)*

After acquiring the adjacency matrix A, one can obtain the neighbor-rule based communication topology $\mathbf{G}$ of AUVs, and then we represent the number of links in $\mathbf{G}$ as $\digamma_d$. Since Algorithm 5 requires the deletion of redundant links in $\mathbf{G}$ to generate the optimally persistent graph $\mathbf{G}^*$, the number of edges in the resulting optimally persistent graph $\mathbf{G}^*$ (denoted by $\digamma_d^*$) satisfies $\digamma_d^* < \digamma_d$. According to Definition 4.6, the number of edges in graph $\mathbf{G}^*$ can be calculated by

$$\digamma_d^* = 3N - N(N+1)/2 = 3N - 6. \tag{4.111}$$

Based on Definition 4.5, the rigid matrix $M(\mathbf{G}^*)$ of $\mathbf{G}^*$ has a row count of $\digamma_d^*$, with the rank satisfying $\mathrm{rank}(M) = 3N - 6$. In view of this, the communication links of AUVs cannot be deleted on the condition of rigidity retaining, i.e., the number of communication links for AUVs is minimum. According to Definition 4.5, the optimally persistent graph has the minimal sum of edge lengths, through which the total communication energy consumption of AUVs can be minimized.

Part 2. Proof of Corollary 4.3*(b)*

According to Definitions 4.4 and 4.6, we know that the optimally rigid graph $\mathcal{G}^*$ and the optimally persistent graph $\mathbf{G}^*$ have the same vertices and edges. Consider the topology of AUVs as $\mathcal{G}^*$, i.e., each AUV requires to communicate with its all neighbors. We assume that in one communication cycle, each AUV communicates once with its neighbors, and the total number of communications is

$$q_{sum} = |d_1| + |d_2| + \cdots + |d_N|. \tag{4.112}$$

Based on (4.112), we define the communication consumption of AUVs in graph $\mathcal{G}^*$ as $q_{sum}E_{\mathrm{con}}$. Consider the topology of AUVs as $\mathbf{G}^*$, thus the communications can only occur in one direction with the AUV pointed to by the link. To this end, the total number of communications is described as

$$\mathbf{q}_{sum} = |d_1^+| + |d_2^+| + \cdots + |d_N^+|. \tag{4.113}$$

We define the communication consumption of AUVs in graph $\mathbf{G}^*$ as $\mathbf{q}_{sum}E_{\mathrm{con}}$. Apparently, $q_{sum} > \mathbf{q}_{sum}$, i.e., $q_{sum}E_{\mathrm{con}} > \mathbf{q}_{sum}E_{\mathrm{con}}$. That completes the proof.

Part 3. Proof of Corollary 4.3*(c)*

Consider an initial topology $\mathcal{G}$ with N vertices. Each vertex in $\mathcal{G}$ needs to communicate with its neighbors, so the number of communication links in topology $\mathcal{G}$ is represented as

$$\digamma_b = N(N-1)/2 = \frac{N^2}{2} - \frac{N}{2}. \tag{4.114}$$

By using graph contraction and direction on topology $\mathcal{G}$, it can be optimized into the optimally persistent topology $\mathbf{G}^*$. From Corollary 4.3(a), the number of communication links in $\mathbf{G}^*$ is $3N-6$, then one has

$$\digamma_b - \digamma_d^* = \frac{N^2}{2} - \frac{7N}{2} + 6. \tag{4.115}$$

Clearly, $\digamma_b - \digamma_d^* \geq 0$ since $N \geq 3$. Therefore, the graph contraction and direction reduce communication redundancy.

In the following, we prove that the graph contraction and direction can ensure topology connectivity. The connectivity (connectivity degree) of each AUV is not uniform in $\mathbf{G}^*$. However, the sum of connectivity for all AUVs equals twice the number of edges, denoted as $2|\mathbf{E}|$. This is because each edge connects two AUVs, contributing one outdegree and one indegree to each AUV. Let $\bar{D}$ represent the average connectivity of each AUV. Based on the Definition 4.4 and Definition 4.6, the total connectivity in $\mathbf{G}^*$ is expressed as

$$2|\mathbf{E}| = 2(3N-6) = 6N - 12. \tag{4.116}$$

Then the average connectivity of each AUV is

$$\bar{D} = \lim_{N\to\infty} \frac{2|\mathbf{E}|}{N} = \lim_{N\to\infty} \frac{6N-12}{N} = 6. \tag{4.117}$$

Based on the above analysis, graph contraction and direction can ensure topology connectivity. That completes the proof. □

4.3.2 Formation Controller for Multiple Underwater Vehicles

With the optimized network topology, this section designs an IRL-based formation controller for AUVs. To be specific, the leader-follower structure is employed, and then the cost function of leader AUV 1 is provided as

$$\bar{g}_1(\eta_1, \tau_1) = \mathbf{e}_1^{\mathrm{T}} \bar{Q}_1 \mathbf{e}_1 + \tau_1^{\mathrm{T}} \bar{R}_1 \tau_1, \tag{4.118}$$

where $\mathbf{e}_1 = \eta_1 - \eta_{\text{tar}}$ is the tracking error of AUV 1. The state weight matrix $\bar{Q}_1$ and control weight matrix $\bar{R}_1$ are positive.

The cost function of follower AUV $i \in \{2, \ldots, N\}$ is

$$\bar{g}_i(\eta_i, \tau_i) = \mathbf{e}_i^{\text{T}} \bar{Q}_i \mathbf{e}_i + \tau_i^{\text{T}} \bar{R}_i \tau_i, \tag{4.119}$$

where $\mathbf{e}_i = \eta_i - \eta_j - \eta_{i,j}$ denotes the formation error between AUVs i and $j \in \mathcal{N}_i$.

From (4.118) and (4.119), the value function for AUV i is

$$\bar{V}_i(\mathbf{X}_i(t)) = \int_t^{t+T} \bar{g}_i\,(\eta_i, \tau_i)\,\mathrm{d}\tau + \bar{V}_i(\mathbf{X}_i(t+T)). \tag{4.120}$$

Based on (4.120), we present the following optimal policy

$$\tau_i^* = \operatorname{argmin}_{\tau_i} \int_t^{t+T} \bar{g}_i(\eta_i, \tau_i)\mathrm{d}\tau + \bar{V}_i(\mathbf{X}_i(t+T)). \tag{4.121}$$

To relax the model dependence, one has

$$\dot{\mathbf{X}}_i = \mathbf{A}_i\mathbf{X}_i + \mathbf{B}_i\tau_i^{(l)} + \mathbf{B}_i(\tau_i - \tau_i^{(l)}) + \mathbf{G}_i, \tag{4.122}$$

where $\tau_i^{(l)}$ is the updated control policy in the l-th iteration, and τ_i is a fixed control policy for the learning procedure.

Note that, the RL-based solutions in Cao et al. (2023), Wu et al. (2019) and Wang and Zhang (2023) require a pre-specified reward function, which cannot adapt the uncertainty of the environmental changes in complex underwater environment. In view of this, we propose the following IRL formation solution. By combining (4.120) and (4.122), the derivative of $\bar{V}_i(\mathbf{X}_i(t))$ in the $l-$th iteration is

$$\begin{aligned}
&\dot{\bar{V}}_i^{(l)}(\mathbf{X}_i(t)) \\
&= (\nabla\bar{V}_i^{(l)})^{\text{T}}(\mathbf{A}_i\mathbf{X}_i + \mathbf{B}_i\tau_i^{(l)} + \mathbf{G}_i) \\
&\quad -(-\bar{R}_i^{-1}\mathbf{B}_i^{\text{T}}\nabla\bar{V}_i^{(l)})\bar{R}_i(\tau_i - \tau_i^{(l)}) \\
&= -\bar{g}_i(\eta_i, \tau_i) - 2(\tau_i^{(l+1)})^{\text{T}}\bar{R}_i(\tau_i - \tau_i^{(l)}),
\end{aligned} \tag{4.123}$$

where $\nabla\bar{V}_i^{(l)} = \partial\bar{V}_i(\mathbf{X}_i(t))/\partial\mathbf{X}_i(t)$, and $\tau_i^{(l+1)} = -\frac{1}{2}\bar{R}_i^{-1}\mathbf{B}_i^{\text{T}}\nabla\bar{V}_i^{(l)}$.

Next, we employ IRL to acquire the desired matrix $\bar{Q}_i$, where $i \in \{1, \ldots, N\}$. To this end, one has

$$\dot{\mathbf{X}}_i = \mathbf{A}_i\mathbf{X}_i + \mathbf{B}_i\tau_{ie}^* + \mathbf{B}_i(\tau_i^{(l)} - \tau_{ie}^*) + \mathbf{G}_i, \tag{4.124}$$

where τ_{ie}^* is the expert input policy.

Substituting (4.124) into the derivative of $\bar{V}_i^{(l)}(\mathbf{X}_i(t))$, one has

$$\begin{aligned}
&\dot{\bar{V}}_i^{(l)}(\mathbf{X}_i(t)) \\
&= (\nabla \bar{V}_i^{(l)})^{\mathrm{T}}(\mathbf{A}_i\mathbf{X}_i + \mathbf{B}_i\tau_{ie}^* + \mathbf{G}_i) \\
&\quad -(-\bar{R}_i^{-1}\mathbf{B}_i^{\mathrm{T}}\nabla\bar{V}_i^{(l)})\bar{R}_i(\tau_i^{(l)} - \tau_{ie}^*) \\
&= -\bar{g}_i(\eta_i, \tau_{ie}^*) - 2\tau_{ie}^{*\mathrm{T}}\bar{R}_i(\tau_i^{(l)} - \tau_{ie}^*).
\end{aligned} \tag{4.125}$$

By integrating both sides of (4.125), one obtains

$$\begin{aligned}
&V_i^{(l)}(\mathbf{X}_i(t+T)) - V_i^{(l)}(\mathbf{X}_i(t)) \\
&= -\int_t^{t+T} \bar{g}_i(\eta_i, \tau_{ie}^*)\mathrm{d}\tau \\
&\quad -\int_t^{t+T} 2\tau_{ie}^{*\mathrm{T}}\bar{R}_i(\tau_i^{(l)} - \tau_{ie}^*)\mathrm{d}\tau.
\end{aligned} \tag{4.126}$$

Substituting (4.119) into (4.126), then $\bar{Q}_i$ can be solved by

$$\begin{aligned}
&\int_t^{t+T} \mathbf{e}_i^{\mathrm{T}}\bar{Q}_i\mathbf{e}_i\mathrm{d}\tau \\
&= \int_t^{t+T} \left((\tau_{ie}^*)^{\mathrm{T}}\bar{R}_i\tau_{ie}^* - 2(\tau_{ie}^*)^{\mathrm{T}}\bar{R}_i\tau_i^{(l)}\right)\mathrm{d}\tau \\
&\quad -\bar{V}_i^{(l)}(\mathbf{X}_i(t+T)) + \bar{V}_i^{(l)}(\mathbf{X}_i(t)).
\end{aligned} \tag{4.127}$$

By solving (4.127), the control policy of AUV i can be improved to the same level as the expert policy, i.e., $\tau_i^{(l)} = \tau_{ie}^*$. Since the choice of different $\bar{R}_i$ can affect the value of $\bar{Q}_i$, we know that the convergence of $\bar{Q}_i$ depends on the selection of $\bar{R}_i$. Furthermore, the convergence of $\tau_i^{(l)}$ to τ_{ie}^* is not affected by the selection of $\bar{R}_i$, because the different cost functions can generate the same optimal behavior. Based on this, Algorithm 6 is designed to acquire the optimal control input τ_i^*.

1. *Initialization*. Set the update procedure $k = 0$ and the iteration step $l = 0$. Based on this, we set $\bar{Q}_i^{(k)} > 0$, $\bar{V}_i^{(k,l)} = 0$ and $\tau_i^{(k,l)} = 0$.
2. *Policy Evaluation*. For each iteration $l = 0, 1, 2, \ldots$, and then one calculates the following function

$$\begin{aligned}
&\bar{V}_i^{(k,l)}(\mathbf{X}_i(t+T)) - \bar{V}_i^{(k,l)}(\mathbf{X}_i(t)) \\
&= -\int_t^{t+T} \bar{g}_i(\eta_i, \tau_i^{(k,l)})\mathrm{d}\tau \\
&\quad -\int_t^{t+T} 2(\tau_i^{(k,l+1)})^{\mathrm{T}}\bar{R}_i(\tau_i - \tau_i^{(k,l)})\mathrm{d}\tau.
\end{aligned} \tag{4.128}$$

3. *Policy Improvement*. Find an updated policy $\tau_i^{(k,l+1)}$ through the following optimal problem:

$$\tau_i^{(k,l+1)} = \mathrm{argmin}_{\tau_i^{(k,l)}} \int_t^{t+T} e(\tau_i^{(k,l)})\mathrm{d}\tau, \tag{4.129}$$

Algorithm 6: IRL-based formation controller

Input: A fixed control policy τ_i;

Initialize $k = 0$, $l = 0$, $\bar{Q}_i^{(k)} > 0$, $\bar{V}_i^{(k,l)} = 0$, $\tau_i^{(k,l)} = 0$, small threshold ϱ_3 and ϱ_4;

repeat

  $k \leftarrow k + 1$;

  repeat

    Solve the value function (4.128);

    Update $\tau_i^{(k,l+1)}$ through (4.130);

  until $\|\tau_i^{(k,l+1)} - \tau_i^{(k,l)}\| \leq \varrho_3$

  Substitute $\tau_i^{(k)} = \tau_i^{(k,l+1)}$ and $\bar{V}_i^k = \bar{V}_i^{(k,l)}$ into (4.127);

  Update $\bar{Q}_i^{(k+1)}$ through (4.131);

  Set $\tau_i^{(k,l)} = \tau_i^{(k)}$;

until $\|\bar{Q}_i^{(k+1)} - \bar{Q}_i^{(k)}\| \leq \varrho_4$

Output: The optimal control policy τ_i^*.

---

with

$$
\begin{aligned}
e(\tau_i^{(k,l)}) = {} & \bar{V}_i^{(k,l)}(\mathbf{X}_i(t+T)) - \bar{V}_i^{(k,l)}(\mathbf{X}_i(t)) \\
& + \int_t^{t+T} \bar{g}_i(\eta_i, \tau_i^{(k,l)}) \mathrm{d}\tau \\
& + \int_t^{t+T} 2(\tau_i^{(k,l+1)})^{\mathrm{T}} \bar{R}_i (\tau_i - \tau_i^{(k,l)}) \mathrm{d}\tau.
\end{aligned} \tag{4.130}
$$

4. *Cost function improvement.* Substituting $\tau^{(k)} = \tau^{(k,l+1)}$ and $\bar{V}^{(k)} = \bar{V}^{(k,l)}$ into (4.127), one has

$$
\begin{aligned}
& \int_t^{t+T} \mathbf{e}_i^{\mathrm{T}} \bar{Q}_i^{(k+1)} \mathbf{e}_i \mathrm{d}\tau \\
& = \int_t^{t+T} \left((\tau_{ie}^*)^{\mathrm{T}} \bar{R}_i \tau_{ie}^* - 2(\tau_{ie}^*)^{\mathrm{T}} \bar{R}_i \tau_i^{(k)} \right) \mathrm{d}\tau \\
& \quad - \bar{V}_i^{(k)}(\mathbf{X}_i(t+T)) + \bar{V}_i^{(k)}(\mathbf{X}_i(t)).
\end{aligned} \tag{4.131}
$$

then, one can obtain the matrix $\bar{Q}_i^{(k+1)}$. Followed by this, one substitutes $\bar{Q}_i^{(k)} = \bar{Q}_i^{(k+1)}$ and $\tau_i^{(k,l)} = \tau_i^{(k)}$ into the Step 2) for conducting the next update procedure k.

The corresponding pseudo code is presented in Algorithm 6, which can handle the impacts of model uncertainties and underwater environmental changes (see Fig. 4.14). Specifically, the control policies of AUVs can be obtained through (4.130), rather than directly calculated through $\tau_i^{(l+1)} = -\frac{1}{2}\bar{R}_i^{-1}\mathbf{B}_i^{\mathrm{T}}\nabla\bar{V}_i^{(l)}$. Hence, the model parameter $\mathbf{B}_i$ can be eliminated to solve the model uncertainty. When $\bar{R}_i$ is fixed, the convergence of $\tau_i^{(l)}$ depends on the value of $\bar{Q}_i$. For different expert policy $\tau_{i,e}^*$, Algorithm 6 can iteratively update $\bar{Q}_i$ through (4.131), which allows AUVs to respond different underwater environments.

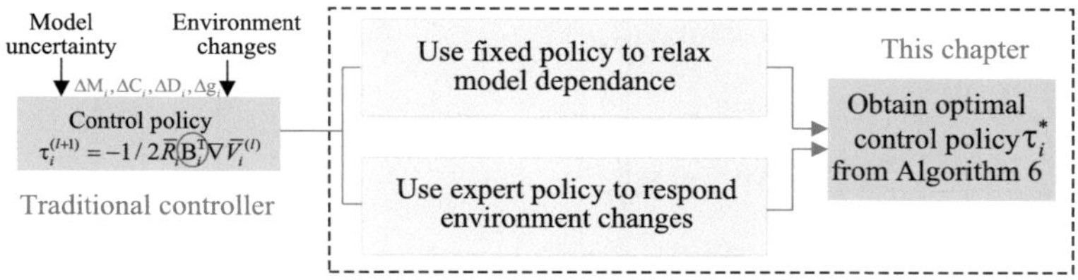

Fig. 4.14 Strategy to handle uncertainty and environment changes

The analytical forms of $\bar{V}_i^{(k,l)}$ and $\tau_i^{(k,l+1)}$ are unknown in (4.128). In view of this, two NN approximators are designed to estimate $\bar{V}_i^{(k,l)}$ and $\tau_i^{(k,l+1)}$, which can be shown as follows

$$\begin{aligned}\bar{V}_i^{(k,l)} &= (\mathbf{W}_i^{(k,l)})^{\mathrm{T}}\phi_i(t)\\ \tau_i^{(k,l+1)} &= [(\mathbf{H}_{i,1}^{(k,l)})^{\mathrm{T}}\psi_{i,1}(t);\, (\mathbf{H}_{i,2}^{(k,l)})^{\mathrm{T}}\psi_{i,2}(t);\\ &\quad (\mathbf{H}_{i,3}^{(k,l)})^{\mathrm{T}}\psi_{i,3}(t);\, (\mathbf{H}_{i,4}^{(k,l)})^{\mathrm{T}}\psi_{i,4}(t)],\end{aligned} \tag{4.132}$$

where $\phi_i(t) \in \mathcal{R}^{M_i}$ and $\psi_{i,\mathbf{i}}(t) \in \mathcal{R}^{m_i}$ are the basis function vectors for $\mathbf{W}_i^{(k,l)}$ and $\mathbf{H}_{i,\mathbf{i}}^{(k,l)}$, respectively. Of note, $\mathbf{i} \in \{1, ..., 4\}$. Based on (4.131) and (4.132), one yields

$$\begin{aligned}& e(\tau_i^{(k,l)})\\ &= (\mathbf{W}_i^{(k,l)})^{\mathrm{T}}(\phi_i(t+T) - \phi_i(t))\\ &\quad +2\textstyle\sum_{\mathbf{i}=1}^{4} \bar{R}_{i,\mathbf{i}} \int_t^{t+T} \psi_{i,1}(t)^{\mathrm{T}}\mathbf{H}_{i,\mathbf{i}}^{(k,l)}(\tau_i - \tau_i^{(k,l)})\mathrm{d}\tau\\ &\quad + \int_t^{t+T} \bar{g}_i(\eta_i, \tau_i^{(k,l)})\mathrm{d}\tau,\end{aligned} \tag{4.133}$$

where $\bar{R}_{i,\mathbf{i}}$ represents the $\mathbf{i}$-th diagonal element in $\bar{R}_i$.

To solve (4.133), one can have

$$\Phi_i^{(k,l)} = \begin{bmatrix} \rho_{i_1} \\ \vdots \\ \rho_{i_\varsigma} \end{bmatrix},\ \Xi_i^{(k,l)} = \begin{bmatrix} \lambda_{i,1}^{(\eta)} & \lambda_{i,1}^{(\tau_1)} & \cdots & \lambda_{i,1}^{(\tau_4)} \\ \vdots & \vdots & \ddots & \vdots \\ \lambda_{i,\varsigma}^{(\eta)} & \lambda_{i,\varsigma}^{(\tau_1)} & \cdots & \lambda_{i,\varsigma}^{(\tau_4)} \end{bmatrix}, \tag{4.134}$$

with

$$\begin{aligned}\rho_{i_\varsigma} &= \int_{t+\varsigma T-T}^{t+\varsigma T} \bar{g}_i(\eta_i, \tau_i^{(k,l)})\mathrm{d}\tau,\\ \lambda_{i,\varsigma}^{(\eta)} &= \phi_i(t+\varsigma T - T) - \phi_i(t+\varsigma T),\\ \lambda_{i,\varsigma}^{(\tau_1)} &= -2\bar{R}_{i,1}\int_{t+\varsigma T-T}^{t+\varsigma T} \psi_{i,1}(t)^{\mathrm{T}}(\tau_i - \tau_i^{(k,l)})\mathrm{d}\tau,\\ \lambda_{i,\varsigma}^{(\tau_4)} &= -2\bar{R}_{i,4}\int_{t+\varsigma T-T}^{t+\varsigma T} \psi_{i,4}(t)^{\mathrm{T}}(\tau_i - \tau_i^{(k,l)})\mathrm{d}\tau.\end{aligned} \tag{4.135}$$

According to the given formulas (4.133)-(4.135), the $\mathbf{W}_i^{(k,l)} \to \mathbf{W}_i^{(k)}$ and $\mathbf{H}_{i,\mathbf{i}}^{(k,l)} \to \mathbf{H}_{i,\mathbf{i}}^{(k)}$ can be solved by

$$
\begin{aligned}
&[\mathbf{W}_i^{(k,l)}, \mathbf{H}_{i,1}^{(k,l)}, \ldots, \mathbf{H}_{i,4}^{(k,l)}]^{\mathrm{T}} \\
&= \left((\Xi_i^{(k,l)})^{\mathrm{T}}(\Xi_i^{(k,l)})\right)^{-1} (\Xi_i^{(k,l)})^{\mathrm{T}}\Phi_i^{(k,l)}.
\end{aligned} \tag{4.136}
$$

The uniqueness of $\mathbf{W}_i^{(k)}$ and $\mathbf{H}_{i,\mathbf{i}}^{(k)}$ is guaranteed by holding the full rank of $(\Xi_i^{(k,l)})^{\mathrm{T}}(\Xi_i^{(k,l)})$, which can be achieved by satisfying $\varsigma \geq M_i + 4m_i$. Of note, $M_i + 4m_i$ is the number of unknown parameters in (4.133).

Theorem 4.4 *Given an initial admissible policy $\tau_i^{(k,0)}$ at procedure k with $\bar{Q}_i^{(k)} > 0$. The policy iteration in (4.129) can make $\tau_i^{(k,l+1)}$ converge to the optimal policy τ_i^*.*

Proof Given an admissible policy $\tau_i^{(k,0)}$ with system trajectory of $\dot{\mathbf{X}}_i = \mathbf{A}_i\mathbf{X}_i + \mathbf{B}_i\tau_i^{(l)} + \mathbf{B}_i(\tau_i - \tau_i^{(l)}) + \mathbf{G}_i$ in (4.122). The model-based Bellman equation is described as

$$
\begin{aligned}
&\bar{H}(\bar{V}_i^{(k,l)}, \tau_i^{(k,l)}) \\
&= \nabla\bar{V}_{\mathbf{X}_i}^{(k,l)\mathrm{T}}(\mathbf{A}_i\mathbf{X}_i + \mathbf{B}_i\tau_i^{(l)} + \mathbf{G}_i) + \bar{g}_i(\eta_i, \tau_i^{(k,l)}) \\
&= 0.
\end{aligned} \tag{4.137}
$$

Based on (4.122), the model-free Bellman equation is

$$
\begin{aligned}
&\bar{H}(\bar{V}_i^{(k,l)}, \tau_i^{(k,l)}) \\
&= \nabla\bar{V}_{\mathbf{X}_i}^{(k,l)\mathrm{T}}(\mathbf{A}_i\mathbf{X}_i + \mathbf{B}_i\tau_i^{(l)} + \mathbf{B}_i(\tau_i - \tau_i^{(l)}) \\
&\quad + \mathbf{G}_i) + \bar{g}_i(\eta_i, \tau_i^{(k,l)}) + 2\tau_i^{(k,l+1)\mathrm{T}}\bar{R}(\tau_i - \tau_i^{(k,l)}) \\
&= 0.
\end{aligned} \tag{4.138}
$$

By substituting $\tau_i^{(k,l+1)} = -\frac{1}{2}\bar{R}_i\mathbf{B}_i\nabla\bar{V}_{\mathbf{X}_i}^{(k,l)}$ into (4.138), it can acquire the same Bellman equation in (4.137). That is, the solution τ_i^* to the Bellman function in model-free equation is the same as the Bellman function in model-based equation. Of note, the model-based Bellman equation has been proven in Theorem 1. To this end, the iteration policy $\tau_i^{(k,l+1)}$ can converge to τ_i^* by solving (4.128) and (4.129).

Next, based on (4.119) and (4.120), we can easily obtain

$$
\begin{aligned}
&\bar{V}_i^{(k)}(\mathbf{X}_i(t+T)) - \bar{V}_i^{(k)}(\mathbf{X}_i(t)) \\
&= -\int_t^{t+T} \left(\mathbf{e}_i^{\mathrm{T}}\bar{Q}_i^{(k)}\mathbf{e}_i + \tau^{(k)}{}_i^{\mathrm{T}}\bar{R}_i\tau_i^{(k)}\right) \mathrm{d}\tau.
\end{aligned} \tag{4.139}
$$

Substituting (4.139) into (4.131), one yields

$$\begin{aligned}
&\int_t^{t+T} \mathbf{e}_i^{\mathrm{T}} \bar{Q}_i^{(k+1)} \mathbf{e}_i \mathrm{d}\tau \\
&= \int_t^{t+T} (\tau_{ie}^* - \tau_i^{(k)})^{\mathrm{T}} \bar{R}_i (\tau_{ie}^* - \tau_i^{(k)}) \mathrm{d}\tau \\
&\quad + \int_t^{t+T} \mathbf{e}_i^{\mathrm{T}} \bar{Q}_i^{(k)} \mathbf{e}_i \mathrm{d}\tau.
\end{aligned} \tag{4.140}$$

With the initial $\bar{Q}_i^{(0)} > 0$, one has

$$0 < \bar{Q}_i^{(0)} \leq \cdots \leq \bar{Q}_i^{(k)} \leq \bar{Q}_i^{(k+1)}. \tag{4.141}$$

Define $\Theta_i^{(k)} \triangleq (\tau_{ie}^* - \tau_i^{(k)})^{\mathrm{T}} \bar{R}_i (\tau_{ie}^* - \tau_i^{(k)})$, thus one has

$$\begin{aligned}
&\int_t^{t+T} \mathbf{e}_i^{\mathrm{T}} \bar{Q}_i^{(k)} \mathbf{e}_i \mathrm{d}\tau \\
&= \int_t^{t+T} (\Theta_i^{k-1} + \cdots + \Theta_i^0 + \mathbf{e}_i^T \bar{Q}_i^{(0)} \mathbf{e}_i) \mathrm{d}\tau.
\end{aligned} \tag{4.142}$$

From (4.142), $\bar{Q}_i^{(k)}$ is increased to the vicinity of $\breve{Q}_i$, i.e., there exists a small threshold ϵ that holds $\left\| \breve{Q}_i - \bar{Q}_i^{(k)} \right\| \leq \epsilon$. Then, the uniform approximate solution of τ_{ie}^* can be acquired, i.e., $\left\| \tau_i^{(k)} - \tau_{ie}^* \right\| \leq \varepsilon_i$, where a small threshold ε_i is used to avoid affecting the positive definiteness of $\bar{Q}_i^{(k)}$. We make the algorithm converge to procedure k, and then one has

$$\begin{aligned}
&\left\| \int_t^{t+T} \mathbf{e}_i^{\mathrm{T}} \bar{Q}_i^{(k+1)} \mathbf{e}_i \mathrm{d}\tau \right\| \\
&\geq \left\| \int_t^{t+T} (k\varepsilon_i^2 \bar{R}_i + \mathbf{e}_i^{\mathrm{T}} \bar{Q}_i^{(0)} \mathbf{e}_i) \mathrm{d}\tau \right\| \\
&\geq k\varepsilon_i^2 T \lambda_{\min}(\bar{R}_i) - \left\| \int_t^{t+T} \mathbf{e}_i^{\mathrm{T}} \bar{Q}_i^{(0)} \mathbf{e}_i \mathrm{d}\tau \right\|,
\end{aligned} \tag{4.143}$$

which implies

$$k \leq \frac{\left\| \int_t^{t+T} \mathbf{e}_i^{\mathrm{T}} \bar{Q}_i^{(k+1)} \mathbf{e}_i \mathrm{d}\tau \right\| + \left\| \int_t^{t+T} \mathbf{e}_i^{\mathrm{T}} \bar{Q}_i^{(0)} \mathbf{e}_i \mathrm{d}\tau \right\|}{\varepsilon_i^2 T \lambda_{\min}(\bar{R}_i)} \equiv k_1. \tag{4.144}$$

The iterative algorithm convergences after a maximum of procedure k_1. Thereby, AUV $i \in \{1, 2, \ldots, N\}$ can converge to the optimal policy τ_i^*. That completes the proof. □

4.4 Technology Practice

4.4.1 Case Study I: Collision-Free Motion Planning for a Single Underwater Vehicle

In this section, simulation results are presented to verify the effectiveness. Specifically, the positions of source sensor node and destination sensor node are set as $[-30, -30, -8]^{\mathrm{T}}$ and $[30, 30, -8]^{\mathrm{T}}$, respectively. The initial position and orientation vectors of underwater vehicles R_1 and R_2 are set as $\eta_1 = [29.5, 29.5, 0, 1]^{\mathrm{T}}$ and $\eta_2 = [29.5, 29.5, 0, 2]^{\mathrm{T}}$, respectively. In addition, the parameters of steepness and repulsive range are given as $c = 10$, $v = 10$, $\nu_1 = 1$, $c_1 = 10$.

(1) *SNR prediction in fading channel*

We first verify the effectiveness of the SNR estimator as proposed in Sect. 4.2.1, where the actual channel parameters are set as $\theta_{R_i} = [-60, 5]^{\mathrm{T}}$, $\xi_{R_i} = 0.07$, $\varphi_{R_i} = 3.51$ and $\rho_{R_i} = 0.2864$. Besides that, $\mathbf{Q}_1$ =diag([5, 5, 5, 5]), $\lambda_1 = 0.5$, $\lambda_2 = 0.01$, and $T = 0.1$. Accordingly, the SNR estimator is adopted, and hence, the estimated path loss, shadowing and multipath fading parameters are shown in Fig. 4.15a–c, respectively. On the basis of this, the link capacity in underwater area can be shown in Fig. 4.15d. The estimated parameters converge to the true values, which verifies the effectiveness of the proposed SNR estimator in this chapter.

In Yan and Mostofi (2012), the least square estimator is adopted to estimate the SNR parameters, but the least square estimator can make the parameters fall into local optimum. To show the above phenomenon, we assume that the SNR measurement process is polluted by external noise, where the external noise is set as $\digamma_{\#} = 5000\cos^2(0.25K_{\mathrm{dB},1} + 5.3) + 2000\cos^2(0.5n_{\mathrm{PL},1} + 0.3)$. Take Part 3 as an example, the optimization problem can be updated as $\hat{\chi}^*_{R_i} = \mathrm{argmin}\{\text{Part 3}+\digamma_{\#}\}$. With the IRL-based estimator in this chapter, the predicted link capacity by using 3500 different sampling points is shown in Fig. 4.15e. Meanwhile, the cost comparison by using the least square estimator (e.g., Yan and Mostofi (2012)) and the IRL-based estimator in this chapter is provided by Fig. 4.15f. We find that the parameters estimation result of IRL-based estimator proposed in this chapter is closer to the true value than that of the least squares estimator.

(2) *Motion planning of a single underwater vehicle*

With the predicted SNR information, we consider a simple motion planning scenario, i.e., a single underwater vehicle is deployed to relay the data from source sensor node to destination sensor node. Along with this, the value and policy basis functions of underwater vehicle R_1 can be expressed as $\phi_2^1 = [e^2_{1,u_1}, e^2_{2,v_1}, e^2_{1,w_1}, e^2_{1,r_1}]^{\mathrm{T}}$, $\phi^1_{4,1} = [2e_{1,u_1}]^{\mathrm{T}}$, $\phi^1_{4,2} = [2e_{1,v_1}]^{\mathrm{T}}$, $\phi^1_{4,3} = [2e_{1,w_1}]^{\mathrm{T}}$, and $\phi^1_{4,4} = [2e_{1,r_1}]^{\mathrm{T}}$, respectively. Of note, e_{1,u_1}, e_{1,v_1}, e_{1,w_1} and e_{1,r_1} denote the error components of underwater vehicle R_1 on surge, sway, depth and yaw, respectively. Meanwhile, $\beta_1 = 100$, $\beta_2 = 1$, $\rho_{o1} = 6$, $\rho_{o2} = 6$, and $\bar{\mathbf{R}}_1$ =diag([0.1, 0.08, 1, 1.2]). Thereby, the trajectory

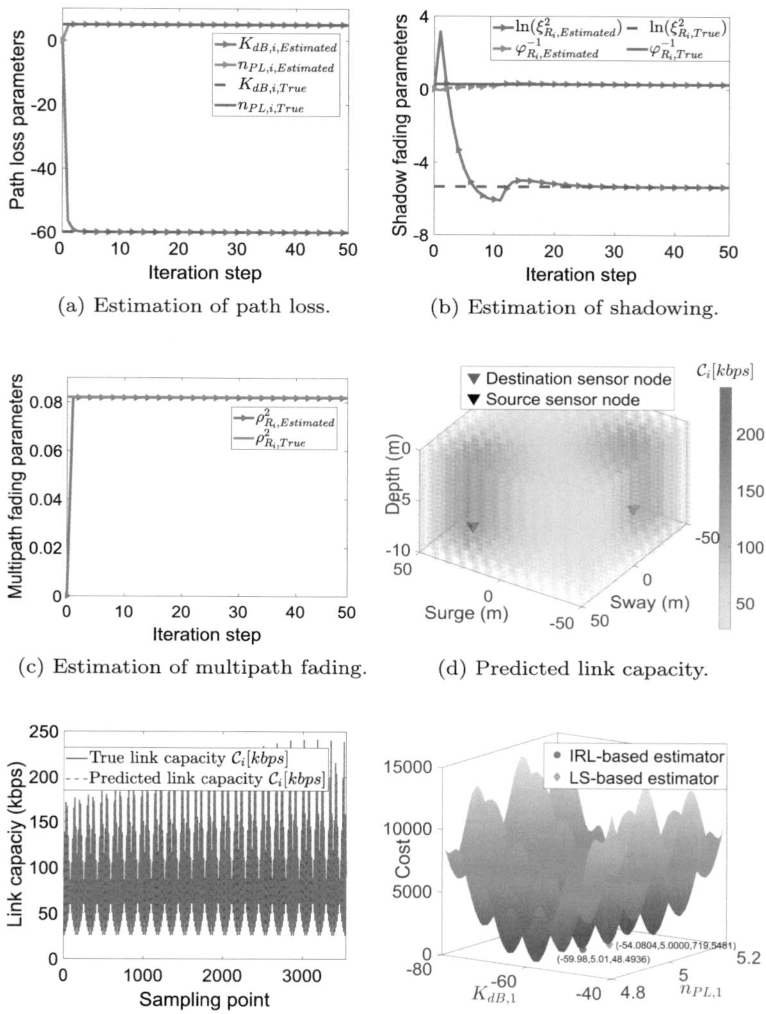

(a) Estimation of path loss. (b) Estimation of shadowing.

(c) Estimation of multipath fading. (d) Predicted link capacity.

(e) Link capacity with noise. (f) Comparison with [29].

Fig. 4.15 Simulation results for the RL-based online SNR prediction

of underwater vehicle R_1 is presented in Fig. 4.16a, whose position and orientation are shown in Fig. 4.16b. Correspondingly, the optimal policy of underwater vehicle R_1 is shown in Fig. 4.16c, where the learned weights are presented in Fig. 4.16d–e. Based on this, the link capacity of underwater vehicle R_1 and the segmented link capacity $c_{0,1}$ and $c_{1,2}$ are shown in Fig. 4.16f. Clearly, the collision avoidance can be guaranteed. Meanwhile, at the beginning, the link capacity of the underwater vehicle is very poor. Through the motion planning procedure, the link capacity gradually

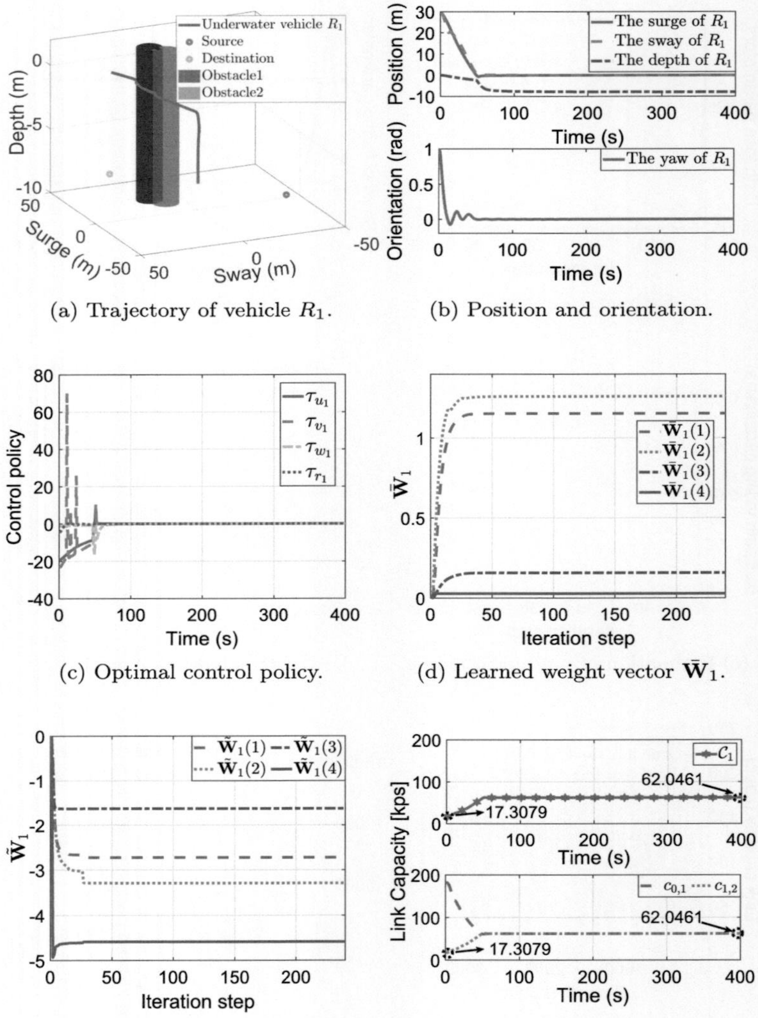

(a) Trajectory of vehicle R_1. (b) Position and orientation.

(c) Optimal control policy. (d) Learned weight vector $\bar{\mathbf{W}}_1$.

(e) Learned weight vector $\tilde{\mathbf{W}}_1$. (f) Link capacity.

Fig. 4.16 Simulation results for motion planning of a single vehicle

increases, wherein $c_{0,1}$ and $c_{1,2}$ become equal and reach stability after $t = 50$ s. Once $\mathcal{C}_1$ is maximized, underwater vehicle R_1 can hover at $(0, 0, -8)$. Overall, the link capacity is increased by 258.48% than the initial value, i.e., from 17.3079 to 62.0461.

(3) *Comparison with the other motion planning solutions*

Note that a Lyapunov guidance vector field (LGVF) based motion planning algorithm was provided in Dixon and Frew (2012), where the spiral forward was performed by vehicle. Clearly, the spiral forward can increase the path length of vehicle, which may reduce the lifetime of vehicle. By ignoring the the underwater obstacle, the LGVF-based motion planning algorithm is adopted here, and hence, the trajectories of the underwater vehicle R_1 by using the above two algorithms are shown in Fig. 4.17a. The positions on surge, sway and depth are given in Fig. 4.17b. The comparison of the path length required by the two algorithms is shown in Fig. 4.17c. Meanwhile, the link capacities of the two algorithms are shown in Fig. 4.17d. From Fig. 4.17a–d, we can see that the link capacities by using the above two algorithms can both be improved, however, the path length required in this chapter is less than the one in Dixon and Frew (2012) since the spiral forward is not required in this chapter.

Another important characteristic of our solution is the independence of system model, i.e., it is not necessary to know the nominal value of the underwater vehicle in advance. This characteristic is of great practical significance to ocean monitoring because it is difficult to obtain the nominal value in harsh underwater environment. With respect to this, the model-based learning controller (e.g., Yan et al. (2022)) is adopted by the underwater vehicle. Then, the following two cases are considered: (1) the model matrix $\mathbf{M}_1$ can be accurately obtained; (2) the model matrix $\mathbf{M}_1$ cannot be accurately obtained due to environment noise and model uncertainty. By employing model-based learning approach, the motion trajectories of underwater vehicle R_1 in Case 1 and Case 2 are shown in Fig. 4.17e and f, respectively. Correspondingly, the link capacities in Case 1 and Case 2 are presented in Fig. 4.17g and h, respectively. Clearly, the underwater vehicle can achieve the motion planning task when the model information is accurate, and meanwhile, the link capacity is significantly improved. However, when the model information is inaccurate, the motion planning task of underwater vehicle cannot be well achieved, which can result in the failure of capacity improvement. The above results demonstrate that the model-free motion planning solution developed in this chapter is meaningful and necessary for underwater vehicles.

This section presents the experimental results. As depicted in Fig. 4.16a, the desired relay position for a single underwater vehicle is on the midpoint between the source sensor node and the destination sensor node. With regard to this, three nodes including an underwater vehicle, a source sensor node and a destination sensor node are considered, as shown in Fig. 4.18. The work frequency band of the wireless communication system is within 21 kHz to 27 kHz, and it adopts the OFDM (Orthogonal Frequency Division Multiple Access) mode. In order to overcome the multipath and Doppler effects, the cyclic prefix is extended and the frequency interval is enlarged. Due to this, our communication system has a stable communication rate of 300 bps and a maximum communication rate of 2000 bps. Different from the terrestrial environment, the signal-to-noise ratio $S_{i-1,i}$ is affected by path loss, shadow fading, multipath fading and various external underwater disturbance, espe-

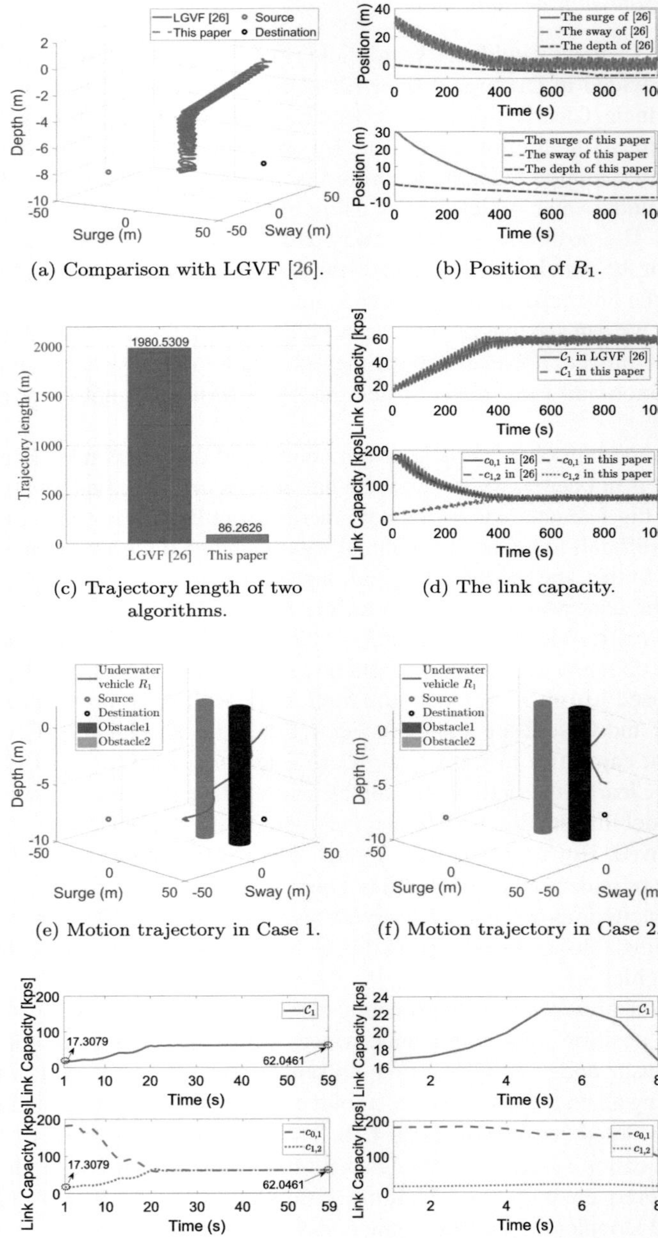

(a) Comparison with LGVF [26]. (b) Position of R_1.

(c) Trajectory length of two algorithms. (d) The link capacity.

(e) Motion trajectory in Case 1. (f) Motion trajectory in Case 2.

(g) The link capacity in Case 1. (h) The link capacity in Case 2.

Fig. 4.17 Comparison for the IRL-based motion planning solution with the other existing solutions

Fig. 4.18 Experiment deployment, where an underwater vehicle, a source sensor node and a destination sensor node is included

cially in shallow water near the shore. In the experiment, with the change of distance, the $S_{i-1,i}$ value is about 0.15, so the communication rate is 300 bps. For underwater vehicle, the BlueROV from Blue Robotics is adopted, which features six thrusters, a flight controller, a wireless communication unit and a Raspberry Pi.

In the following, the underwater vehicle patrols in different positions to relay the data of source sensor node to the destination sensor node, whose motion trajectory is shown in Fig. 4.19a. For clear description, the relative distance between underwater vehicle and source (or destination) sensor node is provided in Fig. 4.19b. Correspondingly, the end-to-end probability of successful data transmission (EPSDT) is shown in Fig. 4.19c, which is defined as the successful data transmission of the worst link. We find that the successful data transmission for each communication link is increased with the increase of relative distance. Meanwhile, the successful data transmission can reach to the same value (i.e., 90%) when the relative distances for the two links are the same (i.e., 14.5803 m). Based on the definition of EPSDT, we can know that the EPSDT is the maximum when the underwater vehicle is on the midpoint between the source sensor node and the destination sensor node. It is clear that these results are consistent with the simulation results. Similarly, the results for multiple underwater vehicles can also be obtained, and this part is omitted here due to page limitation.

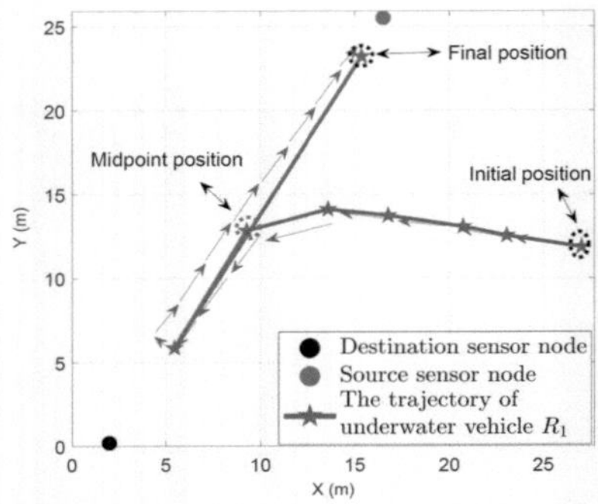

(a) Motion trajectory of underwater vehicle.

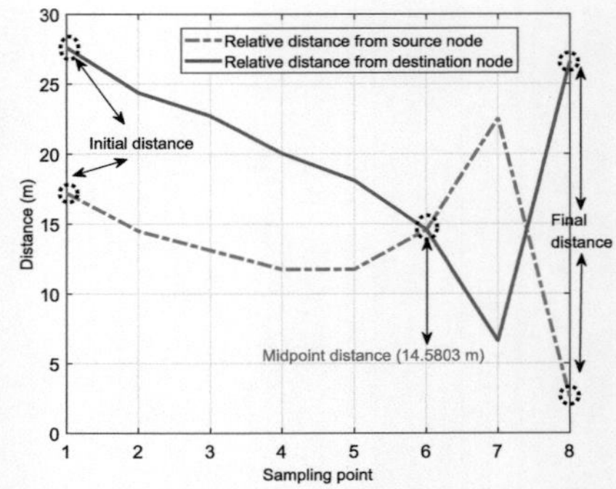

(b) Relative distances with source and destination sensor nodes.

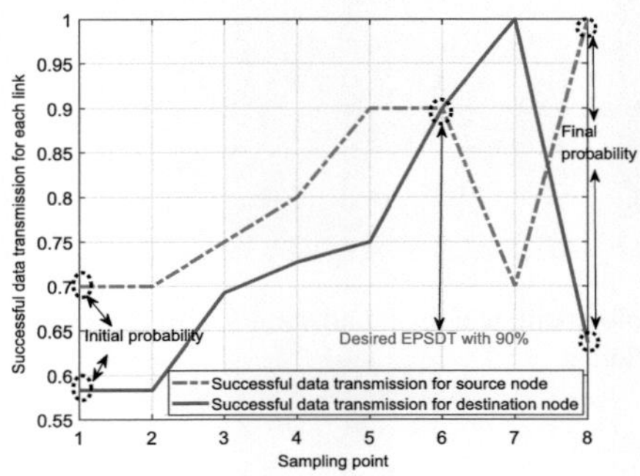

(c) Probability of successful data transmission for each link.

Fig. 4.19 Experimental results for deployment of underwater vehicle

4.4.2 Case Study II: Topology-Aware Formation for Multiple Underwater Vehicles

(1) *Effectiveness of the proposed formation solution*

In this section, five AUVs are employed to achieve the formation task. AUV 1 plays the role of leader, and its observation radius is $R_{\mathrm{f}} = 6$m. The destination and the desired relative state vectors are set as: $\eta_{\mathrm{tar}} = [10\mathrm{m}, 10\mathrm{m}, -5\mathrm{m}, 0\mathrm{rad}]^{\mathrm{T}}$, $\eta_{1,2} = [3\mathrm{m}, 3\mathrm{m}, 3\mathrm{m}, 0\mathrm{rad}]^{\mathrm{T}}$, $\eta_{1,3} = [4.5\mathrm{m}, 0\mathrm{m}, 3\mathrm{m}, 0\mathrm{rad}]^{\mathrm{T}}$, $\eta_{1,4} = [0\mathrm{m}, 4.5\mathrm{m}, 3\mathrm{m}, 0\mathrm{rad}]^{\mathrm{T}}$ and $\eta_{1,5} = [6\mathrm{m}, 6\mathrm{m}, 0\mathrm{m}, 0\mathrm{rad}]^{\mathrm{T}}$.

We first verify the effectiveness of the topology inference estimator in Sect. 4.3.1. Similar to Neu and Szepesvari (2007), the expert data can be indirectly obtained by selecting policies that minimize the topology estimation errors. For $i, j \in \{1, \ldots, 5\}$, we denote l_{ij} as the i-th row and j-th column of Laplacian matrix L, whose estimation process and error of l_{ij} are shown in Fig. 4.20a and b, respectively. From Fig. 4.20a–b, we find L can converge to the expected value. The actual topology of AUVs is presented in Fig. 4.20c, which is unknown to AUVs. For that reason, the leader AUV 1 patrols the entire region, where its constant speed is $\mathbf{v} = [0.2\text{m/s}, 0.2\text{m/s}, 0.3\text{m/s}]^{\mathrm{T}}$. The IRL-based topology estimator is conducted on the leader AUV, through which the procedure of topology inference is shown in Fig. 4.20d–f. Clearly, the IRL-based topology estimator can accurately estimate the topology of AUVs.

With the inferred topology, a topology optimization algorithm is provided in Sect. 4.3.1 to balance the connectivity and energy efficiency. To verify its significance, we utilize Algorithm 5 on the inferred topology, through which the optimally rigid graph is shown in Fig. 4.20g. Followed by this, the optimally persistent graph is shown in Fig. 4.20h. From Fig. 4.20g–h, it can be observed that Algorithm 5 can reduce redundant edges for the original topology.

Based on the optimized topology relationship, an IRL-based formation controller is employed to drive the motions of AUVs, as presented in Sect. 4.3.2. Note that the expert data can be obtained using optimal policies generated by RL. For AUV $i \in \{1, \ldots, 5\}$, the basis function of critic-network can be designed as $\phi_i = [\mathbf{e}_{i,1}^2, \mathbf{e}_{i,2}^2, \mathbf{e}_{i,3}^2, \mathbf{e}_{i,4}^2]^{\mathrm{T}}$, where $\mathbf{e}_{i,\mathbf{i}\in\{1,2,3,4\}}$ represents the $\mathbf{i}$-th row of $\mathbf{e}_i$. In addition, the basis functions of actor-network are designed as $\psi_{i,1} = [2\mathbf{e}_{i,1}]^{\mathrm{T}}$, $\psi_{i,2} = [2\mathbf{e}_{i,2}]^{\mathrm{T}}$, $\psi_{i,3} = [2\mathbf{e}_{i,3}]^{\mathrm{T}}$ and $\psi_{i,4} = [2\mathbf{e}_{i,4}]^{\mathrm{T}}$. Based on this, the trajectories of AUVs are presented in Fig. 4.21a. It can be seen that the trajectories of AUVs converge to the desired expert trajectories. In order to show more clearly, the formation errors, i.e., $E_{ij} = \left\|\eta_i - \eta_j - \eta_{i,j}\right\|$ for AUVs $i, j \in \{1, \ldots, 5\}, i \neq j$ are depicted in Fig. 4.21b. The tracking error of leader AUV 1, i.e., $\bar{E} = \|\eta_1 - \eta_{\text{tar}}\|$, is shown in Fig. 4.21c. From Fig. 4.21b–c, we can find the that the formation task can be achieved, since the formation and tracking errors converge to zero. In addition, the tracking state errors of $\mathbf{X}_i$ with respect to the expert system are depicted in Fig. 4.21d, while the tracking policy errors with respect to the expert system are given in Fig. 4.21e. We can find that learned state and policy of AUVs can well capture the ones with expert system. Besides that, the weight matrix Q_i and policy weight $\mathbf{H}_{i,\mathbf{i}\in\{1,\ldots,4\}}$ are depicted in Fig. 4.21f–g, respectively. The control inputs of AUVs are presented in Fig. 4.21h. It is obvious that the weights and inputs can arrive at the optimization values. Overall, the above results demonstrate the effectiveness of the IRL-based formation controller.

(2) *Comparison for the topology inference estimator*

In Li et al. (2023), a least squares-based estimator is adopted to infer the topology relationship. However, the least squares estimator can make the estimation fall into local optimum. In view of this, we assume that the state collection process is affected by external Gaussian noise, e.g., $\omega_i \sim (0, \sigma^2 I)$ where $\sigma = 0.05$. Based on this, the collected information of AUV $i \in \{1, \ldots, 5\}$ can be described by $\tilde{\Delta}_i = \Delta_i + \omega_i$. By using the least squares-based estimator and the IRL-based estimator in this chapter, the estimation of $L_{HF,1}$ is presented in Fig. 4.22a. The actual value of $L_{HF,1}$ is $L_{HF,1} = [0, 0, 0, 0; -1, 3, -1, -1; -1, 0, 2, -1]$. From Fig. 4.22a, we know the IRL-based estimator in this chapter can acquire the global optimization value.

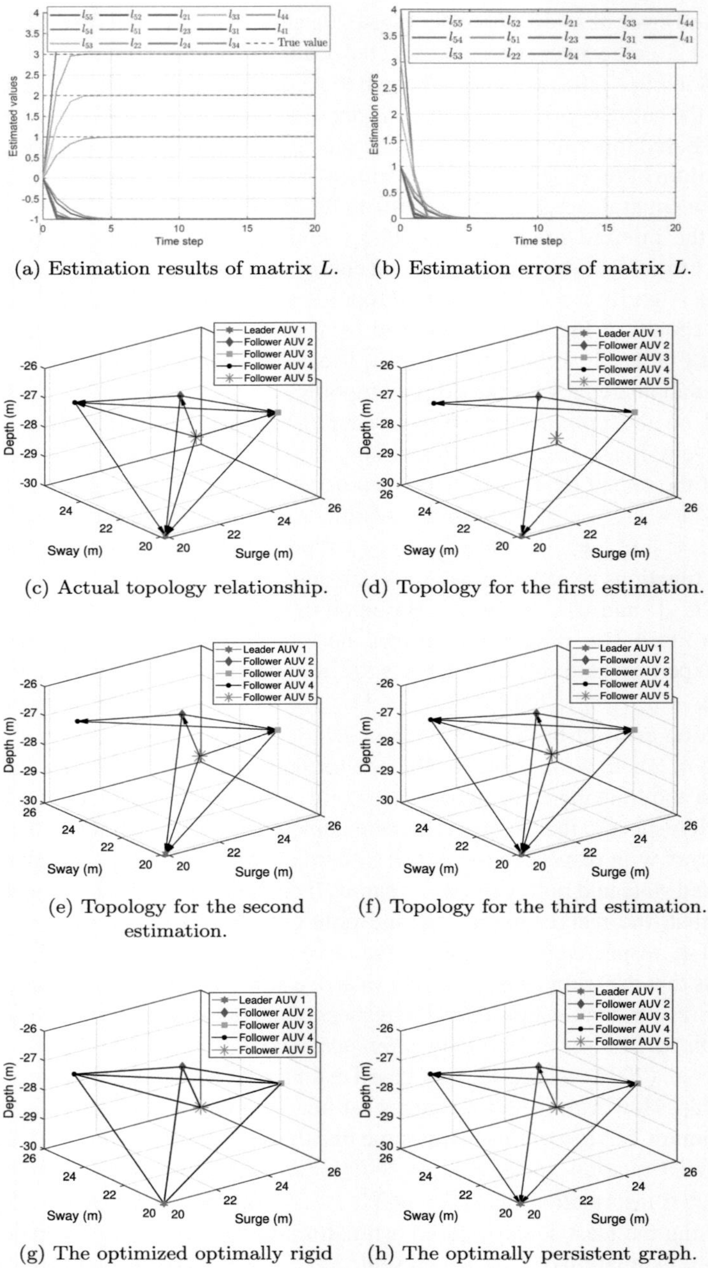

(a) Estimation results of matrix L. (b) Estimation errors of matrix L.

(c) Actual topology relationship. (d) Topology for the first estimation.

(e) Topology for the second estimation. (f) Topology for the third estimation.

(g) The optimized optimally rigid graph. (h) The optimally persistent graph.

Fig. 4.20 Simulation results for topology inference and optimization

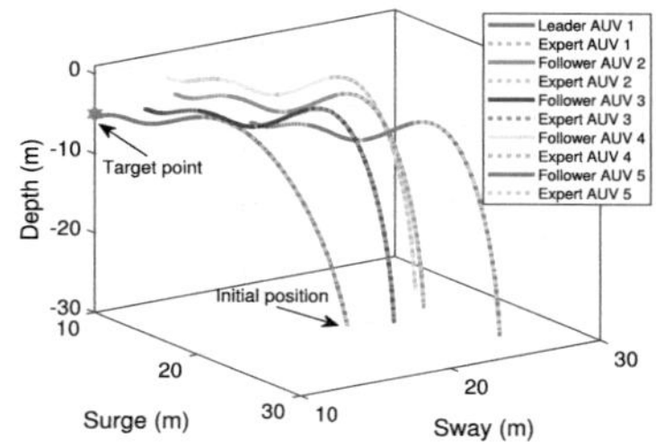

(a) Formation trajectories.

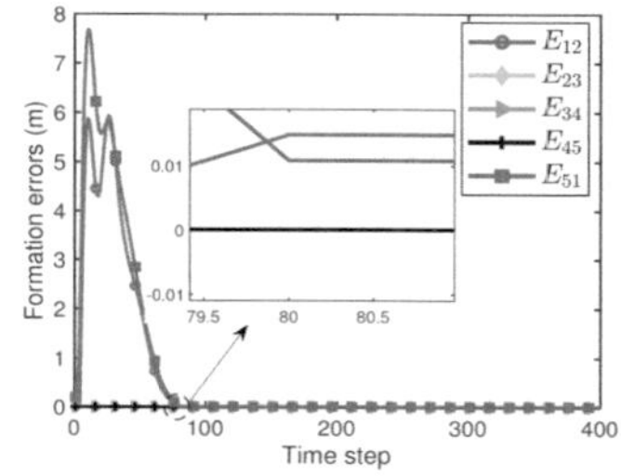

(b) Formation errors.

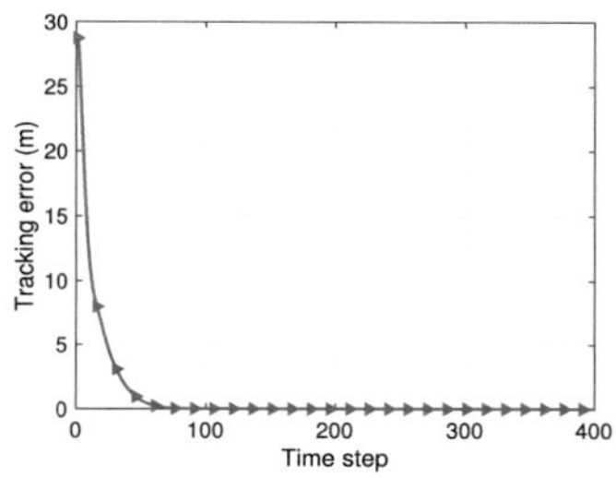

(c) Tracking errors.

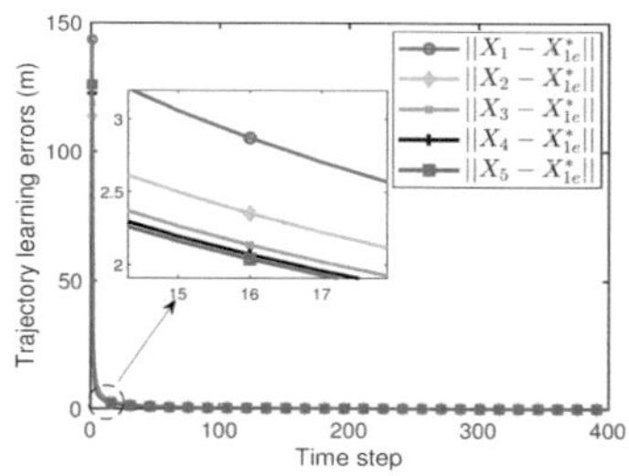

(d) Trajectory learning errors.

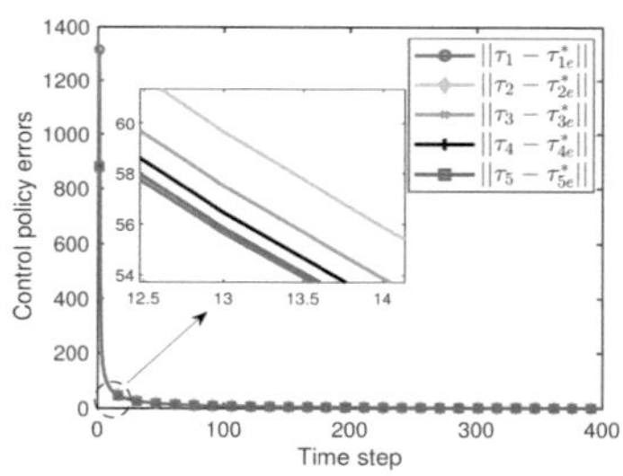

(e) Control policy errors.

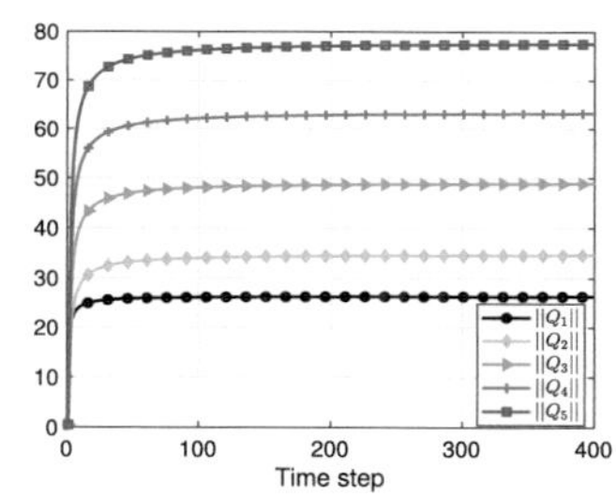

(f) Convergence of Q_i.

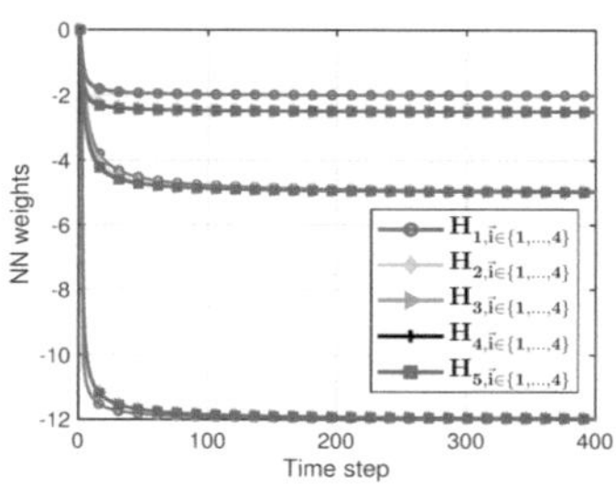

(g) Weight convergence of control policy.

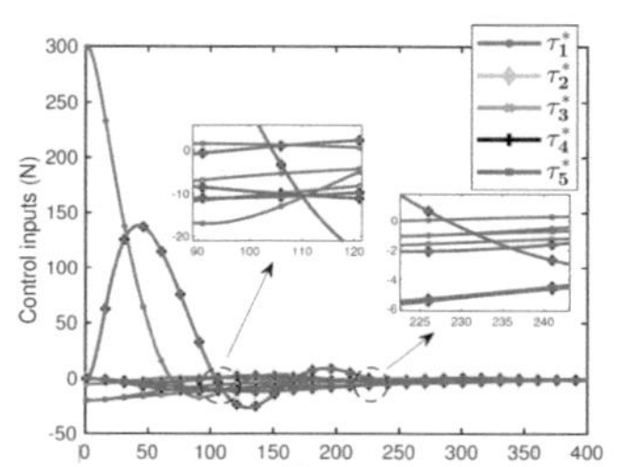

(h) Optimal control inputs of AUVs.

Fig. 4.21 Simulation results of IRL-based formation controller

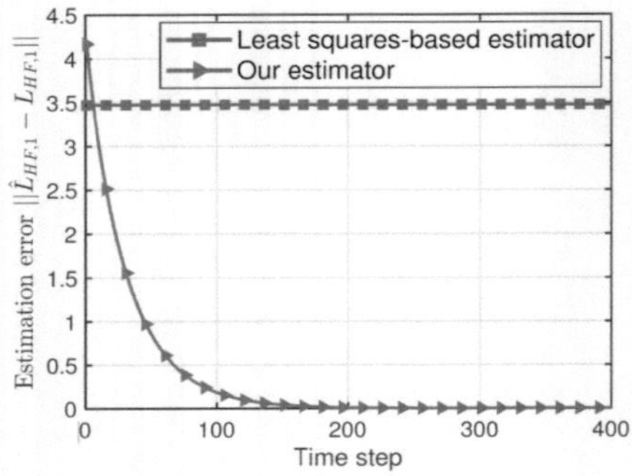

(a) The estimation of $L_{HF,1}$.

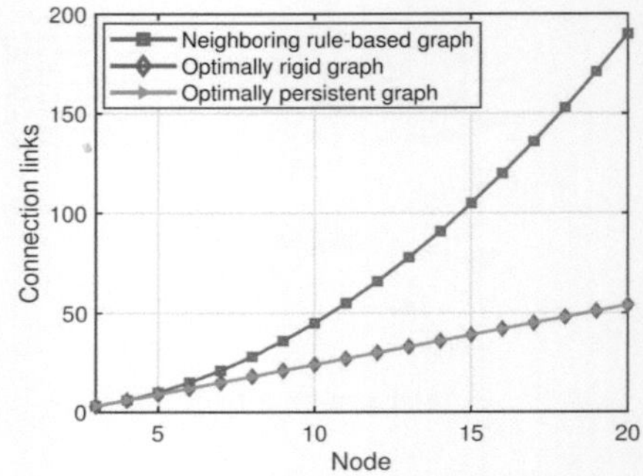

(b) Comparison of communication links.

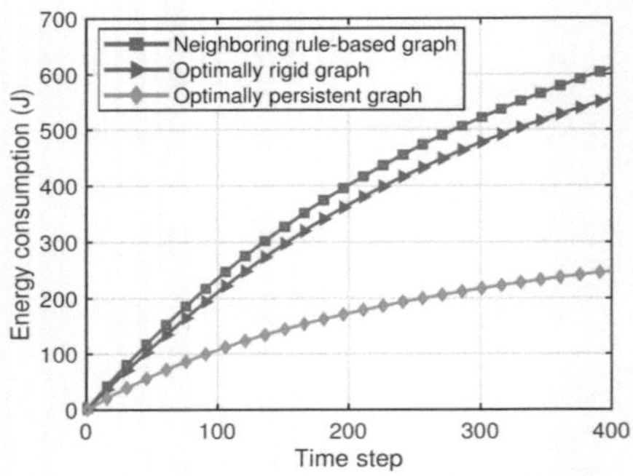

(c) Comparison of communication energy.

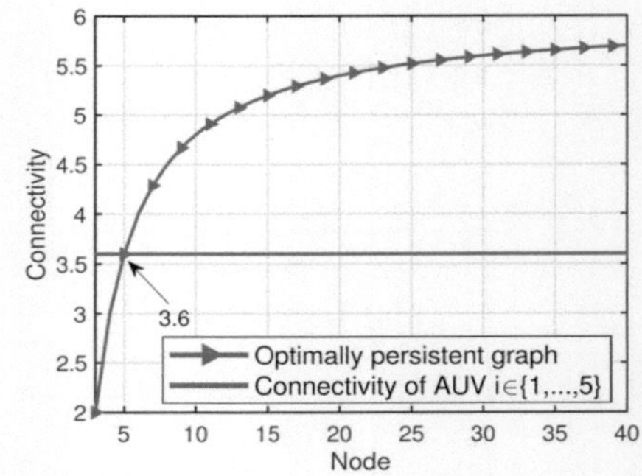

(d) Average connectivity.

Fig. 4.22 Comparison of topology inference and optimization

(3) *Comparison for the topology optimization algorithm*

Note that some existing works (e.g., Xu et al. (2023), Mao and Yan (2018) and Chen et al. (2022)) adopt the neighboring rule-based algorithm to ensure the connectivity. However, it cannot make a balance between topology connectivity and energy efficiency. Alternatively, the optimally persistent graph-based solution in Sect. 4.3.1 can bridge this gap. In view of this, the number of connection links and the communication energy consumption by using the neighboring rule-based algorithm, the optimally rigid graph (e.g., Yan et al. (2024) and Yan et al. (2018b)) and the optimally persistent graph in this chapter are presented by Fig. 4.22b–c, respectively. It can be seen that AUVs under the optimally persistent graph in this chapter have the minimum communication energy consumption, and the topology of AUVs constructed based on the neighboring rule-based has the highest communication energy consumption. On the other hand, the connectivity of AUVs converges to 6 in the optimally persistent graph, as shown in Fig. 4.22d. From the above results, we find the optimally persistent graph in this chapter can ensure the connectivity of AUVs and reduce the communication energy consumption.

(4) *Comparison for the IRL-based formation controller*

To demonstrate the environmental adaptability of the IRL-based formation controller, we compare the formation stability with the model-based RL and model-free RL controllers. For the model-based controller (e.g. Shou et al. (2021)), we first train a control policy that enables AUVs to reach their destinations. Followed by this, we introduce a periodic noise into the system to demonstrate the environment change, i.e., $\dot{\mathbf{X}}_i = \mathbf{A}_i\mathbf{X}_i + \mathbf{B}_i\tau_i + \mathbf{G}_i + \omega_2$, where the environment change is depicted by $\omega_2 = 10\cos(0.34x_i)[1; 1; 0; 0; 0; 0; 0; 0]$. Along with this, the trajectories of AUVs under the model-based RL formation controller are shown in Fig. 4.23a. Correspondingly, Fig. 4.23b provides the formation errors of AUVs. In addition, we utilize the model-based RL control policy $\tau_i = -1/2\bar{R}_i\partial\bar{V}(\mathbf{X}_i)/\partial\mathbf{X}_i$ to collect state data, through which we use model-free RL to train the control policy. Under the periodic environment change, the trajectories of AUVs are shown in Fig. 4.23c, and the formation errors are shown in Fig. 4.23d. For the IRL-based formation controller in this chapter, we use the expert policy τ_{ie}^* that performs well in the disturbance environment to learn the optimal control policy for AUVs. Based on this, the trajectories of AUVs are shown in Fig. 4.23e, and f presents the formation errors. In addition, the comparison of average errors are also shown in Fig. 4.23f. From Fig. 4.23b, d and f, we find that the IRL-based controller in this chapter has the smallest formation errors, i.e., it has better environment adaptability by learning from expert demonstrations. Therefore, it is necessary and meaningful to design an IRL-based formation controller for AUVs to enhance the environmental adaptability. The results in this chapter can be extended to the other types of rigid formations where less sensor measurements are needed, such as bearing (or angle) rigid formations Chen et al. (2022) and Chen et al. (2021).

Experiment in the pool environment: As shown in Fig. 4.24, the hardware system mainly consists of the following units: (1) *AUVs*. Three AUVs with STM32 and NVIDIA are employed to achieve the formation task, where AUV 1 plays the role of leader; (2) *Communication unit*. The NRF24L01 transceivers are installed on each AUV for data transmission; (3) *Localization unit*. The ultra-wideband (UWB) and ultra-short baseline (USBL) units are both utilized to achieve localization function for AUVs.

The position vector of destination is set as $\eta_{\text{tar}} = [7.1\text{m}, 4.6\text{m}, 0\text{m}, 0\text{rad}]^{\text{T}}$. The desired relative state vectors of AUVs are $\eta_{1,2} = [0.4\text{m}, -1.4\text{m}, 0\text{m}, 0\text{rad}]^{\text{T}}$, $\eta_{1,3} = [0.33\text{m}, -0.43\text{m}, 0.5\text{m}, 0\text{rad}]^{\text{T}}$ and $\eta_{2,3} = [-0.1\text{m}, 0.95\text{m}, 0.5\text{m}, 0\text{rad}]^{\text{T}}$. Before performing the formation controller, we conduct the following steps to train the control policy: (1) *Date collection*. Proportional controllers are used as expert policies to collect the data information of AUVs; (2) *Train the network weights*. Construct data sets from the collected data, and thus the actor networks are trained; (3) *Output the torque for AUVs*. The pulse-width modulations (PWMs) are generated to drive the motions of AUVs. To ensure the training effectiveness, 3000 sets of expert data are used in the experiment.

By using the IRL-based formation controller, the trajectories of AUVs are presented in Fig. 4.25a. It can be seen that the trajectories of AUVs converge to the trajectories by using expert policy. In addition, the control inputs of AUVs during

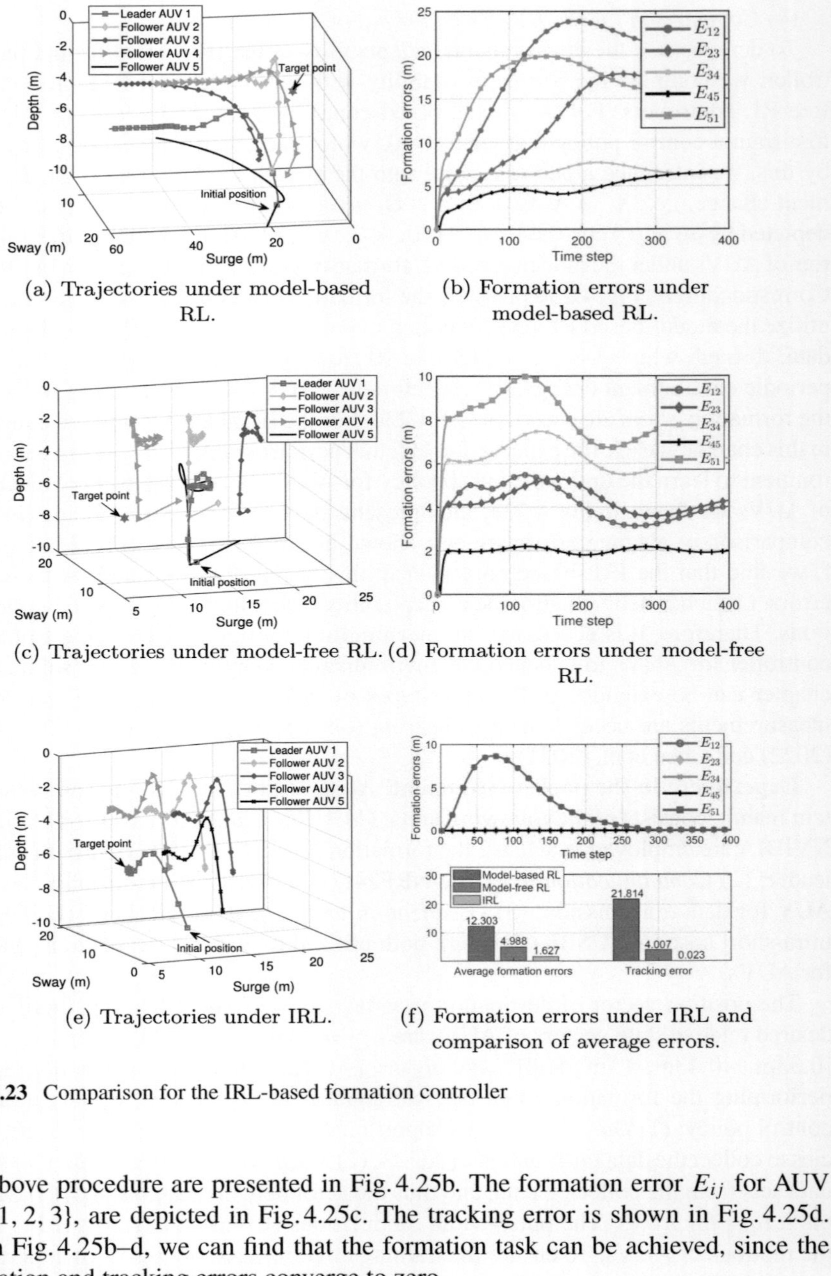

(a) Trajectories under model-based RL.
(b) Formation errors under model-based RL.
(c) Trajectories under model-free RL.
(d) Formation errors under model-free RL.
(e) Trajectories under IRL.
(f) Formation errors under IRL and comparison of average errors.

Fig. 4.23 Comparison for the IRL-based formation controller

the above procedure are presented in Fig. 4.25b. The formation error E_{ij} for AUV $i \in \{1, 2, 3\}$, are depicted in Fig. 4.25c. The tracking error is shown in Fig. 4.25d. From Fig. 4.25b–d, we can find that the formation task can be achieved, since the formation and tracking errors converge to zero.

It should be emphasized that our formation solution has environment adaptability. To verify the correctness of this judgment, we compare our approach with the RL-based solutions (e.g. Wang et al. (2022)). Specifically, we set the same destination

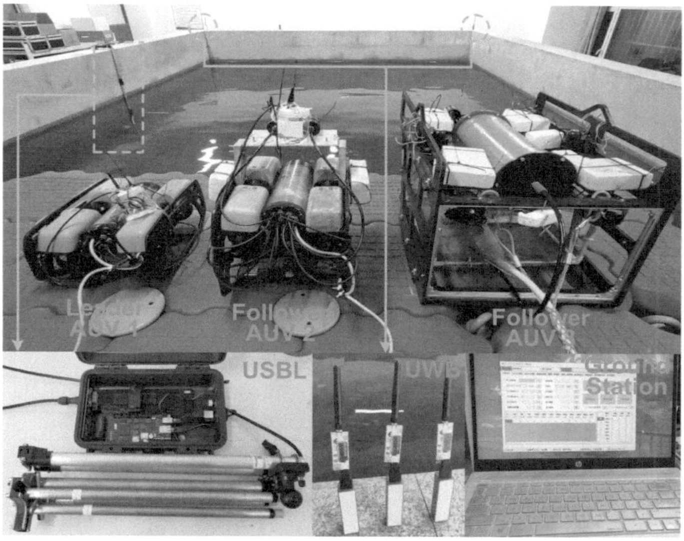

Fig. 4.24 Experiment setup in the pool environment

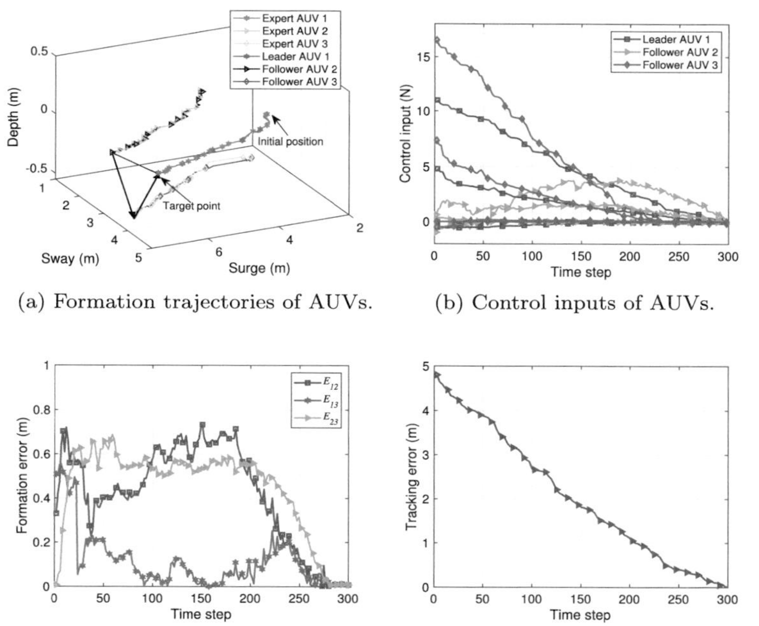

(a) Formation trajectories of AUVs.

(b) Control inputs of AUVs.

(c) Formation errors.

(d) Tracking error for the leader AUV.

Fig. 4.25 Experimental results in the pool environment

and formation shape in the experiment. Meanwhile, the mass of AUV $i \in \{1, 2\}$ is reduced by 2.5kg to simulate environmental effects. Based on this, the trajectories of AUV by using the RL controller are shown in Fig. 4.26a, and the formation errors are provided in Fig. 4.26b. On the other hand, the trajectories of AUV by using the IRL controller in this chapter are shown by Fig. 4.26c, whose formation errors are provided in Fig. 4.26d. To further enhance the experimental design and data analysis, we set up an initial topology where only AUV 1 communicates with AUV 2 and AUV 3, respectively, while the communication situation between AUV 2 and AUV 3 is unknown to AUV 1. Based on this, the trajectories by using the RL controller under topology inference are shown in Fig. 4.26e, and and the formation errors are provided in Fig. 4.26f. The trajectories by using the IRL controller under topology inference are shown in Fig. 4.26g, whose formation errors are provided in Fig. 4.26h. Apparently, the formation errors by using IRL controller is smaller than the RL-based controller, i.e., our method has better control performance by learning from expert demonstrations.

Experiment in the lake environment: We further give the experimental results in lake, as depicted by Fig. 4.27. The position vector of destination is set as $\eta_{\text{tar}} = [3.95\text{m}, 9.35\text{m}, -0.25\text{m}, 0\text{rad}]^{\text{T}}$. The desired relative state vectors of AUVs are $\eta_{1,2} = [0.75\text{m}, -1.35\text{m}, 0.25\text{m}, 0\text{rad}]^{\text{T}}$, $\eta_{1,3} = [-0.55\text{m}, -1.4\text{m}, 0.25\text{m}, 0\text{rad}]^{\text{T}}$ and $\eta_{2,3} = [-1.3\text{m}, 1.95\text{m}, 0\text{m}, 0\text{rad}]^{\text{T}}$. By using the IRL-based formation controller, the trajectories of AUVs are presented in Fig. 4.28a. In addition, the control inputs of AUVs during the above procedure are presented in Fig. 4.28b. The formation error E_{ij} for AUV $i \in \{1, 2, 3\}$, are depicted in Fig. 4.28c. The tracking error is shown in Fig. 4.28d. From Fig. 4.28c–d, we can find that the formation task can be achieved, since the formation and tracking errors converge to zero.

Similar to Sect. 4.4.2, we verify the environment adaptability of our formation controller. Specifically, the mass of AUV $i \in \{1, 2\}$ is reduced by 2.5kg, and hence, the trajectories of AUV by using the RL controller are shown in Fig. 4.29a. The formation errors are provided in Fig. 4.29b. With the same environment changes, the trajectories of AUV by using the IRL controller in this chapter are shown by Fig. 4.29c, and the formation errors are provided in Fig. 4.29d. It can be seen that our method has better environment adaptability as compared to the RL-based formation controller. To further enhance the experimental design and data analysis, we set up an initial topology where only AUV 1 communicates with AUV 2 and AUV 3, respectively, while the communication situation between AUV 2 and AUV 3 is unknown to AUV 1. Based on this, the trajectories by using the RL controller under topology inference are shown in Fig. 4.29e, and the formation errors are provided in Fig. 4.29f. The trajectories by using the IRL controller under topology inference are shown in Fig. 4.29g, whose formation errors are provided in Fig. 4.29h. Apparently, the formation errors by using IRL controller is smaller than the RL-based controller, i.e., our method has better control performance.

Experiment in the ocean environment: To further validate our approach, we conduct a marine experiment in the Bohai Sea in China, as shown in Fig. 4.30. The vector of destination is set as $\eta_{\text{tar}} = [8\text{m}, 2\text{m}, 0\text{m}, 0\text{rad}]^{\text{T}}$. The desired relative state vectors are set as $\eta_{1,2} = [1.7\text{m}, 0.4\text{m}, 0\text{m}, 0\text{rad}]^{\text{T}}$, $\eta_{1,3} = [1.5\text{m}, -0.55\text{m}, 0.1\text{m}, 0\text{rad}]^{\text{T}}$

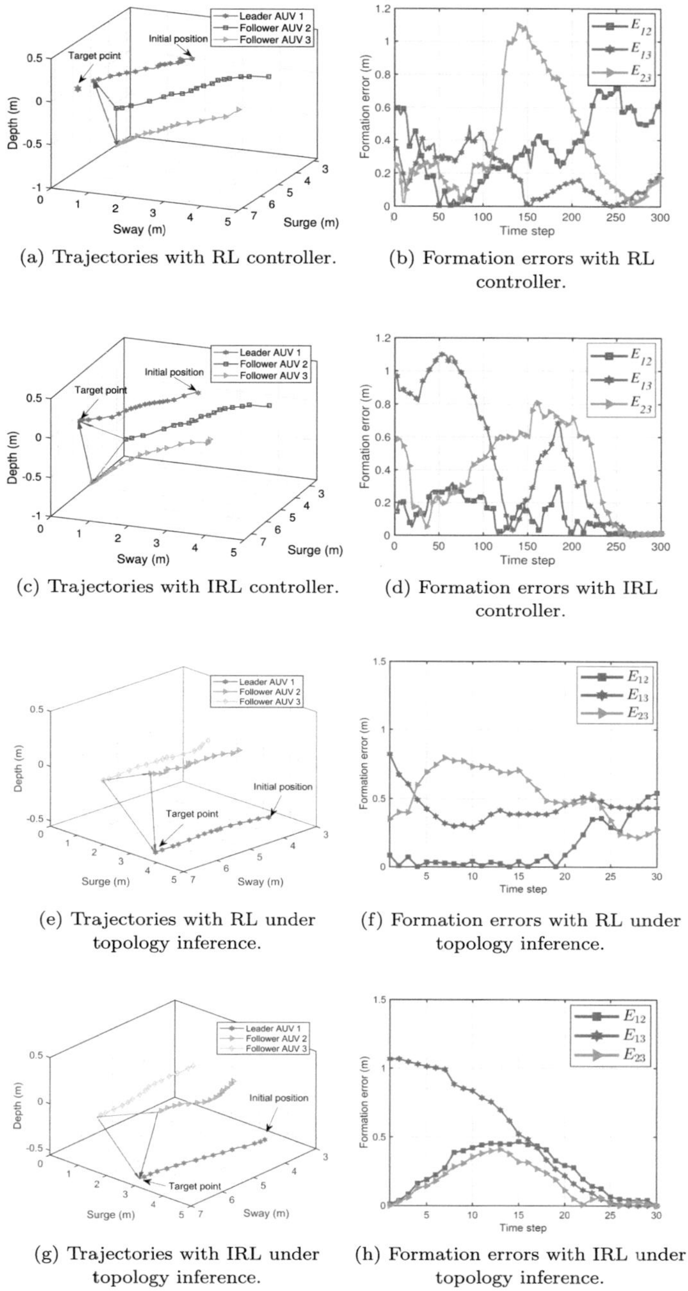

(a) Trajectories with RL controller.

(b) Formation errors with RL controller.

(c) Trajectories with IRL controller.

(d) Formation errors with IRL controller.

(e) Trajectories with RL under topology inference.

(f) Formation errors with RL under topology inference.

(g) Trajectories with IRL under topology inference.

(h) Formation errors with IRL under topology inference.

Fig. 4.26 Comparison for the IRL-based solution in the pool

Fig. 4.27 Experiment setup in lake environment

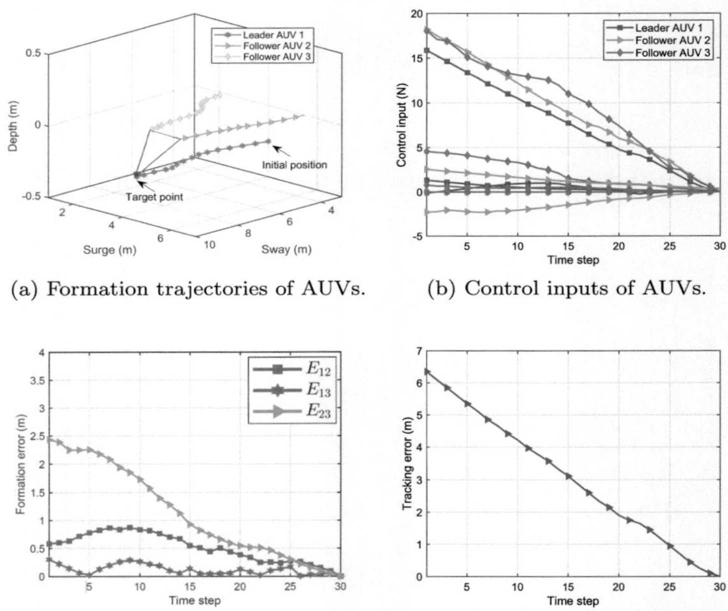

(a) Formation trajectories of AUVs. (b) Control inputs of AUVs.

(c) Formation errors. (d) Tracking error for the leader AUV.

Fig. 4.28 Experimental results in the lake environment

and $\eta_{2,3} = [0.15\text{m}, -0.95\text{m}, 0.1\text{m}, 0\text{rad}]^{\text{T}}$. By using the IRL-based controller, the trajectories of AUVs are presented in Fig. 4.31a. The control inputs of AUVs are presented in Fig. 4.31b. Besides that, the formation and tracking errors are provided in Fig. 4.31c–d. It is clear that the task can be accomplished by using the IRL-based controller in this chapter.

Similar to Sect. 4.4.2, we verify the environment adaptability of our formation controller. Specifically, the mass of AUV $i \in \{1, 2\}$ is reduced by 2.5kg, and hence,

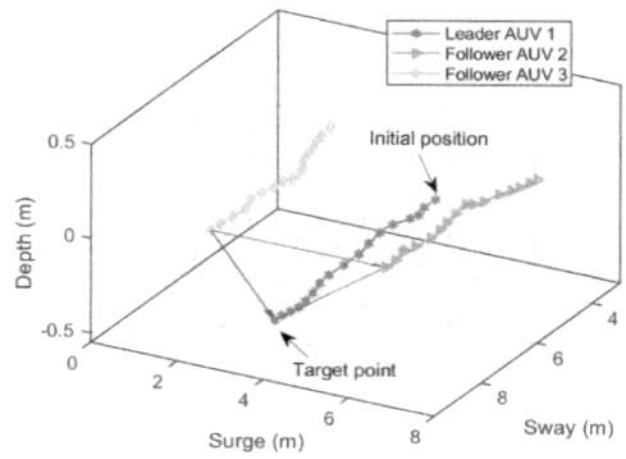

(a) Trajectories with RL controller.

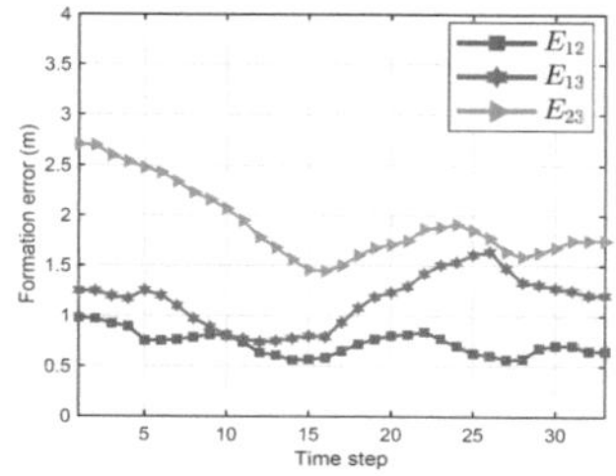

(b) Formation errors with RL controller.

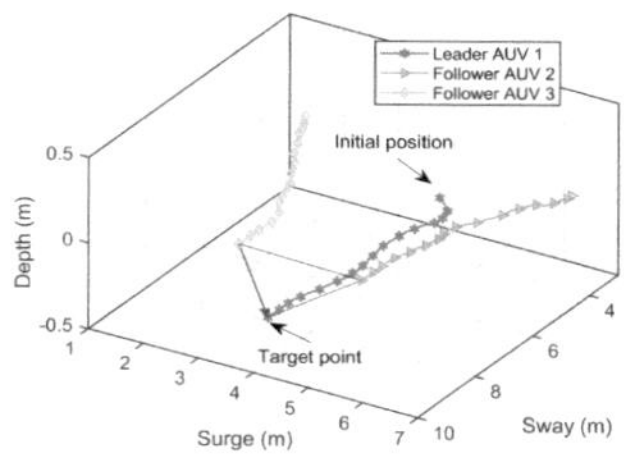

(c) Trajectories with IRL controller.

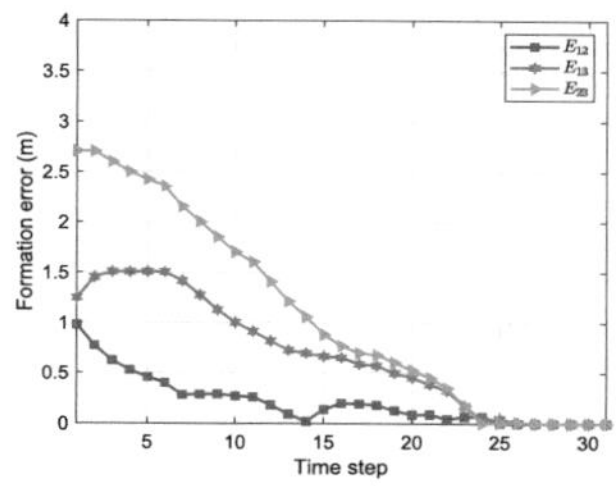

(d) Formation errors with IRL controller.

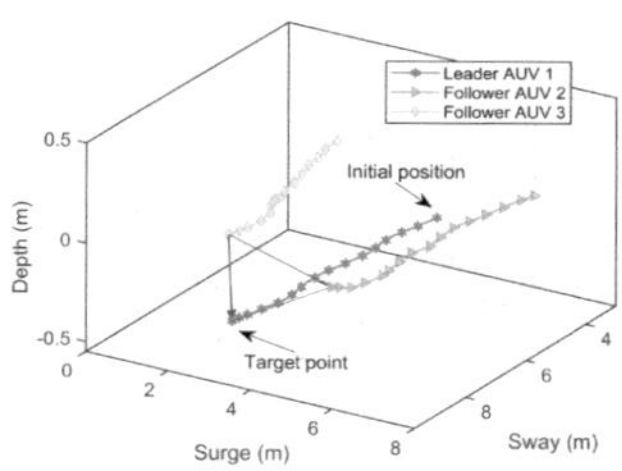

(e) Trajectories with RL under topology inference.

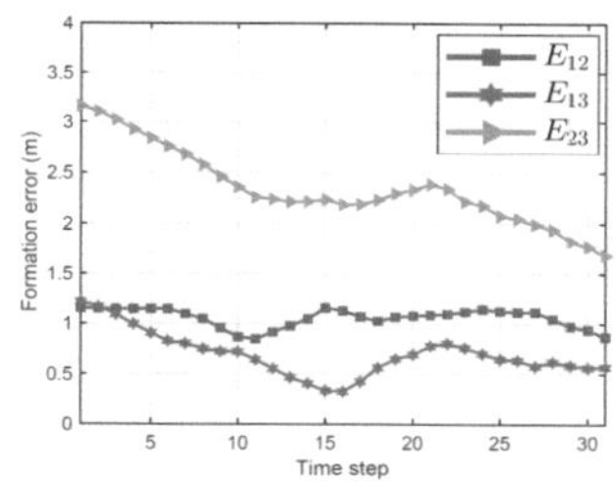

(f) Formation errors with RL under topology inference.

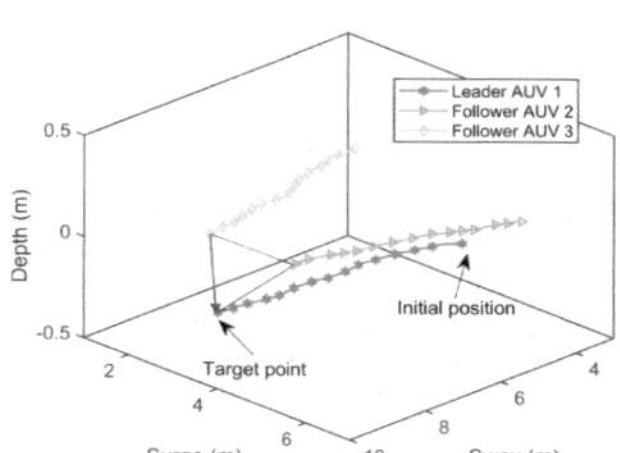

(g) Trajectories with IRL under topology inference.

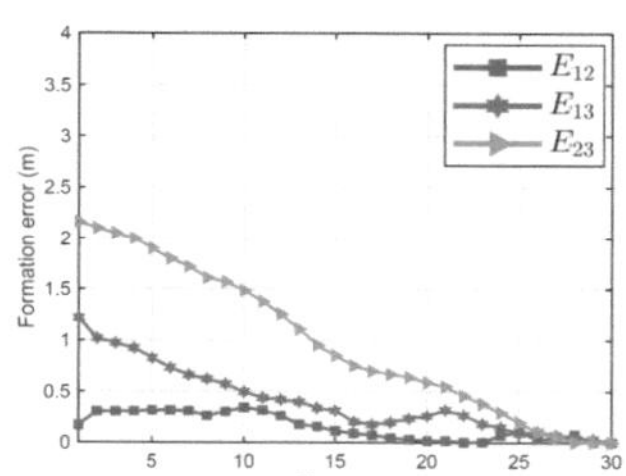

(h) Formation errors with IRL under topology inference.

Fig. 4.29 Comparison for the IRL-based solution in the lake

Fig. 4.30 Experiment setup in ocean environment

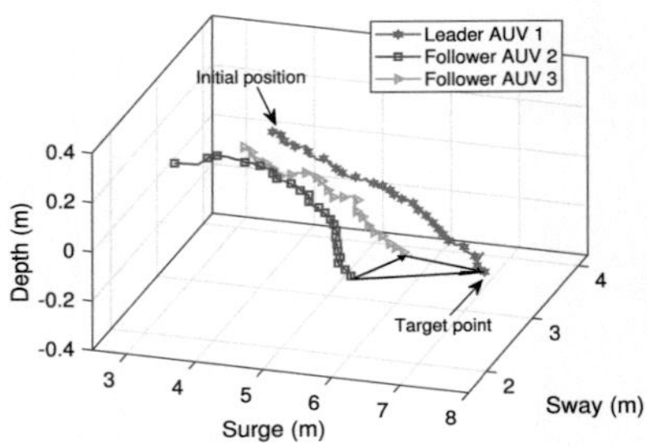

(a) Formation trajectories.

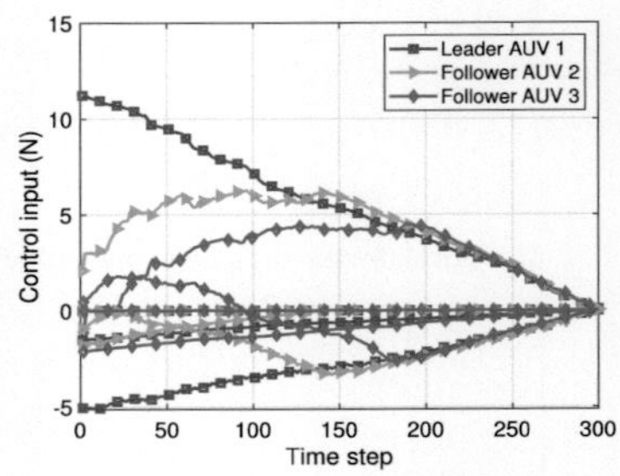

(b) Control inputs of AUVs.

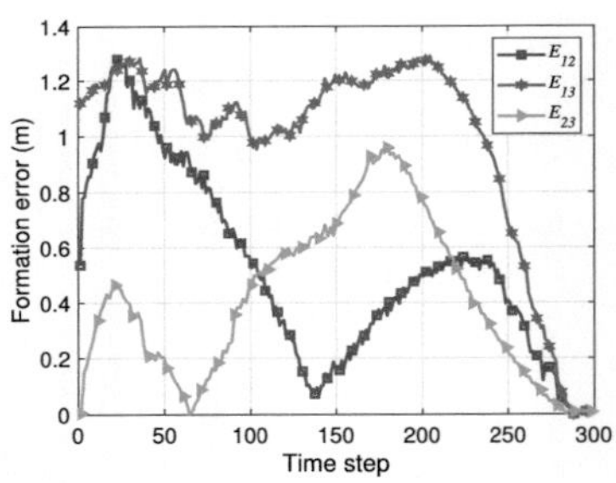

(c) formation error.

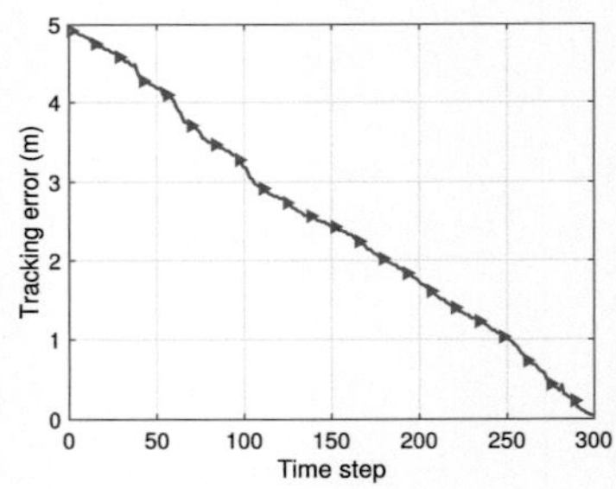

(d) Destination error.

Fig. 4.31 Experiment results in the ocean environment

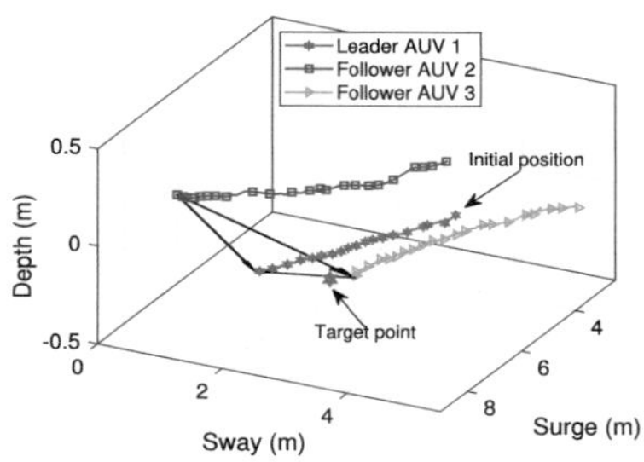

(a) Trajectories with RL controller.

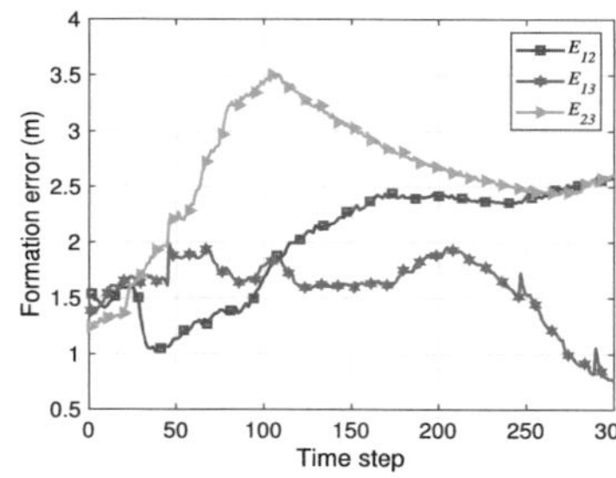

(b) Formation errors with RL controller.

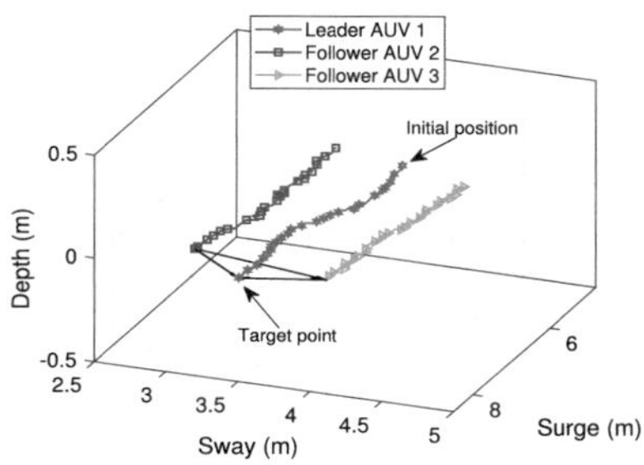

(c) Trajectories with IRL controller.

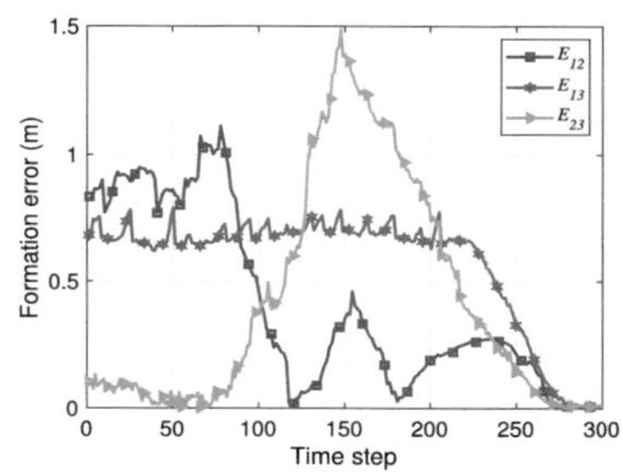

(d) Formation errors with IRL controller.

Fig. 4.32 Comparison for the IRL-based solution in the ocean

the trajectories of AUV by using the RL controller are shown in Fig. 4.32a. The formation errors are provided in Fig. 4.32b. With the same environment changes, the trajectories of AUV by using the IRL controller in this chapter are shown by Fig. 4.32c, and the formation errors are provided in Fig. 4.32d. It can be seen that our method has better environment adaptability as compared to the RL-based formation controller.

4.5 Summary

This chapter studies two fundamental tasks in maritime unmanned systems: communication-efficient and collision-free motion planning, and optimally persistent formation control with unknown interaction topology. For the first task, a model-based IRL estimator was designed to predict SNR in unvisited positions by capturing unknown shadowing and multipath parameters. Based on this, a model-free IRL motion planning algorithm was developed to drive underwater vehicles to desired positions while maximizing communication capacity and avoiding collisions. This approach successfully integrates realistic underwater acoustic channels with vehicle dynamics. For the second task, an IRL-based topology estimator was designed to infer the unknown interaction topology of multi-DOF AUVs. An optimally persistent graph generation algorithm was then developed to balance connectivity and energy efficiency. Furthermore, an IRL-based formation controller was designed to enhance environmental adaptability by learning from expert demonstrations without requiring pre-specified reward functions or accurate model parameters. Both simulation and experimental results have validated the effectiveness of the proposed approaches, demonstrating superior performance in communication capacity, collision avoidance, topology optimization, and formation control under complex marine conditions.

References

Ali, U., Cai, H., Mostofi, Y., & Wardi, Y. (2019, June). Motion-communication co-optimization with cooperative load transfer in mobile robotics: An optimal control perspective. *IEEE Transactions on Control of Network Systems*, *6*(2), 621–632.

Bhat, S., Panteli, C., Stenius, I., & Dimarogonas, D. (2023, October). Nonlinear model predictive control for hydrobatics: Experiments with an underactuated AUV. *Journal of Field Robotics*, *40*(7), 1840–1859.

Boban, M., Vinhoza, T., Ferreira, M., Barros, J., & Tonguz, O. (2011, January). Impact of vehicles as obstacles in vehicular ad hoc networks. *IEEE Journal on Selected Areas in Communications*, *29*(1), 15–28.

Cao, W., Yan, J., Yang, X., Luo, X., & Guan, X. (2020). Communication-aware formation control of AUVs with model uncertainty and fading channel via integral reinforcement learning. *IEEE/CAA Journal of Automatica Sinica*, JAS-2022-0608.R1.

Cao, W., Yan, J., Yang, X., Luo, X., & Guan, X. (2023, January). Communication-aware formation control of AUVs with model uncertainty and fading channel via integral reinforcement learning. *IEEE-CAA Journal of Automatica Sinica*, *10*(1), 159–176.

Cao, X., Yang, F., Gan, X., Liu, J., Qian, L., Tian, X., & Wang, X. (2013). Joint estimation of clock skew and offset in pairwise broadcast synchronization mechanism. *IEEE Transactions on Communications*, *61*(6), 2508–2521.

Cao, X., Zhang, J., & Poor, H. (2018, August). Joint energy procurement and demand response towards optimal deployment of renewables. *IEEE Journal of Selected Topics in Signal Processing*, *12*(4), 657–672.

Chen, L., Cao, M., & Li, C. (2021, August). Angle rigidity and its usage to stabilize multiagent formations in 2-D. *IEEE Transactions on Automatic Control*, *66*(8), 3667–3681.

Chen, M., Gündüz, D., Huang, K., Saad, W., Bennis, M., Feljan, A., & Poor, H. (2021, December). Distributed learning in wireless networks: recent progress and future challenges. *IEEE Journal on Selected Areas in Communications*, *39*(12), 3579–3605.

Chen, L., Han, T., Xiao, B., & Yan, H. (2023, November). Prescribed-time time-varying multi-group formation tracking control of NMSVs via estimator-based hierarchical control algorithm. *IEEE Transactions on Intelligent Vehicles*, *8*(11), 1–6.

Chen, L., Xie, L., Li, X., Fang, X., & Feroskhan, M. (2022, December). Simultaneous localization and formation using angle-only measurements in 2D. *Automatica*, *146*, Article 110605.

Chen, Y., Zhu, J., Wan, L., Fang, X., Tong, F., & Xu, X. (2022, January). Routing failure prediction and repairing for AUV-assisted underwater acoustic sensor networks in uncertain ocean environments. *Applied Acoustics*, *186*, Article 108479.

Cui, R., Chen, L., Yang, C., & Chen, M. (2017, August). Extended state observer-based integral sliding mode control for an underwater robot with unknown disturbances and uncertain nonlinearities. *IEEE Transactions on Industrial Electronics*, *64*(8), 6785–6795.

Cui, R., Yang, C., Li, Y., & Sharma, S. (2017, June). Adaptive neural network control of AUVs with control input nonlinearities using reinforcement learning. *IEEE Transactions on Systems Man and Cybernetics Systems*, *47*(6), 1019–1029.

Cui, R., Yang, C., Li, Y., & Sharma, S. (2017). Adaptive neural network control of AUVs with control input nonlinearities using reinforcement learning. *IEEE Transactions on Systems Man and Cybernetics Systems, 47*(6), 1019–1029.

Diamant, R., Francescon, R., & Zorzi, M. (2018, October). Topology-efficient discovery: A topology discovery algorithm for underwater acoustic networks. *IEEE Journal of Oceanic Engineering*, *43*(4), 1200–1214.

Dixon, C., & Frew, E. (2012, June). Optimizing cascaded chains of unmanned aircraft acting as communication relays. *IEEE Journal on Selected Areas in Communications*, *30*(5), 883–898.

Goldsmith, A. (2005). *Wireless communication*. Cambridge University Press.

Guo, G., & Zhang, Q. (2023, October). Finite-time command-filtered lateral-longitudinal formation control of vehicles on rough roads. *IEEE Transactions on Intelligent Vehicles*, *8*(10), 4331–4340.

Guo, Y., You, C., Yin, C., & Zhang, R. (2021, November). UAV trajectory and communication co-design: flexible path discretization and path compression. *IEEE Journal on Selected Areas in Communications*, *39*(11), 3506–3523.

He, Y., Han, G., Tang, Z., Martíez-García, M., & Peng, Y. (2022, April). State prediction-based data collection algorithm in underwater acoustic sensor networks. *IEEE Transactions on Wireless Communications*, *21*(4), 2830–2842.

He, C., Wan, Y., Gu, Y., & Lewis, F. (2021, July). Integral reinforcement learning-based multi-robot minimum time-energy path planning subject to collision avoidance and unknown environmental disturbances. *IEEE Control Systems Letters*, *5*(3), 1–1.

Hendrickx, J., Anderson, B., Delvenne, J., & Blondel, V. (2007, July). Directed graphs for the analysis of rigidity and persistence in autonomous agent systems. *International Journal of Robust and Nonlinear Control*, *17*(10–11), 960–981.

Heshmati-Alamdari, S., Nikou, A., & Dimarogonas, D. (2021, July). Robust trajectory tracking control for underactuated autonomous underwater vehicles in uncertain environments. *IEEE Transactions on Automation Science and Engineering*, *18*(3), 1288–1301.

Huang, S., Sun, D., & Zhao, M. (2024). Distributed MPC-based hierarchical cooperative control for mixed vehicle groups with T-CPS in the vicinity of traffic signal light. *IEEE Transactions on Intelligent Transportation Systems, 25*(7), 8003–8016.

Hurst, W., Cai, H., & Mostofi, Y. (2021). Communication-aware RRT*: Path planning for robotic communication operation in obstacle environments. In *IEEE International Conference on Communications (ICC)* (pp. 1–6).

Jawhar, I., Mohamed, N., Al-Jaroodi, J., & Zhang, S. (2019, March). An architecture for using autonomous underwater vehicles in wireless sensor networks for underwater pipeline monitoring. *IEEE Transactions on Industrial Informatics*, *15*(3), 1329–1340.

Jiang, J., Han, G., Guo, H., Shu, L., & Rodrigues, J. (2016, January). Geographic multipath routing based on geospatial division in duty-cycled underwater wireless sensor networks. *Journal of Network and Computer Applications*, *1*(1), 4–13.

Jiang, L., Huang, H., & Ding, Z. (2020, July). Path planning for intelligent robots based on deep Q-learning with experience replay and heuristic knowledge. *IEEE/CAA Journal of Automatica Sinica*, *7*(4), 1179–1189.

Jiang, J., Wang, S., Han, G., & Wang, H. (2023, May). Reinforcement-learning-based adaptive neighbor discovery algorithm for directional transmission-enabled internet of underwater things. *IEEE Internet of Things Journal*, *10*(10), 9038–9048.

Khan, M., Jembre, Y., Ahmed, S., Seo, J., & Kim, D. (2019). Data freshness based AUV path planning for UWSN in the internet of underwater things. *IEEE Global Communications Conference (GLOBECOM)* (pp. 1–6).

Kontoudis, G., & Vamvoudakis, K. (2019, December). Kinodynamic motion planning with continuous-time Q-learning: an online, model-Free, and safe navigation framework. *IEEE Transactions on Neural Networks and Learning Systems*, *30*(12), 3803–3817.

Li, K., & Li, Y. (2024, February). Fuzzy adaptive optimization prescribed performance control for nonlinear vehicle platoon. *IEEE Transactions on Fuzzy Systems*, *32*(2), 360–372.

Li, Y., & Tong, S. (2023, November). Bumpless transfer distributed adaptive backstepping control of nonlinear multi-agent systems with circular filtering under DoS attacks. *Automatica*, *157*, Article 111250.

Li, Y., & Tong, S. (2024, May). Adaptive backstepping control for uncertain nonlinear strict-feedback systems with full state triggering. *Automatica*, *163*, Article 111574.

Li, Y., He, J., Cai, L., & Guan, X. (2023, November). Local topology inference of mobile robotic networks under formation control. *IEEE Transactions on Automatic Control*, *68*(11), 6450–6465.

Lian, B., Kartal, Y., Lewis, F., Mikulski, D., Hudas, G., Wan, Y., & Davoudi, A. (2023, July). Anomaly detection and correction of optimizing autonomous systems with inverse reinforcement learning. *IEEE Transactions on Cybernetics*, *53*(7), 4555–4566.

Licea, D., Bonilla, M., Ghogho, M., Lasaulce, S., & Varma, V. (2020, April). Communication-aware energy efficient trajectory planning with limited channel knowledge. *IEEE Transactions on Robotics*, *36*(2), 431–442.

Lin, C., Han, G., Du, J., Bi, Y., Shu, L., & Fan, K. (2020). A path planning scheme for AUV flock-based internet-of-underwater-things systems to enable transparent and smart ocean. *IEEE Internet of Things Journal, 7*(10), 9760–9772.

Liu, Y., & Bucknall, R. (2018, July). A survey of formation control and motion planning of multiple unmanned vehicles. *Robotica*, *36*(7), 1019–1047.

Liu, M., Wan, Y., Lewis, F., & Lopez, V. (2020, December). Adaptive optimal control for stochastic multiplayer differential games using on-policy and off-policy reinforcement learning. *IEEE Transactions on Neural Networks and Learning Systems*, *31*(12), 5522–5533.

Liu, Y., Wang, H., Cai, L., Shen, X., & Zhao, R. (2021, October). Fundamentals and advancements of topology discovery in underwater acoustic sensor networks: A review. *IEEE Sensors Journal*, *21*(19), 21159–21174.

Liu, Y., & Zhao, R. (2018). *Collision-free topology discovery protocol based on node ID for underwater acoustic networks* (pp. 1–4). OCEANS-MTS/IEEE Kobe Techno-Oceans: OCEANS-Kobe.

Lucani, D., Medard, M., & Stojanovic, M. (2008, December). Underwater acoustic networks: channel models and network coding based lower bound to transmission power for multicast. *IEEE Journal on Selected Areas in Communications*, *26*(9), 1708–1719.

Luo, X., Li, S., & Guan, X. (2009, November). Automatic generation of min-weighted persistent formations. *Chinese Physics*, *18*(8), 3104–3114.

Mahmoudzadeh, S., Powers, D., & Atyabi, A. (2019, August). UUV's hierarchical de-based motion planning in a semi dynamic underwater wireless sensor network. *IEEE Transactions on Cybernetics*, *49*(8), 2992–3005.

Mao, Z., & Yan, S. (2018, January). Design and analysis of the thermal-stress coupled topology optimization of the battery rack in an AUV. *Ocean Engineering*, *148*(1), 401–411.

Mendíez Váquez, B., & Barca, J. (2018). Network topology inference in swarm robotics. In *Proceedings of the IEEE International Conference on Robotics and Automation* (pp. 7660–7666).

Neu, G., & Szepesvari, C. (2007). Apprenticeship learning using inverse reinforcement learning and gradient methods. In *Proceedings of the Conference on Uncertainty in Artificial Intelligence* (pp. 295–302).

Ng, A., & Russell, S. (2000). Algorithms for inverse reinforcement learning. *Proceedings of the SICM* 663–670.

Pang, W., Zhu, D., Chu, Z., & Chen, Q. (2023, June). Distributed adaptive formation reconfiguration control for multiple AUVs based on affine transformation in three-dimensional ocean environments. *IEEE Transactions on Vehicular Technology*, *72*(6), 7338–7350.

Peng, Z., Wang, D., & Wang, J. (2021, December). Data-driven adaptive disturbance observers for model-free trajectory tracking control of maritime autonomous surface ships. *IEEE Transactions on Neural Networks and Learning Systems*, *32*(12), 5584–5594.

Rappaport, T. (2002). *Wireless communications, principles and practice*. Prentice Hall PTR, Press.

Ren, R., Zhang, Y., Luo, X., & Bao, S. (2010, November). Automatic generation of optimally rigid formations using decentralized methods. *International Journal of Automation and Computing*, *7*(1), 557–564.

Sahu, B., & Subudhi, B. (2018, October). Flocking control of multiple AUVs based on fuzzy potential functions. *IEEE Transactions on Fuzzy Systems*, *26*(5), 2539–2551.

Shi, X., & Wu, J. (2018, September). To hide private position information in localization using time difference of arrival. *Transactions on Signal Processing*, *66*(18), 4946–4956.

Shou, Y., Xu, B., Lu, H., Zhang, A., & Mei, T. (2022, March). Finite-time formation control and obstacle avoidance of multi-agent system with application. *International Journal of Robust and Nonlinear Control*, *32*(5), 2883–2901.

Shou, Y., Xu, B., Zhang, A., & Mei, T. (2021, December). Virtual guidance-based coordinated tracking control of multi-autonomous underwater vehicles using composite neural learning. *IEEE Transactions on Neural Networks and Learning Systems*, *32*(12), 5565–5574.

Smith, B., Egerstedt, M., & Howard, A. (2009, January). Automatic generation of persistent formations for multi-agent networks under range constricts. *Mobile Networks and Applications*, *14*(1), 322–335.

Song, Z., Lipinski, D., & Mohseni, K. (2017, September). Multi-vehicle cooperation and nearly fuel-optimal flock guidance in strong background flows. *Ocean Engineering*, *141*(1), 388–404.

Stefanov, A., & Stojanovic, M. (2011, December). Design and performance analysis of underwater acoustic networks. *IEEE Journal on Selected Areas in Communications*, *29*(10), 2012–2021.

Stojanovic, M., & Preisig, J. (2009, January). Underwater acoustic communication channels: Propagation models and statistical characterization. *IEEE Communications Magazine*, *47*(1), 84–89.

Su, R., Zhang, D., Li, C., Gong, Z., Venkatesan, R., & Jiang, F. (2019, November). Localization and data collection in AUV-aided underwater sensor networks: challenges and opportunities. *IEEE Network*, *33*(6), 86–93.

Vasilijević, A., Nad, D., Mandić, F., Mišković, N., & Vukić, Z. (2017, June). Coordinated navigation of surface and underwater marine robotic vehicles for ocean sampling and environmental monitoring. *IEEE/ASME Transactions on Mechatronics*, *22*(3), 1174–1184.

Wang, Z., & Zhang, L. (2023, July). Distributed optimal formation tracking control based on reinforcement learning for underactuated AUVs with asymmetric constraints. *Ocean Engineering*, *280*, Article 114491.

Wang, C., Cai, W., Lu, J., Ding, X., & Yang, J. (2022, October). Design, modeling, control, and experiments for multiple AUVs formation. *IEEE Transactions on Automation Science and Engineering*, *19*(4), 2776–2787.

Wang, Y., Ji, Y., Woo, H., Tamura, Y., Tsuchiya, H., Yamashita, A., & Asama, H. (2021, July). Acoustic camera-based pose graph SLAM for dense 3-D mapping in underwater Environments. *IEEE Journal of Oceanic Engineering*, *46*(3), 829–847.

Wang, J., Jiang, Q., & Wang, H. (2025). Generalized disturbance estimation based continuous integral terminal sliding mode control for magnetic levitation systems. *IEEE Transactions on Automation Science and Engineering, 22*(1), 1725–1737.

Wang, J., Wu, Z., Tan, M., & Yu, J. (2021). 3-D path planning with multiple motions for a gliding robotic dolphin. *IEEE Transactions on Systems Man and Cybernetics Systems, 51*(5), 2904–2915.

Wu, H., Song, S., You, K., & Wu, C. (2019, December). Depth control of model-free AUVs via reinforcement learning. *IEEE Transactions on Systems Man and Cybernetics Systems*, *49*(12), 2499–2510.

Wu, H., Yang, Z., Cao, J., & Lai, L. (2020, July). TRiForm: Formation control for underwater sensor networks with measurement errors. *IEEE Transactions on Vehicular Technology*, *69*(7), 7679–7691.

Wu, Q., Zeng, Y., & Zhang, R. (2018, March). Joint trajectory and communication design for multi-UAV enabled wireless networks. *IEEE Transactions on Wireless Communications*, *17*(3), 2109–2121.

Xia, N., Ou, Y., Wang, S., Zheng, R., Du, H., & Xu, C. (2017, April). Localizability judgment in UWSNs based on skeleton and rigidity theory. *IEEE Transactions on Mobile Computing*, *16*(4), 980–989.

Xu, B., Fei, Y., Wang, X., Tang, J., & Razzaqi, A. (2023, March). Optimal topology design of multi-target AUVs for 3D cooperative localization formation based on angle of arrival measurement. *Ocean Engineering*, *271*, Article 113758.

Xue, W., Kolaric, P., Fan, J., Lian, B., Chai, T., & Lewis, F. (2022, October). Inverse reinforcement learning in tracking control based on inverse optimal control. *IEEE Transactions on Cybernetics*, *52*(10), 10570–10581.

Yan, Y., & Mostofi, Y. (2012, August). Robotic router formation in realistic communication environments. *IEEE Transactions on Robotics*, *28*(4), 810–827.

Yan, J., Lin, J., Yang, X., et al. (2025, February). Cooperation detection and tracking of underwater target via aerial-surface-underwater vehicle. *IEEE Transactions on Automatic Control*, *70*(2), 1068–1083.

Yan, J., Meng, Y., Yang, X., Luo, X., & Guan, X. (2021, January). Privacy-preserving localization for underwater sensor networks via deep reinforcement learning. *IEEE Transactions on Information Forensics and Security*, *16*(1), 1880–1895.

Yan, J., Wang, H., Yang, X., Luo, X., & Guan, X. (2024, September). Optimal rigid graph-based cooperative formation control of AUVs in anchor-free environments. *IEEE Transactions on Intelligent Vehicles*, *9*(9), 5922–5938.

Yan, J., Li, X., Yang, X., Luo, X., Hua, C., & Guan, X. (2022). Integrated localization and tracking for AUV with model uncertainties via scalable sampling-based reinforcement learning approach. *IEEE Transactions on Systems Man and Cybernetics Systems, 52*(11), 6952–6967.

Yan, J., Yang, X., Luo, X., & Chen, C. (2018). Energy-efficient data collection over AUV-assisted underwater acoustic sensor network. *IEEE Systems Journal, 12*(4), 3519–3530.

Yetkin, H., Lutz, C., & Stilwell, D. (2020, May). A decision-theoretic approach to acquire environmental information for improved subsea search performance. *Ocean Engineering*, *204*(1), 1–12.

Yordanova, V., & Gips, B. (2020, July). Coverage path planning with track spacing adaptation for autonomous underwater vehicles. *IEEE Robotics and Automation Letters*, *5*(3), 4774–4780.

Yu, X., Chen, W., Gu, T., Yuan, H., Zhang, H., & Zhang, J. (2019, August). ACO-A*: Ant colony optimization plus A* for 3-D traveling in environments with dense obstacles. *IEEE Transactions on Evolutionary Computation*, *23*(4), 617–631.

Yuan, C., Licht, S., & He, H. (2018, October). Formation learning control of multiple autonomous underwater vehicles with heterogeneous nonlinear uncertain dynamics. *IEEE Transactions on Cybernetics*, *48*(10), 2920–2934.

Zhang, J., Sha, J., Han, G., Liu, J., & Qian, Y. (2021, March). A cooperative-control-based underwater target escorting mechanism with multiple autonomous underwater vehicles for underwater internet of things. *IEEE Internet of Things Journal*, *8*(6), 4403–4416.

Zhang, Y., Zhang, Z., Chen, L., & Wang, X. (2021, March). Reinforcement learning-based opportunistic routing protocol for underwater acoustic sensor networks. *IEEE Transactions on Vehicular Technology*, *70*(3), 2756–2770.

Chapter 5
Integrated Design of Detection and Control

5.1 Introduction

The aim of detection and tracking is to detect the existence of underwater target before it is close enough to attack, through which unmanned underwater vehicles (UUVs) can be scheduled to track the real-time trajectory of target (Biggs & Holderbaum, 2009). It is of great importance to improve the marine sensing ability in various applications, such as object recognition, intrusion surveillance, subsea exploration, map construction, marine rescue and so on. Particularly, a typical example is the deployment of Bluefin-21 UUV for the search of Malaysia MH370 airplane in 2014 (McNutt, 2014). During the search mission, the Bluefin-21 UUV clocked in 370 h of search time and covered 250 square miles. Although the lost plane is not detected eventually, it has shown the potential of a growing application value in future marine search and rescue.

Till now, there have been many studies on the topic of UUV-based detection for underwater target. For instance, Cui et al. (2020) employed the binocular vision system to develop a deep learning-based detection algorithm for UUV. Followed by this, a YOLO-T (you only look twice) network based underwater detection algorithm was designed in Li et al. (2023) to reduce the time cost. It should be emphasized note, the above vision-based algorithms have the characteristics of high resolution ability and low cost, which are very suitable for the short-distance underwater detection. However, the sensing radius of vision system relies heavily on the illumination changes, which results in the narrow spatio-temporal range for long-distance underwater detection.

In view of this, some researchers attempt to employ the sonar system to enlarge the spatio-temporal detection range. For example, the forward-looking sonar was utilized in Williams (2015) to design an unsupervised strategy for the detection of underwater target. In Munaf et al. (2019), a statistical-based unsupervised algorithm with synthetic aperture sonar imagery detection was proposed, which can provide a higher target update rate as compared with the pulsed active sonar. Besides that, Tian

J. Yan and X. Guan, *Maritime Unmanned Systems*, Cognitive Intelligence and Robotics 1, https://doi.org/10.1007/978-981-95-6907-6_5

et al. (2022) analyzed the properties of different kinds of transient/detection signals, through which a multi-target passive detection algorithm was provided. In Hongwei et al. (2022), a multi-scale change entropy method was designed to achieve remote passive sonar detection in chaotic characteristics. Our previous work (Yan et al., 2023) proposed an active-passive sonar fusion algorithm to improve the detection accuracy and reduce the communication load. Note that the detection mission in the above works is accomplished by a single type of platform, however the single type of platform may be incapable to achieve complex detection task. For that reason, it is necessary to employ multiple types of platforms to construct a heterogeneous detection network for multiple areas, such as air, surface and underwater (Luo et al., 2022). Inspired by this, a cooperation detection framework of UUV and unmanned aerial vehicle (UAV) was constructed in Ke and Chen (2022), while a framework of UUV and unmanned surface vehicle (USV) was developed in McCammon et al. (2021). Followed by this, a joint design of UUV, UAV and USV networks for underwater target detection was presented in Wu et al. (2020). Nevertheless, the works in Ke and Chen (2022), McCammon et al. (2021), Wu et al. (2020) focus on the design of network architecture, while they donot theoretically analyze the influences of motion trajectory and formation shape on the detection probability. With regard to this, how to employ the UAV, USV and UUV to construct a heterogeneous detection mode that can both determine the optimal motion trajectory and formation shape is still an unsolved issue.

After the target is detected, the next urgent issue to be addressed is to track the trajectory of target. For instance, a consensus formation controller was developed in Yang et al. (2020) for the trajectory tracking of UUV, UAV and USV networks within obstacle environment, where the block kronecker product was adopted to describe the communication topology. Followed by this, an artificial potential field-based underwater tracking controller was presented in Zhang et al. (2021) to achieve target monitoring via the cooperation of USVs and UUVs. In Lindsay et al. (2022), a collaborative tracking system with UAV, USV and UUV was constructed to achieve multidomain sensing on partially submerged target. In Ke and Chen (2022b), an improved particle swarm optimization (PSO) based tracking algorithm was provided to address the air-sea path planning issue with obstacles. Notice that the motion model of USV or UUV in Yang et al. (2020), Zhang et al. (2021), Lindsay et al. (2022), Ke and Chen (2022b) treats each vehicle as a point mass, such that a second-order linear/nonlinear equation can be conducted to describe its kinematics. However, the kinematics cannot capture the low-level interactions during the implementation of tracking controller on actual vehicles. As such, a model predictive control based tracking solution was proposed in Jia et al. (2022), where the kinematics and dynamics models were both utilized to capture the motion of USV or UUV. In Li et al. (2023), a bearing-only tracking controller was designed for USV and UUV in the presence of unknown velocity measurement. In Enayat and Khorasani (2017), the dynamic surface control technique was conducted to ensure the tracking stabilities of USV and UUV with Lagrangian dynamics. Nevertheless, the above tracking controllers in Jia et al. (2022), Li et al. (2023), Enayat and Khorasani (2017) rely on full or partial knowledge of vehicle dynamic model. Due to the rigorous ocean conditions, it is difficult or impossible to acquire the accurate dynamics model of USV or UUV. Along with this, Wei et al. (2023) employed the Markov decision process to model air-sea tracking problem, such that a deep Q-network (DQN) based model-free tracking algorithm was designed for UAV, USV and UUV. However, the tracking algorithm in Wei et al. (2023) only considers the kinematics model of vehicle, where the dynamics model is ignored. Meanwhile, the DQN cannot be directly applied to continuous state space since it finds action that maximizes the action-value function (Lillicrap et al., 2016), and hence, it suffers from the difficulty of handling continuous underwater complex environment. In our previous works (Cao et al., 2023; Yan et al., 2022), the reinforcement learning (RL) was employed to reduce the dependency on model parameters, however they are not developed in the context of tracking control of UAV, USV and UUV. Due to the complex dynamics and strong-coupling relationship of UAV-USV-UUV networks, the issue of how to design a model-free tracking algorithm for the UAV-USV-UUV networks is largely unexplored.

In this chapter, we provide a heterogeneous network architecture, which includes UAV, USV and UUV. Specifically, UAV makes air detection by visual sensors, while USV and UUV utilize sonars for underwater detection and tracking. In addition, USV also relays the data between UAV and UUV. Based on this, a heterogeneous detection mode is constructed to maximize the detection probability by adjusting the motion trajectory and formation shape. Then, a deep deterministic policy gradient (DDPG) based tracking algorithm is designed for UUV to track the trajectory of target. Particularly, main features of this chapter are summarized as follows.

(1) ***Construction of the detection mode to maximize the detection probability.*** We utilize the measure theory to construct a heterogeneous detection mode for UAV, USV and UUV. Along with this, the coverage and detection probabilities for underwater target are given in theory. Compared with the single type of detection platform as developed in Williams (2015), Munaf et al. (2019), Tian et al. (2022), Hongwei et al. (2022), Yan et al. (2023), the multi-platform detection developed in this chapter can improve the detection range. Additionally, the detection probability is given analytically in our work, which is lacked in the exiting heterogeneous detection networks, e.g., Ke and Chen (2022), McCammon et al. (2021), Wu et al. (2020).

(2) ***DDPG-based tracking algorithm to achieve persistent monitoring.*** After the target is detected by MUSs, we design a DDPG-based tracking algorithm to achieve persist monitoring, where a multi-step location prediction strategy is incorporated to guarantee the communication connectivity. Compared with the solutions in Jia et al. (2022), Li et al. (2023), Enayat and Khorasani (2017), the model parameters are not required in our work. Different from the DQN-based tracking controller (Wei et al., 2023), the DDPG-based solution in this chapter is more efficient for handling continuous complex action space.

5.2 Model Construction and Problem Formulation

As depicted by Fig. 5.1, we consider a heterogeneous underwater detection system, which includes an UAV, an USV and an UUV. Particularly, UAV is equipped with visual sensor for air detection, and it is also outfitted with radio communication equipment to relay the data between control center and USV. The USV is equipped with sonar and visual sensors for detecting both underwater and surface targets, while it also serves as relay node between UAV and UUV. The UUV is installed with acoustic unit for communicating with USV, and with sonar unit for underwater target detection.

The IBF and BRF are utilized to depict the motion. Without loss of generality, we assume that UAV operates at the same height h. Then, the dynamics of UAV can be given as Zhao et al. (2022)

$$\begin{aligned} m_q \ddot{\mathbf{p}}_a &= \mathbf{R}_q \mathbf{F}_q - m_q g \mathbf{e}_3 + \mathbf{A}_q, \\ \mathbf{J}_q \ddot{\eta}_a &= -\mathbf{C}_q(\eta_a, \dot{\eta}_a)\dot{\eta}_a + \tau_\eta, \end{aligned} \tag{5.1}$$

where m_q is the UAV mass, g is the gravity coefficient and $\mathbf{e}_3 = [0, 0, 1]^T$. $\mathbf{A}_q$ is the external disturbance vector. $\mathbf{p}_a = [p_{a,x}, p_{a,y}, p_{a,z}]^T \in \mathcal{R}^3$ is the position vector in IRF, where $p_{a,x}$, $p_{a,y}$ and $p_{a,z}$ are the positions on X, Y and Z axes, respectively. $\eta_a = [\phi_a, \theta_a, \psi_a]^T \in \mathcal{R}^3$ is the angle vector in IRF, where ϕ_a, θ_a and ψ_a are the roll, pitch and yaw angles, respectively. $\mathbf{R}_q \in \mathcal{R}^{3\times3}$ is rotation matrix. $\mathbf{F}_q \in \mathcal{R}^3$ and $\tau_\eta \in \mathcal{R}^3$ are the control force and torque, respectively. $\mathbf{J}_q$ is the inertial scalar and $\mathbf{C}_q(\eta_a, \dot{\eta}_a)$ is the Coriolisa term.

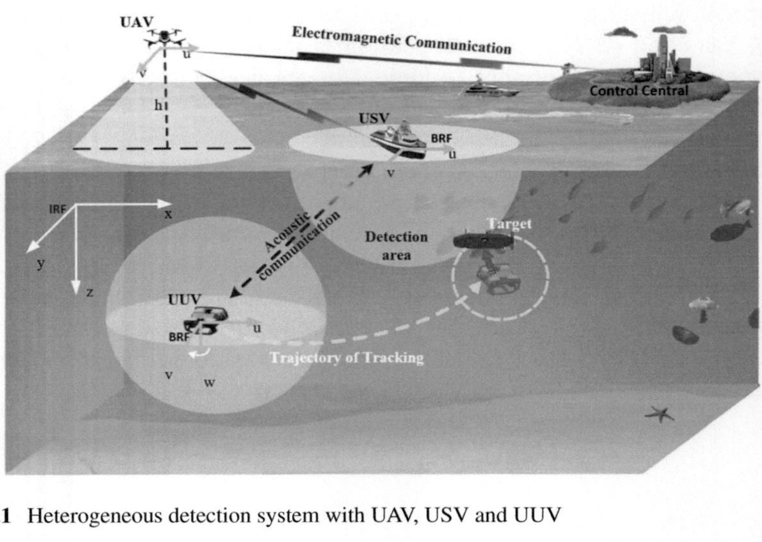

Fig. 5.1 Heterogeneous detection system with UAV, USV and UUV

An underactuated model is considered for USV. Note that USV usually sails on the water surface, and hence we ignore the roll, pitch and heave motions, as similar to the assumption in Basso et al. (2023), Er et al. (2023). That means the height of USV in IRF is regarded to be zero, where the disturbances on heave direction are ignored. This assumption is reasonable in practical scenarios, since the mean of ocean current disturbances on height direction of USV can be set as zero (Caruso et al., 2008). Based on this, the position and attitude vector in IRF is denoted as $\eta_{\mathrm{s}} = [x_s, y_s, \phi_s]^{\mathrm{T}}$, where x_s and y_s are the positions on X and Y axes, respectively. ϕ_s is the yaw angle. The linear and angular velocity vector in BRF is expressed as $\mathbf{v}_{\mathrm{s}} = [u_s, \upsilon_s, r_s]^{\mathrm{T}}$, where u_s and υ_s are the linear velocities on surge and sway, respectively. In addition, r_s is the angle velocity on yaw. Correspondingly, the dynamics of USV can be represented as

$$\begin{aligned} &\dot{\eta}_{\mathrm{s}} = \mathbf{J}_{\mathrm{s}}(\eta_{\mathrm{s}})\mathbf{v}_{\mathrm{s}}, \\ &\mathbf{M}_{\mathrm{s}}\dot{\mathbf{v}}_{\mathrm{s}} + \mathbf{C}_{\mathrm{s}}(\mathbf{v}_{\mathrm{s}})\mathbf{v}_{\mathrm{s}} + \mathbf{D}_{\mathrm{s}}(\mathbf{v}_{\mathrm{s}})\mathbf{v}_{\mathrm{s}} = \tau_{\mathrm{s}}, \end{aligned} \tag{5.2}$$

where $\mathbf{J}_{\mathrm{s}}(\eta_{\mathrm{s}}) \in \mathcal{R}^{3\times 3}$ is the rotation matrix. $\mathbf{M}_{\mathrm{s}}$ =diag$\{m_{11}, m_{22}, m_{44}\} \in \mathcal{R}^{3\times 3}$ is the inertia matrix. $\mathbf{C}_{\mathrm{s}}(\mathbf{v}_{\mathrm{s}}) \in \mathcal{R}^{3\times 3}$ represents the Coriolis-centripetal matrix. $\mathbf{D}_{\mathrm{s}}(\mathbf{v}_{\mathrm{s}}) \in \mathcal{R}^{3\times 3}$ represents the damping matrix. The control input vector for USV is denoted by $\tau_{\mathrm{s}} = [\tau_u^s, 0, \tau_r^s]^{\mathrm{T}}$, where τ_u^s and τ_r^s are the surge force and the yaw force, respectively.

With regard to the UUV, a fully actuated model is considered in this chapter to ensure the mobility performance. The position and attitude vector of UUV in IRF is denoted as $\eta_{\mathrm{u}} = [x_u, y_u, z_u, \phi_u]^{\mathrm{T}}$. The linear and angular velocity vector in BRF is expressed as $\mathbf{v}_{\mathrm{u}} = [u_u, \upsilon_u, w_u, r_u]^{\mathrm{T}}$. Hence, the motion model of UUV can be expressed as Yan et al. (2022)

$$\begin{aligned} &\dot{\eta}_{\mathrm{u}} = \mathbf{J}_{\mathrm{u}}(\eta_{\mathrm{u}})\mathbf{v}_{\mathrm{u}}, \\ &\mathbf{M}_{\mathrm{u}}\dot{\mathbf{v}}_{\mathrm{u}} + \mathbf{C}_{\mathrm{u}}(\mathbf{v}_{\mathrm{u}})\mathbf{v}_{\mathrm{u}} + \mathbf{D}_{\mathrm{u}}(\mathbf{v}_{\mathrm{u}})\mathbf{v}_{\mathrm{u}} + \mathbf{g}_{\mathrm{u}}(\eta_{\mathrm{u}}) = \tau_{\mathrm{u}}, \end{aligned} \tag{5.3}$$

where $\tau_{\mathrm{u}} = [\tau_u^u, \tau_\upsilon^u, \tau_w^u, \tau_r^u]^{\mathrm{T}}$ denotes the control input. $\mathbf{M}_{\mathrm{u}} \in \mathcal{R}^{4\times 4}$ denotes the inertia matrix. $\mathbf{J}_{\mathrm{u}}(\eta_{\mathrm{u}}) \in \mathcal{R}^{4\times 4}$ is the rotation matrix. $\mathbf{C}_{\mathrm{u}}(\mathbf{v}_{\mathrm{u}}) \in \mathcal{R}^{4\times 4}$ is the matrix of Coriolis-centripetal term. Besides that, $\mathbf{D}_{\mathrm{u}}(\mathbf{v}_{\mathrm{u}}) \in \mathcal{R}^{4\times 4}$ represents the drag matrix, and $\mathbf{g}_{\mathrm{u}}(\eta_{\mathrm{u}})$ is the hydrostatic force vector.

Remark 5.1 Note that most UAVs (or UUVs) can provide the lateral propulsion, longitudinal propulsion and rudder moment for maneuverability control. In contrast, most USVs rely only on rudder moment and longitudinal propulsion for maneuverability control (Liang et al., 2017). For that reason, the fully actuated models are adopted by UAV and UUV, while the underactuated model is adopted by USV. The design of maneuverability controller for underactuated USV is harder than the ones for fully actuated vehicles, due to the lack of dimensions for controller.

With the above motion models, we consider a rectangular detection space, whose length, width and height are a, b and c, respectively. Without loss of generality, the detection areas of USV and UUV are depicted by hemispherical and spherical regions with radius R_d, respectively. Then, the following definition is given to explain the detection effectiveness.

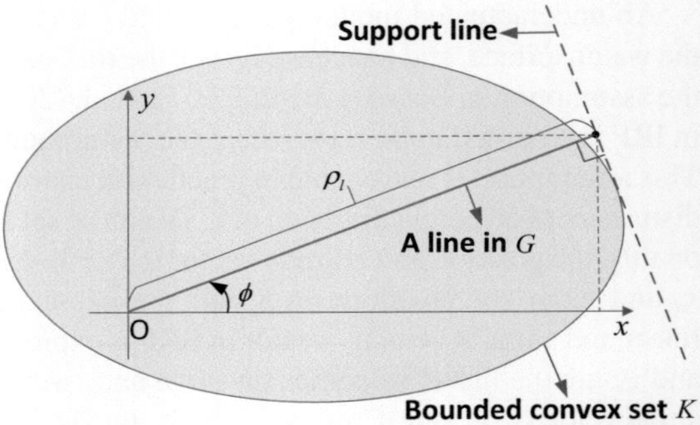

Fig. 5.2 Description of the line set on the plane

Definition 5.1 (*Detection effectiveness*) Target is regarded to be detected by the UAV-USV-UUV networks if it is within one of the scanning areas of UUV, UAV and USV.

In order to quantitatively evaluate the detection performance, we calculate the detection probability, which can be indirectly expressed by the *measure*. Next, the preliminary knowledge of the *measure theory* (Santalo, 2004) is presented.

Definition 5.2 (*Measure of line set*) For a bounded convex set K and origin $O \in K$, one defines $G : x \cos \phi + y \sin \phi - \rho_l = 0$ as the whole line set on the plane, where (x, y) denotes the straight-line coordinate of an arbitrary point. In addition, ϕ is the angle of normal with the X axis, as depicted by Fig. 5.2. Thereby, for a specified line set $X = \{G : G \cap K \neq \varnothing\}$, its measure is defined as (see Eq. (3.12) in Santalo, 2004)

$$m(X) = \int_{G \cap K \neq \emptyset} \mathrm{d}\rho_l \wedge \mathrm{d}\phi = \int_0^{2\pi} \rho_l \mathrm{d}\phi = L, \tag{5.4}$$

where L represents the perimeter of convex set K.

If line set G has intersected the convex set K, one can conclude that the probability of $G \cap K_\mathrm{m}$ is $P_\mathrm{m} = L_\mathrm{m}/L$ by using Definition 5.2, where L_m is the perimeter of convex set $K_\mathrm{m} \subset K$. Based on this, we extend the measure of line set to the strip set of width d_l. As depicted by Fig. 5.3, the strip set of width d_l is a closed set formed by two parallel lines. The position of the strip set is determined by its mid-parallel line, which is defined as the line G_H located at a distance of $d_l/2$ from the parallel lines. In view of this, the detection space and its boundary of a vehicle can be defined by a strip set. Correspondingly, the strip sets for UAV, USV and UUV are denoted by S_A, S_S and S_U, respectively. The density of strip set S_ϱ for $\varrho \subset \{\mathrm{A,S,U}\}$ is given as (see Eq. (3.5) in Santalo, 2004)

$$\mathrm{d}S_\varrho = \mathrm{d}G_\mathrm{H} = \mathrm{d}\rho_l \wedge \mathrm{d}\varphi, \tag{5.5}$$

whose role is to calculate the measure for a strip set.

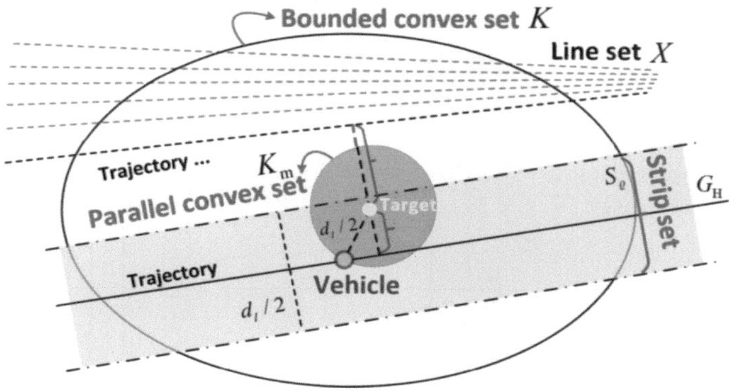

Fig. 5.3 Line set G passing through convex set K_m

Next, the measure of strip set is to calculate the detection probability. Specifically, a circle centered at the target point T_p with radius $d_l/2$ is regarded as the parallel convex set T_r with width d_l. The union of all circles with radius $d_l/2$ and center at any point in the detection space is regarded as the parallel convex set for S_I, which can be denoted by S_r. Of note, S_I is the bounded convex set for the detection space. The boundary curve of S_r is denoted by ∂S_r, and the perimeter of S_r is $L_r = L_I + \pi d_l$, where L_I is the perimeter of S_I. Then, the measure of S_ϱ with width d_l and center point T_p can be acquired by determining the measure of the intersection between G_H and T_r, i.e., for $\varrho \subset \{A,S,U\}$

$$m(S_\varrho; T_p \in S_\varrho) = \int_{T_p \in S_\varrho} dS_\varrho = \int_{G_H \cap K \neq \varnothing} dG_H = \pi d_l. \tag{5.6}$$

Constraint 5.1 (*UAV-USV connectivity*) In order to guarantee the cooperation between UAV and USV, the distance between USV and UAV needs to satisfy

$$h \leq \|\bar{\mathbf{p}}_a - \bar{\mathbf{p}}_s\| \leq C_{AS}, \tag{5.7}$$

where $\bar{\mathbf{p}}_a = [p_{a,x}, p_{a,y}, h]^T$ and $\bar{\mathbf{p}}_s = [x_s, y_s, 0]^T$ are the three-dimensional position vectors of UAV and USV in IRF, respectively. $C_{AS} > h$ is the threshold of reliable communication distances between UAV and USV. Since the UAV operates at the same height h, we know the distance between USV and UAV is larger than or equal to h, as shown in Fig. 5.4.

Constraint 5.2 (*USV-UUV connectivity*) Acoustic communication is conducted by USV and UUV. For connectivity maintenance, the distance between USV and UUV satisfies

$$0 < \|\bar{\mathbf{p}}_s - \bar{\mathbf{p}}_u\| \leq C_{US}, \tag{5.8}$$

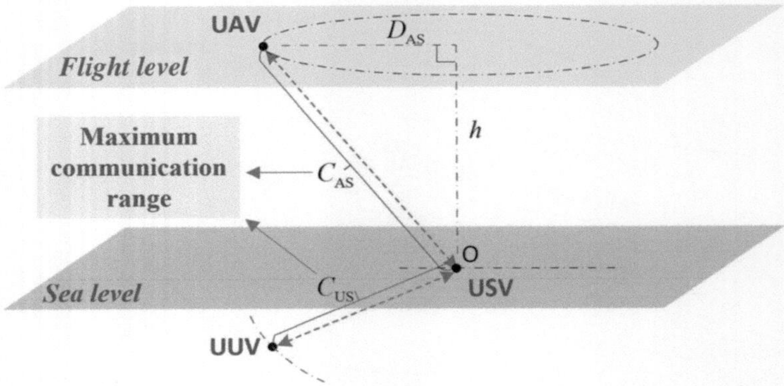

Fig. 5.4 Communication connectivity of UAV-USV-UUV

where $\bar{\mathbf{p}}_u = [x_u, y_u, z_u]^T$ is the three-dimensional position vector of UUV in IRF. $C_{US} > 0$ is the threshold of the reliable communication distances between USV and UUV.

In summary, the cooperation detection and tracking missions in this chapter are detailed as the following two problems.

Problem 1 (*Construction of the detection mode*) We aim to adopt UAV, USV and UUV to construct a heterogeneous detection mode. This problem is reduced to analyze the effects of the angle and distance of the USV-UUV formation on the target detection probability with the consideration of Constraints 5.1–5.2, such that the target detection probability can be maximized by adjusting the formation shape.

Problem 2 (*Design of tracking algorithm*) The complex dynamics and strong-coupling relationship make it difficult to achieve persistently target monitoring. We aim to design a model-free tracking algorithm for UUV to track the trajectory of target. This problem is reduced to design a DDPG-based tracking algorithm for UAV-USV-UUV networks, with considerations of dynamics models in (5.1), (5.2) and (5.3).

5.3 Strategy Design for Detection and Tracking Control

5.3.1 Heterogeneous Detection Mode

As mentioned above, UAV and USV cooperatively detect the surface target by visual sensors, while USV and UUV cooperatively detect the underwater target by sonar sensors. In view of this, the heterogeneous detection mode can be divided into the

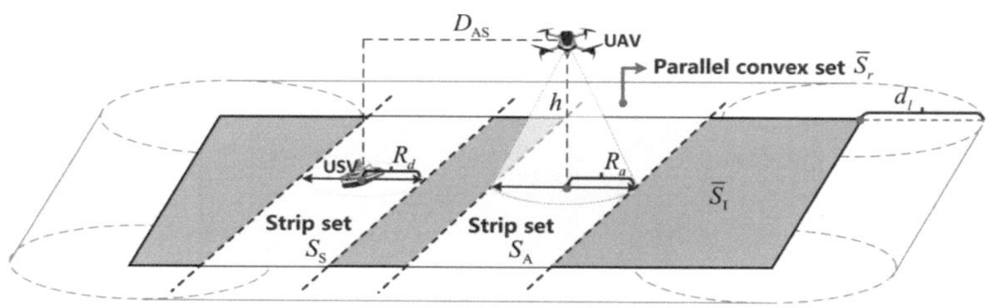

Fig. 5.5 UAV-USV detection networks, where $d_l = R_a + R_d$

following two parts: (1) detection mode between UAV and USV; (2) detection mode between USV and UUV.

For the first part, the detection performance can be evaluated by the scanning area. As shown in Fig. 5.5, the scanning area is related with the motion direction of UAV (or USV), however we cannot calculate all the possible directions of scanning area. In view of this, the strip set, which is a set of all points between any two parallel straight lines (Santalo, 2004), is employed to describe the scanning area. Specifically, the scanning area of UAV on water surface is denoted by the strip set S_{A} with width $2R_a$, where $R_a > 0$ represents the scanning radius of UAV on water surface. Meanwhile, the scanning area of USV on water surface is denoted by strip set S_{S} with width $2R_d$, where $R_d > 0$ is the scanning radius of USV on water surface. Based on this, one needs to maximize the measure of set $S_{\mathrm{A}} \cup S_{\mathrm{S}}$, such that the scanning area can be maximized. To this end, we calculate the measure of UAV-USV detection system. Define D_{AS} as the distance between UAV and USV in horizontal direction. Then, if the scanning areas of UAV and USV overlap, i.e., $D_{\mathrm{AS}} < (R_a + R_d)$, the set $S_{\mathrm{A}} \cup S_{\mathrm{S}}$ is equivalent to the strip set $\bar{E}_{\mathrm{AS}}$ with the width $D_{\mathrm{AS}} + R_a + R_d$, where $\mathrm{d}\bar{E}_{\mathrm{AS}} = \mathrm{d}G_{\mathrm{H}}^{\mathrm{AS}}$ is the density of the strip set $\bar{E}_{\mathrm{AS}}$. Meanwhile, $G_{\mathrm{H}}^{\mathrm{AS}}$ is defined as the mid-parallel line of strip set $\bar{E}_{\mathrm{AS}}$. According to (5.6), the measure of target T_{p} within the strip set $\bar{E}_{\mathrm{AS}} = S_{\mathrm{A}} \cup S_{\mathrm{S}}$ can be represented by

$$
\begin{aligned}
m(\bar{E}_{\mathrm{AS}}; T_{\mathrm{p}} \in \ \bar{E}_{\mathrm{AS}}) & \\
&= \int_{T_{\mathrm{p}} \in \bar{E}_{\mathrm{AS}}} \mathrm{d}\bar{E}_{\mathrm{AS}} = \int_{G_{\mathrm{H}}^{\mathrm{AS}} \cap K \neq \varnothing} \mathrm{d}G_{\mathrm{H}}^{\mathrm{AS}} \\
&= \pi(D_{\mathrm{AS}} + R_a + R_d).
\end{aligned} \tag{5.9}
$$

On the other hand, if the scanning areas of UAV and USV do not overlap, i.e., $D_{\mathrm{AS}} \geq (R_a + R_d)$, the measure of target T_{p} within the strip set $\bar{E}_{\mathrm{AS}}$ can be denoted by

$$
\begin{aligned}
m(\bar{E}_{\mathrm{AS}}; T_{\mathrm{p}} \in \ \bar{E}_{\mathrm{AS}}) & \\
&= m(S_{\mathrm{A}}; T_{\mathrm{p}} \in S_{\mathrm{A}}) + m(S_{\mathrm{S}}; T_{\mathrm{p}} \in S_{\mathrm{S}}) \\
&= 2\pi(R_a + R_d).
\end{aligned} \tag{5.10}
$$

From (5.9) and (5.10), we have the following conclusions: (1) if $D_{\mathrm{AS}} < (R_a + R_d)$, the measure of $\bar{E}_{\mathrm{AS}}$ increases with D_{AS}; (2) if $D_{\mathrm{AS}} \geq (R_a + R_d)$, the measure of $\bar{E}_{\mathrm{AS}}$ is a constant; (3) the value in (5.10) is greater than the one in (5.9), which means the scanning areas of UAV and USV should avoid overlapping for seeking the maximization of measure. According to the above conclusions, we further investigate the target detection probability of UAV-USV system when there is no overlapping. For that purpose, we need to acquire the measure of the strip set intersecting the detection space $\bar{S}_{\mathrm{I}}$, such that the measure of the strip set intersecting the detection space $\bar{S}_{\mathrm{I}}$ can be indirectly converted to the measure of a straight line G intersecting a parallel convex set $\bar{S}_r$. Based on this, the measure of $\bar{S}_r$ is calculated by

$$m(\bar{S}_r; G \cap \bar{S}_r \neq \varnothing) = 2\pi(R_a + R_d) + 2a + 2b. \tag{5.11}$$

Through geometric probability principle and (5.9)–(5.11), we calculate the target detection probability when target intersects $\bar{E}_{\mathrm{AS}}$ in $\bar{S}_{\mathrm{I}}$. If the scanning areas of UAV and USV overlap, i.e., $D_{\mathrm{AS}} < (R_a + R_d)$, the detection probability is

$$\begin{aligned} P(T_{\mathrm{p}} \in\ & \bar{E}_{\mathrm{AS}} | T_{\mathrm{p}} \in \bar{S}_{\mathrm{I}}) \\ &= \frac{m(\bar{E}_{\mathrm{AS}}; T_{\mathrm{p}} \in \bar{E}_{\mathrm{AS}})}{m(\bar{S}_r; G \cap \bar{S}_r \neq \varnothing)} = \frac{\pi(D_{\mathrm{AS}} + R_a + R_d)}{\pi(R_a + R_d) + a + b}. \end{aligned} \tag{5.12}$$

If the scanning areas of UAV and USV do not overlap, i.e., $D_{\mathrm{AS}} \geq (R_a + R_d)$, the target detection probability is

$$P(T_{\mathrm{p}} \in \bar{E}_{\mathrm{AS}} | T_{\mathrm{p}} \in \bar{S}_{\mathrm{I}}) = \frac{\pi(R_a + R_d)}{\pi(R_a + R_d) + a + b}. \tag{5.13}$$

Clearly, the detection probability in (5.13) is a constant. Note that the motion monitoring in a linear way can allow unmanned system to maximize the scanning area (Santalo, 2004). Based on this, the detection mode for UAV-USV system can be summarized as follows: (1) UAV and USV should keep a straight formation while avoiding the overlapping of scanning area, i.e. $D_{\mathrm{AS}} \geq (R_a + R_d)$; (2) the connectivity between UAV and USV should be guaranteed, i.e., $D_{\mathrm{AS}} \leq \sqrt{C_{\mathrm{AS}}^2 - h^2}$.

For the second part, we construct the detection mode of USV and UUV by evaluating the scanning areas. As shown in Fig. 5.6, the scanning areas of USV and UUV are described by the three-dimension strip sets, i.e., B_{s} and B_{u} with width d_c. Similar to the first part, the maximization of scanning areas is obtained by maximizing the size of $B_{\mathrm{s}} \cup B_{\mathrm{u}}$. Accordingly, the following two cases are considered. Case 1: the scanning areas have intersection points, i.e., $0 < D_{\mathrm{SU}} < d_c$, where D_{SU} is the distance between USV and UUV; Case 2: the scanning areas are tangent or separated, i.e. $D_{\mathrm{SU}} \geq d_c$. With respect to Case 1, the measure of $E_{\mathrm{SU}} = B_{\mathrm{s}} \cup B_{\mathrm{u}}$ satisfies

$$m(T_{\mathrm{p}} \in E_{\mathrm{SU}}) < m(T_{\mathrm{p}} \in B_{\mathrm{u}}) + m(T_{\mathrm{p}} \in B_{\mathrm{s}}). \tag{5.14}$$

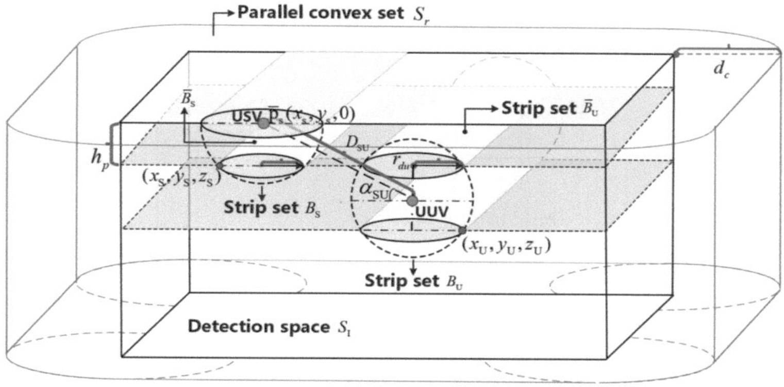

Fig. 5.6 USV-UUV detection networks where $d_c = 2R_d$

For Case 2, one has $B_s \cap B_u = \varnothing$, and hence, the measure of E_{SU} is equal to the sum of the measures of B_u and B_s, i.e.,

$$m(T_p \in E_{SU}) = m(T_p \in B_u) + m(T_p \in B_s). \tag{5.15}$$

From (5.14) and (5.15), we know that the value in (5.15) is greater than the one in (5.14), and thereby, the scanning areas of USV and UUV should avoid intersection for seeking the maximization of the measure. Therefore, we study the detection probability at any depth h_p in Case 2. To this end, we need to calculate the measure of $B_s \cup B_u$ at any depth h_p. Along with this, we set the depth of UUV as $h_u = D_{SU} \sin \alpha_{SU}$, where α_{SU} represents the angle with respect to the horizontal plane (see Fig. 5.6). In order to calculate the scanning radius of B_s and B_u, the scanning area is projected onto X-Z plane, and its projection round can be given as

$$\begin{cases} (x_S - x_s)^2 + (z_S - z_s)^2 = R_d^2, \\ (x_U - x_u)^2 + (z_U - z_u)^2 = R_d^2, \\ y_S = 0, \;\; y_U = 0, \end{cases} \tag{5.16}$$

where x_S, y_S and z_S are the positions along X, Y and Z axes on the scanning boundary of USV, respectively. Besides that, x_U, y_U and z_U are the positions along X, Y and Z axes on the scanning boundary of UUV, respectively.

Based on this, the strip sets of B_s and B_u intersecting the plane of depth h_p are the strip sets $\bar{B}_s$ and $\bar{B}_u$, respectively. For $z_S = h_p$ and $z_U = h_p$, the scanning radius r_{ds} and r_{du} for USV and UUV can be expressed as

$$r_{ds} = \begin{cases} \sqrt{R_d^2 - h_p^2}, \text{ if } 0 \le h_p \le R_d \\ 0, \quad \text{else} \end{cases} \tag{5.17}$$

$$r_{du} = \begin{cases} \triangle, \text{ if } \max(h_u - R_d, 0) \in [h_p, h_u + R_d] \\ 0, \text{ else} \end{cases} \tag{5.18}$$

where $\triangle = \sqrt{R_d^2 - \left(h_p - D_{\mathrm{SU}} \sin(\alpha_{\mathrm{SU}})\right)^2}$.

When $h_u - R_d \le 0$, we employ (5.17) and (5.18) to calculate the measure of $\bar{E}_{\mathrm{US}}$, i.e.,

$$\begin{aligned} & m(\bar{E}_{\mathrm{US}}; T_{\mathrm{p}} \in \bar{E}_{\mathrm{US}}) \\ = & \begin{cases} 2\pi(r_{ds} + r_{du}), \text{ if } 0 < h_p \le R_d \\ 2\pi r_{du}, \quad \text{if } R_d < h_p \le h_u + R_d \\ 0, \quad \text{else} \end{cases} \end{aligned} \tag{5.19}$$

When $0 < h_u - R_d \le R_d$, the measure of $\bar{E}_{\mathrm{US}}$ is

$$\begin{aligned} & m(\bar{E}_{\mathrm{US}}; T_{\mathrm{p}} \in \bar{E}_{\mathrm{US}}) \\ = & \begin{cases} 2\pi r_{ds}, \quad \text{if } 0 \le h_p \le h_u - R_d \\ 2\pi(r_{ds} + r_{du}), \text{ if } h_u - R_d < h_p \le R_d \\ 2\pi r_{du}, \quad \text{if } R_d < h_p \le h_u + R_d \\ 0, \quad \text{else} \end{cases} \end{aligned} \tag{5.20}$$

When $h_u - R_d > R_d$, the measure of $\bar{E}_{\mathrm{US}}$ is

$$\begin{aligned} & m(\bar{E}_{\mathrm{US}}; T_{\mathrm{p}} \in \bar{E}_{\mathrm{US}}) \\ = & \begin{cases} 2\pi r_{ds}, \text{ if } 0 \le h_p \le R_d \\ 2\pi r_{du}, \text{ if } h_u - R_d \le h_p \le h_u + R_d \\ 0, \quad \text{else} \end{cases} \end{aligned} \tag{5.21}$$

By geometric probability principle and (5.19)–(5.21), the target detection probability of $T_{\mathrm{p}} \in \bar{E}_{\mathrm{US}}$ at depth h_p is

$$\begin{aligned} & P_{dect}(h_p, D_{\mathrm{SU}}, \alpha_{\mathrm{SU}}; T_{\mathrm{p}} \in \bar{E}_{\mathrm{US}}) \\ = & \frac{m(\bar{E}_{\mathrm{US}}; T_{\mathrm{p}} \in \bar{E}_{\mathrm{US}})}{m(\bar{S}_p; G \cap \bar{S}_p \ne \varnothing)} \\ = & \begin{cases} \frac{\pi\sqrt{R_d^2 - h_p^2}}{\pi d_c + a + b}, \text{ if } 0 \le h_p < h_u - R_d \text{ and } h_u - R_d > 0 \\ \triangle_1, \quad \text{if } \max(h_u - R_d, 0) \le h_p \le R_d \\ \qquad \text{and } h_u - R_d < R_d \\ \triangle_2, \quad \text{if } \max(R_d, h_u - R_d) < h_p \le h_u + R_d \\ 0 \quad \text{else} \end{cases} \end{aligned} \tag{5.22}$$

with

$$m(\bar{S}_p; G \cap \bar{S}_p \neq \varnothing) = 2\pi d_c + 2a + 2b. \tag{5.23}$$

where $\triangle_1 = \frac{\pi(\sqrt{R_d^2 - h_p^2} + \sqrt{R_d^2 - (h_p - D_{\mathrm{SU}}\sin(\alpha_{\mathrm{SU}}))^2})}{\pi d_c + a + b}$. In addition, $m(\bar{S}_p; G \cap \bar{S}_p \neq \varnothing)$ is the measure of G intersecting parallel convex set $\bar{S}_p$. Meanwhile, $\triangle_2 = \frac{\pi\sqrt{R_d^2 - (h_p - D_{\mathrm{SU}}\sin(\alpha_{\mathrm{SU}}))^2}}{\pi d_c + a + b}$.

From (5.22), we can see the value of $\triangle_1$ is greater than the other ones, i.e., the detection probability is largest when $h_p - D_{\mathrm{SU}}\sin(\alpha_{\mathrm{SU}}) = 0$. Based on this and noting with (5.23), the detection mode for USV-UUV system is summarized as follows: (1) USV and UUV should keep a straight formation while avoiding the overlapping of scanning area, i.e. $D_{\mathrm{SU}} \geq d_c$; (2) formation angle α_{SU} should satisfy the condition $\alpha_{\mathrm{SU}} = \arcsin\frac{h_p}{D_{\mathrm{SU}}}$ or $\alpha_{\mathrm{SU}} = \pi - \arcsin\frac{h_p}{D_{\mathrm{SU}}}$ for maximizing the detection probability; (3) the connectivity between USV and UUV should be guaranteed, i.e. $D_{\mathrm{SU}} \leq C_{\mathrm{US}}$.

5.3.2 DDPG-Based Tracking Algorithm

After target is detected, we assume the position information of target can be acquired by the localization units installed on UAV, USV and UUV. After that, this section designs a DDPG-based tracking algorithm to drive UAV, USV and UUV to pursuit the target. To this end, we define $\eta_t = [x_t, y_t, z_t]^{\mathrm{T}}$ as the position vector of underwater target. Then, the final desired position vectors of UAV, USV and UUV are $\eta_{\mathrm{D,UAV}} = [x_t, y_t, h]^{\mathrm{T}}$, $\eta_{\mathrm{D,USV}} = [x_t, y_t, 0]^{\mathrm{T}}$ and η_t, respectively.

In order to guarantee the communication connectivity among UAV, USV and UUV, a multi-step location prediction strategy is incorporated into the tracking procedure. The main idea of multi-step location prediction strategy is to construct a virtual force for UAV, USV or UUV, through which its possible location can be predicted. For UAV, the relationship among the detection mode, the multi-step location prediction and the tracking algorithm is depicted by Fig. 5.7. To facilitate illustration, the motion vector of the UAV is redefined as $\mathbf{X}_{\mathrm{a}} = [\mathbf{p}_{\mathrm{a}}; \omega_{\mathrm{q}}]$, where $\omega_{\mathrm{q}} = \dot{\eta}_{\mathrm{a}}$ is the angular velocity vector in IRF. Denote $\delta > 0$ as the sampling interval, and hence, model (5.1) at time step k is rearranged as

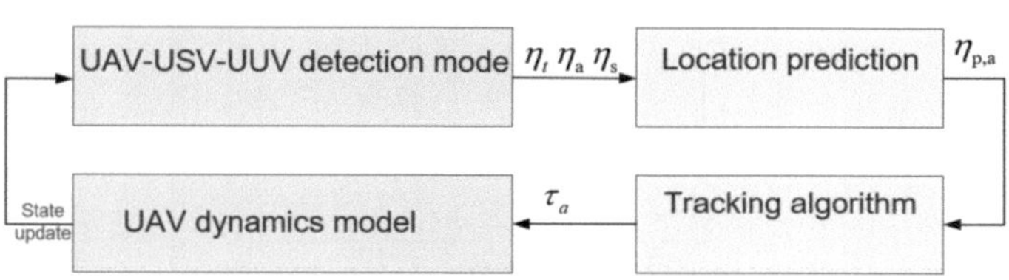

Fig. 5.7 Relationship with the detection mode and location prediction

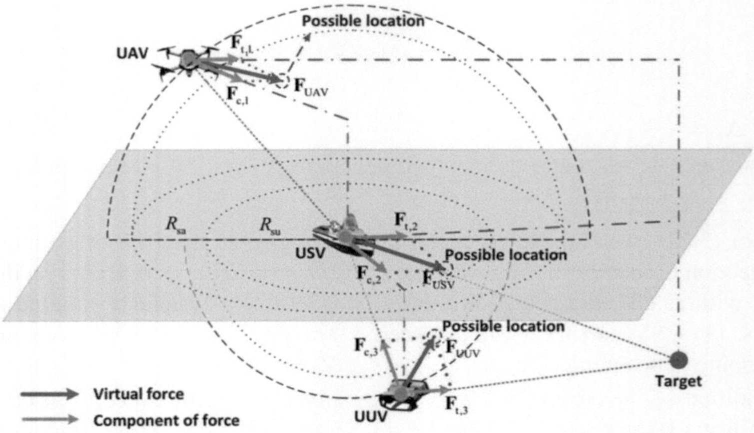

Fig. 5.8 The virtual forces of UAV, USV and UUV

$$\mathbf{X}_{\mathrm{a}}(k+1) = f_{\mathrm{a}}(\mathbf{X}_{\mathrm{a}}(k)) + g_{\mathrm{a}}(\mathbf{X}_{\mathrm{a}}(k))\tau_{\mathrm{a}}(k), \tag{5.24}$$

with

$$f_{\mathrm{a}}(\mathbf{X}_{\mathrm{a}}(k)) = \begin{bmatrix} \mathbf{p}_{\mathrm{a}} + \delta\mathbf{v}_{\mathrm{a}}(k) - \frac{\delta^2 g\mathbf{e}_3}{m_{\mathrm{q}}} + \frac{\delta^2\mathbf{A}_{\mathrm{q}}}{m_{\mathrm{q}}} \\ \omega_{\mathrm{q}} - \delta\mathbf{J}_{\mathrm{q}}^{-1}\mathbf{C}_{\mathrm{q}}(\eta_{\mathrm{a}}, \omega_{\mathrm{q}})\omega_{\mathrm{q}} \end{bmatrix}, \tag{5.25}$$

$$g_{\mathrm{a}}(\mathbf{X}_{\mathrm{a}}(k)) = \begin{bmatrix} \frac{\delta^2\mathbf{R}_{\mathrm{q}}}{m_{\mathrm{q}}} & 0_{3\times 3} \\ 0_{3\times 3} & \delta\mathbf{J}_{\mathrm{q}}^{-1} \end{bmatrix}, \tag{5.26}$$

where $\tau_{\mathrm{a}} = [\mathbf{F}_{\mathrm{q}};\ \tau_\eta]$ denotes the control input vector for UAV. $\mathbf{v}_{\mathrm{a}} = \dot{\mathbf{p}}_{\mathrm{a}}$ denotes the linear velocity vector of UAV in BRF. As shown by Fig. 5.8, the virtual force is used to predict the possible location. The virtual force of UAV can be designed as

$$\mathbf{F}_{\mathrm{UAV}} = \underbrace{\mathbf{F}_{\mathrm{c},1}}_{\text{Communication keeping}} + \underbrace{\mathbf{F}_{\mathrm{t},1}}_{\text{Target tracking}}, \tag{5.27}$$

with

$$\mathbf{F}_{\mathrm{c},1} = \begin{cases} k_{\mathrm{ca}}\left[\frac{x_s - p_{a,x}}{d_{\mathrm{as}}};\ \frac{y_s - p_{a,y}}{d_{\mathrm{as}}}\right], \text{if } d_{\mathrm{as}} \in (R_{\mathrm{sa}}, C_{\mathrm{AS}}] \\ 0, \qquad\qquad\qquad\qquad \text{else} \end{cases} \tag{5.28}$$

$$\mathbf{F}_{\mathrm{t},1} = k_{\mathrm{ta}}\left[\frac{x_t - p_{a,x}}{d_{\mathrm{ta}}};\ \frac{y_t - p_{a,y}}{d_{\mathrm{ta}}}\right], \tag{5.29}$$

$$k_{\mathrm{ca}} = \beta_{\mathrm{a}}(d_{\mathrm{as}} - R_{\mathrm{sa}}),\ \beta_{\mathrm{a}} = k_{\mathrm{ta}}/(V_{\mathrm{a,max}}T), \tag{5.30}$$

where $V_{\text{a,max}} = \left\|\mathbf{v}_{\text{a,max}}\right\|$, and $\mathbf{v}_{\text{a,max}} = \left[\dot{p}_{a,x,\text{max}}, \dot{p}_{a,y,\text{max}}\right]^{\text{T}}$ is the vector of maximum linear velocity of UAV. $k_{\text{ta}} > 0$ is the coefficient. $d_{\text{as}} = \sqrt{(x_s - p_{a,x})^2 + (y_s - p_{a,y})^2}$ is the relative plane distance between UAV and USV, while $R_{\text{sa}} = \sqrt{C_{\text{AS}}^2 - h^2 - 2V_{\text{a,max}}T}$ is the maximum safe communication distance. In addition, $d_{\text{ta}} = \sqrt{(x_t - p_{a,x})^2 + (y_t - p_{a,y})^2}$ is the distance from UAV to target on plane.

Based on (5.24) and (5.27), the possible location of UAV at time step $k + T$ can be designed as

$$\eta_{\text{p,a}}(k + T) = [\mathbf{p}_{\text{a}}(k) + \delta_{\#}\mathbf{F}_{\text{UAV}}(k);\ 0], \tag{5.31}$$

where $n \in \mathcal{Z}^+$ and $\delta_{\#} \in \mathcal{R}^+$. The purpose of $\delta_{\#}$ is to modulate the impact of virtual force, which is dynamically adjusted by the values of $\mathbf{F}_{\text{UAV}}(k)$ and R_{sa}. The larger the $\delta_{\#}$ the lower the prediction accuracy. Moreover, $T = n\delta$ is the prediction window. A small n means the real-time performance is strong, but a high computational cost is required. One can choose n according the the system requirement.

In the following, the multi-step location prediction strategy for USV is designed. The motion vector of the USV is redefined as $\mathbf{X}_{\text{s}} = [\eta_{\text{s}};\ \mathbf{v}_{\text{s}}]$. Similar to the implementation in (5.24), the dynamics of USV can be discretized as

$$\mathbf{X}_{\text{s}}(k + 1) = f_{\text{s}}(\mathbf{X}_{\text{s}}(k)) + h_{\text{s}}(\mathbf{X}_{\text{s}}(k))\tau_{\text{s}}^{\#}(k), \tag{5.32}$$

with

$$\begin{aligned} f_{\text{s}}(\mathbf{X}_{\text{s}}) &= \begin{bmatrix} \eta_{\text{s}} + \delta\mathbf{J}_{\text{s}}(\eta_{\text{s}})\mathbf{v}_{\text{s}} \\ \mathbf{v}_{\text{s}} - \delta\mathbf{M}_{\text{s}}^{-1}[\mathbf{C}_{\text{s}}(\mathbf{v}_{\text{s}})\mathbf{v}_{\text{s}} + \mathbf{D}_{\text{s}}(\mathbf{v}_{\text{s}})\mathbf{v}_{\text{s}}] \end{bmatrix}, \\ h_{\text{s}}(\mathbf{X}_{\text{s}}) &= \begin{bmatrix} 0_{3\times 3} \\ \delta\mathbf{M}_{\text{s}}^{-1} \end{bmatrix}. \end{aligned} \tag{5.33}$$

where $\tau_{\text{s}}^{\#}(k) = [0; 0; 0; \tau_{\text{s}}(k)]$.

Then, the virtual force acting on USV is designed as

$$\mathbf{F}_{\text{USV}} = \underbrace{\mathbf{F}_{\text{c,2}}}_{\text{Communication keeping}} + \underbrace{\mathbf{F}_{\text{t,2}}}_{\text{Target tracking}}, \tag{5.34}$$

with

$$\mathbf{F}_{\text{c,2}} = \begin{cases} k_{\text{cs}}\left[\frac{x_u - x_s}{d_{\text{us}}};\ \frac{y_u - y_s}{d_{\text{us}}}\right], \text{ if } d_{\text{us}} \in (R_{\text{su}},\ C_{\text{SU}}] \\ 0, \qquad\qquad\qquad \text{else} \end{cases} \tag{5.35}$$

$$\mathbf{F}_{\text{t,2}} = k_{\text{ts}}\left[\frac{x_t - x_s}{d_{\text{ts}}};\ \frac{y_t - y_s}{d_{\text{ts}}}\right], \tag{5.36}$$

$$k_{\text{cs}} = \beta_{\text{s}}(d_{\text{us}} - R_{\text{su}}),\ \beta_{\text{s}} = k_{\text{ts}}/(V_{\text{su,max}}T) \tag{5.37}$$

where $k_{\text{ts}} > 0$, $d_{\text{us}} = \sqrt{(x_s - x_u)^2 + (y_s - y_u)^2}$ is the relative plane distance between UUV and USV, and $V_{\text{su,max}} = \max\left(\left\|\mathbf{v}_{s,\text{max}}\right\|, \left\|\mathbf{v}_{u,\text{max}}\right\|\right)$. Of note, $\mathbf{v}_{s,\text{max}} =$

$\left[u_{s,\max}, v_{s,\max}\right]^{\mathrm{T}}$ and $\mathbf{v}_{u,\max} = \left[u_{u,\max}, v_{u,\max}, w_{u,\max}\right]^{\mathrm{T}}$ are the maximum linear velocity vectors of USV and UUV, respectively. $R_{\mathrm{su}} = C_{\mathrm{SU}} - 2V_{\mathrm{su,max}}T$ is the maximum safe communication distance, and $d_{\mathrm{ts}} = \sqrt{(x_t - x_s)^2 + (y_t - y_s)^2}$ is the distance from USV to target on plane.

From (5.32) and (5.34), the possible location of USV at time step $k + T$ is designed as

$$\eta_{\mathrm{p,s}}(k+T) = \eta_{\mathrm{s}}(k) + [\delta_{\#}\mathbf{F}_{\mathrm{USV}}(k);\ 0]. \tag{5.38}$$

Along with this, the multi-step location prediction strategy for UUV is also designed. For UUV, the motion vector of the UUV is redefined as $\mathbf{X}_{\mathrm{u}} = [\eta_{\mathrm{u}};\ \mathbf{v}_{\mathrm{u}}]$ and the dynamics model (5.3) can be rearranged as

$$\mathbf{X}_{\mathrm{u}}(k+1) = f_{\mathrm{u}}(\mathbf{X}_{\mathrm{u}}(k)) + h_{\mathrm{u}}(\mathbf{X}_{\mathrm{u}}(k))\tau_{\mathrm{u}}^{\#}(k), \tag{5.39}$$

with

$$\begin{aligned} f_{\mathrm{u}}(\mathbf{X}_{\mathrm{u}}) &= \begin{bmatrix} \eta_{\mathrm{u}} + \delta\mathbf{J}_{\mathrm{u}}(\eta_{\mathrm{u}})\mathbf{v}_{\mathrm{u}} \\ \mathbf{v}_{\mathrm{u}} - \delta\mathbf{M}_{\mathrm{u}}^{-1}[\mathbf{C}_{\mathrm{u}}(\mathbf{v}_{\mathrm{u}})\mathbf{v}_{\mathrm{u}} + \mathbf{D}_{\mathrm{u}}(\mathbf{v}_{\mathrm{u}})\mathbf{v}_{\mathrm{u}} + \mathbf{g}_{\mathrm{u}}(\eta_{\mathrm{u}})] \end{bmatrix}, \\ h_{\mathrm{u}}(\mathbf{X}_{\mathrm{u}}) &= \begin{bmatrix} 0_{4\times 4} \\ \delta\mathbf{M}_{\mathrm{u}}^{-1} \end{bmatrix}. \end{aligned}$$

where $\tau_{\mathrm{u}}^{\#}(k) = [0;\ 0;\ 0;\ 0;\ \tau_{\mathrm{u}}(k)]$.

The virtual force acting on UUV is designed as

$$\mathbf{F}_{\mathrm{UUV}} = \underbrace{\mathbf{F}_{\mathrm{c},3}}_{\text{Communication keeping}} + \underbrace{\mathbf{F}_{\mathrm{t},3}}_{\text{Target tracking}}, \tag{5.40}$$

with

$$\mathbf{F}_{\mathrm{c},3} = \begin{cases} k_{\mathrm{cu}}\boldsymbol{\sigma}_{\mathrm{su}}, & \text{if } d_{\mathrm{us}} \in (R_{\mathrm{su}},\ C_{\mathrm{SU}}] \\ 0, & \text{else} \end{cases}, \tag{5.41}$$

$$\mathbf{F}_{\mathrm{t},3} = k_{\mathrm{tu}}\left[\frac{x_t - x_u}{d_{\mathrm{tu}}};\ \frac{y_t - y_u}{d_{\mathrm{tu}}};\ \frac{z_t - z_u}{d_{\mathrm{tu}}}\right], \tag{5.42}$$

$$k_{\mathrm{cu}} = \beta_{\mathrm{u}}(d_{\mathrm{us}} - R_{\mathrm{su}}),\ \beta_{\mathrm{u}} = k_{\mathrm{tu}}/(V_{\mathrm{su,max}}T), \tag{5.43}$$

where $\boldsymbol{\sigma}_{\mathrm{su}} = [\frac{x_s - x_u}{d_{\mathrm{us}}};\ \frac{y_s - y_u}{d_{\mathrm{us}}};\ \frac{z_s - z_u}{d_{\mathrm{us}}}]$ is a unit vector. $k_{\mathrm{tu}} > 0$ is the coefficient. $d_{\mathrm{tu}} = \sqrt{(x_t - x_u)^2 + (y_t - y_u)^2 + (z_t - z_u)^2}$ represents the relative distance from UUV to target.

From (5.24) and (5.40), the possible location of UUV at time step $k + T$ is designed as

$$\eta_{\mathrm{p,u}}(k+T) = \eta_{\mathrm{u}}(k) + [\delta_{\#}\mathbf{F}_{\mathrm{UUV}}(k);\ 0]. \tag{5.44}$$

With the predicted location information (i.e., (5.31), (5.38) and (5.44)), our next objective is to control the UAV, USV, and UUV to reach the desired positions, i.e. $\eta_{\mathrm{a}} \rightarrow \eta_{\mathrm{D,UAV}}$, $\eta_{\mathrm{s}} \rightarrow \eta_{\mathrm{D,USV}}$, and $\eta_{\mathrm{u}} \rightarrow \eta_{\mathrm{t}}$ as $t \rightarrow \infty$. To this end, the control process can be effectively modeled as a Markov decision process (MDP). Therefore, the optimal control inputs τ_{a}^*, τ_{s}^* and τ_{u}^* for UAV, USV, and UUV can be obtained by solving the MDP with DDPG. Specifically, the control inputs τ_{a}, τ_{s} and τ_{u} for UAV, USV and UUV are defined as actions. According to Markov property, $\tau_{\mathrm{a}}(k)$, $\tau_{\mathrm{s}}(k)$ and $\tau_{\mathrm{u}}(k)$ at time step k are determined by the states at time step k, respectively. We define the states of UAV, USV and UUV at time step k as

$$\mathbf{s}_{\mathrm{a}}(k) = [\triangle\eta_{\mathrm{p,a}}; \mathbf{v}_{\mathrm{a}}] = [\eta_{\mathrm{p,a}} - \eta_{\mathrm{a}}; \mathbf{v}_{\mathrm{a}}], \tag{5.45}$$

$$\mathbf{s}_{\mathrm{u}}(k) = [\triangle\eta_{\mathrm{p,u}}; \mathbf{v}_{\mathrm{u}}] = [\eta_{\mathrm{p,u}} - \eta_{\mathrm{u}}; \mathbf{v}_{\mathrm{u}}], \tag{5.46}$$

$$\mathbf{s}_{\mathrm{s}}(k) = [\triangle\eta_{\mathrm{p,s}}; \mathbf{v}_{\mathrm{s}}] = [\eta_{\mathrm{p,s}} - \eta_{\mathrm{s}}; \mathbf{v}_{\mathrm{s}}]. \tag{5.47}$$

The value of the action performed in current state $\mathbf{s}_\xi$ is evaluated by cost functions. Thereby, the cost functions of UAV, USV and UUV are

$$c_{\mathrm{a}}(\mathbf{s}_{\mathrm{a}}(k)) = \rho_{\mathrm{a}} \left\|\triangle\eta_{\mathrm{p,a}}\right\|^2, \; c_{\mathrm{s}}(\mathbf{s}_{\mathrm{s}}(k)) = \rho_{\mathrm{s}} \left\|\triangle\eta_{\mathrm{p,s}}\right\|^2, \tag{5.48}$$

$$c_{\mathrm{u}}(\mathbf{s}_{\mathrm{u}}(k)) = \rho_{\mathrm{u}} \left\|\triangle\eta_{\mathrm{p,u}}\right\|^2 \tag{5.49}$$

where $\rho_{\mathrm{a}} > 0$, $\rho_{\mathrm{s}} > 0$ and $\rho_{\mathrm{u}} > 0$ are coefficients.

An actor network $\mu_\xi(\mathbf{s}_\xi|\theta_\xi)$ with weights of θ_ξ is established to get the control inputs of τ_{a}, τ_{s} and τ_{u}, where $\xi = \{\mathrm{a,s,u}\}$. Meanwhile, a critic network $Q(\mathbf{s}_\xi, \mu_\xi|\omega_\xi)$ with weight of ω_ξ is established to evaluate the payoff of the policy. Based on this, the optimal estimated value at k time can be written by using the Bellman equation, i.e.,

$$\begin{aligned} & Q_\xi^*(\mathbf{s}_\xi(k), \tau_\xi(k)) \\ = \; & E[c_\xi(\mathbf{s}_\xi(k+1)) + \gamma \min Q_\xi(\mathbf{s}_\xi(k+1), \tau_\xi(k+1))] \end{aligned} \tag{5.50}$$

where $0 < \gamma < 1$ is a discount rate.

Based on (5.50), the optimal input control $\tau_\xi^*(k)$ for $\xi = \{\mathrm{a,s,u}\}$ can be acquired by minimizing the output of the critic network for the action taken from the current state. Hence, the optimal policy at k time is expressed as

$$\mu_\xi^*(\mathbf{s}_\xi(k)) = \arg\min Q_\xi(\mathbf{s}_\xi(k), \tau_\xi(k)). \tag{5.51}$$

Remark 5.2 In (5.48) and (5.49), the cost functions rather than critic networks are used to evaluate the quality of current actions. The reasons associated with this operation are summarized as follows: (1) the cost function can provide a straightforward way to evaluate actions of immediate consequences, through which they play the role of directly quantifying the desirability associated with each action; (2) the critic networks can evaluate the global reward and guide the decision-making

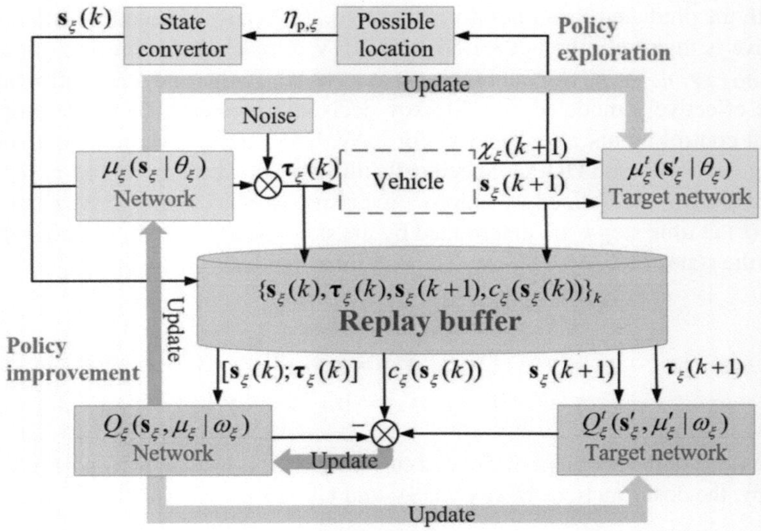

Fig. 5.9 DDPG-based algorithm framework

process, through which they play the role of offering insights into the long-term consequences of actions.

From (5.50) and (5.51), the following steps are conducted to seek $\tau_\xi^*(k)$, as represented by Fig. 5.9.

Step 1. Policy exploration and experience collection.

With $\mathbf{X}_a(k)$, $\mathbf{X}_s(k)$ and $\mathbf{X}_u(k)$, the possible locations $\eta_{p,a}(k)$, $\eta_{p,s}(k)$ and $\eta_{p,u}(k)$ are obtained by (5.31), (5.38) and (5.44), respectively. Then, the states $\mathbf{s}_a(k)$, $\mathbf{s}_s(k)$ and $\mathbf{s}_u(k)$ are converted from $\mathbf{X}_\xi(k)$ and $\eta_{p,\xi}(k)$, through which the exploration policies $\tau_a(k)$, $\tau_s(k)$ and $\tau_u(k)$ are obtained from the actor networks $\mu_a(\mathbf{s}_a(k)|\theta_a)$, $\mu_s(\mathbf{s}_s(k)|\theta_s)$ and $\mu_u(\mathbf{s}_u(k)|\theta_u)$, respectively. To improve the degree of policy exploration, we employ action noise to make the exploration of the action space. The control input can be expressed as

$$\tau_\xi(k) = \mu_\xi(\mathbf{s}_\xi|\theta_\xi) + \mathcal{N}^\#, \tag{5.52}$$

where $\mathcal{N}^\# \backsim \mathcal{N}(0, 1)$ is a gaussian noise. Thereby, the next states $\mathbf{s}_a(k+1)$, $\mathbf{s}_s(k+1)$ and $\mathbf{s}_u(k+1)$ are acquired by executing the current actions $\tau_a(k)$, $\tau_s(k)$ and $\tau_u(k)$, respectively. At time $k+T$, the possible locations $\eta_{p,a}(k+T)$, $\eta_{p,s}(k+T)$ and $\eta_{p,u}(k+T)$ are updated by (5.31), (5.38) and (5.44) after every T times. Repeat the above exploration steps to acquire exploration sequences $\mathbf{s}_\xi(k)$, $\mu_\xi(\mathbf{s}_\xi(k))$, $\mathbf{s}_\xi(k+1)$, $\mu_\xi(\mathbf{s}_\xi(k+1))$. Furthermore, we construct a replay buffer R_o to collect the exploration experience to train the actor and critic networks. An experience at time step k is denoted as

$$\{\mathbf{s}_\xi(k), \tau_\xi(k), \mathbf{s}_\xi(k+1), c_\xi(\mathbf{s}_\xi(k))\}_k. \tag{5.53}$$

Step 2. Policy improvement and network training.
After the policy exploration, the policies are improved by adjusting the policy weights $\theta_{\rm a}$, $\theta_{\rm s}$ and $\theta_{\rm u}$ of the actor networks $\mu_{\rm a}(\mathbf{s}_{\rm a}|\theta_{\rm a})$, $\mu_{\rm s}(\mathbf{s}_{\rm s}|\theta_{\rm s})$ and $\mu_{\rm u}(\mathbf{s}_{\rm u}|\theta_{\rm u})$, respectively. Specifically, $\theta_{\rm a}$, $\theta_{\rm s}$ and $\theta_{\rm u}$ are adjusted by using the policy gradient to minimize the return obtained from the critic networks according to (5.51). Meanwhile, the weights $\omega_{\rm a}$, $\omega_{\rm s}$ and $\omega_{\rm u}$ for critic networks $Q_{\rm a}(\mathbf{s}_{\rm a}, \mu_{\rm a}|\omega_{\rm a})$, $Q_{\rm s}(\mathbf{s}_{\rm s}, \mu_{\rm s}|\omega_{\rm s})$ and $Q_{\rm u}(\mathbf{s}_{\rm u}, \mu_{\rm u}|\omega_{\rm u})$ are updated by using temporal-difference learning, respectively.

To improve the training stability, two networks of the target critic network $Q^t_\xi(\mathbf{s}_\xi, \mu_\xi|\omega^t_\xi)$ with weight ω^t_ξ and target actor network $\mu^t_\xi(\mathbf{s}_\xi(k)|\theta^t_\xi)$ with weight θ^t_ξ are employed. With (5.50), the critic network weight is updated by

$$\omega_{\xi,j+1}=\omega_{\xi,j}+\frac{\alpha_\omega}{N}\sum\nolimits_{i=1}^{N}\delta_{\xi,i}\nabla_\omega Q_\xi(\mathbf{s}_{\xi,i}, \tau_{\xi,i}|\omega_{\xi,j}), \tag{5.54}$$

with

$$\begin{aligned}\delta_{\xi,i} = \ & c_\xi(\mathbf{s}'_{\xi,i}) + \gamma Q^t_\xi(\mathbf{s}'_{\xi,i}, \mu(\mathbf{s}'_{\xi,i})|\omega^t_\xi)\\ & - Q_\xi(\mathbf{s}_{\xi,i}, \tau_{\xi,i}|\omega_{\xi,j})\end{aligned} \tag{5.55}$$

where $\delta_{\xi,i}$ is the error value and j is the iteration number. N is the mini-batch size. $0 < \alpha_\omega < 1$ is the learning rate of critic network. $\nabla_\omega Q_\xi(\mathbf{s}_{\xi,i}, \tau_{\xi,i}|\omega_{\xi,j})$ is the deterministic gradient value of the critic network weights.

Algorithm 1: DDPG-based tracking algorithm

Input: Initial $\chi_{\xi,1}$, $\eta_{{\rm p},\xi}(0)$, ω_ξ, θ_ξ, $\omega^t_{\xi,j+1}$ and $\theta^t_{\xi,j+1}$;
Output: Control input τ_a,τ_s and τ_u.
Convert $\chi_{\xi,1}$ to $\eta_{{\rm p},\xi}(1)$ and state $\mathbf{s}_\xi(1)$;
for *episode* $= 1$ *to* M **do**
  Reset state $\mathbf{s}_\xi(0)$ based on $\chi_{\xi,0}$ and $\chi_{\xi,0}$;
  for $k = 0$ *to* N **do**
    Get $\tau_\xi(k)$ by (5.52);
    Execute $\tau_\xi(k)$ to get $\mathbf{s}_\xi(k+1)$ and $\chi_\xi(k+1)$;
    Calculate the cost $c_\xi(k)$ by (5.48) and (5.49);
    Push $\{\mathbf{s}_\xi(k), \tau_\xi(k), \mathbf{s}_\xi(k+1), c_\xi(k)\}_k$ into the replay buffer $R_{\rm o}$;
    if MOD$(k, T) = 0$ **then**
      Update $\eta_{{\rm p},\xi}(k+T)$ by (5.31), (5.38) and (5.44);
    Update $\mathbf{s}_\xi(k+1)$ by (5.45)–(5.47);
  if *episode* $> N_m$ **then**
    Randomly select N samples from $R_{\rm o}$;
    Calculate the $\delta_{\xi,i}$ by (5.55);
    Update critic networks by (5.54);
    Update actor networks by (5.56);
    Update $\omega^t_{\xi,j+1}$ by (5.57);
    Update $\theta^t_{\xi,j+1}$ by (5.58);

From (5.51), the actor network weight is update by

$$\theta_{\xi,l+1} = \theta_{\xi,l} - \frac{\alpha_\theta}{N} \sum\nolimits_{i=1}^{N} \nabla_\theta Q(\mathbf{s}_{\xi,i}, \mu_\xi(\mathbf{s}_{\xi,i}|\theta_{\xi,l})|\omega_{\xi,j}), \tag{5.56}$$

where $0 < \alpha_\theta < 1$ is the learning rate of critic network and l is the iteration number. $\nabla_\theta Q(s_k, \mu(s_k|\theta)|\omega) = \nabla_\theta \mu(s_k|\theta) \nabla_\mu Q(s_k, a_k|\omega)$ is the deterministic gradient value of weights of the actor network.

Meanwhile, two target networks are updated, where the weights of the target networks are updated gradually towards the weights of the main networks. It expresses as

$$\omega^t_{\xi,j+1} = \alpha^t_w \omega^t_{\xi,j} + (1 - \alpha^t_w)\omega_{\xi,j}, \tag{5.57}$$

$$\theta^t_{\xi,j+1} = \alpha^t_w \theta^t_{\xi,j} + (1 - \alpha^t_w)\theta_{\xi,j}, \tag{5.58}$$

where α^t_w and α^t_w are the learning rates. For clear illustration, the DDPG-based tracking algorithm is shown in Algorithm 1.

5.4 Performance Analysis

5.4.1 Detection Probability

Theorem 5.1 *With the communication constraints in (5.7) and (5.8), the probability of detection mode between UAV and USV is maximized, if parameter D_{AS} satisfies the condition of*

$$D_{AS} = \min(\sqrt{C^2_{AS} - h^2}, R_a + R_d). \tag{5.59}$$

The probability of detection mode between USV and UUV is maximized, if parameters D_{SU} and α_{SU} satisfy

$$D_{SU} = \min(C_{US}, 2R_d) \text{ and } \alpha_{SU} = \arcsin\left(\frac{h_p}{D_{SU}}\right). \tag{5.60}$$

Proof For detection mode between UAV and USV, the target detection probability via measure in (5.9) and (5.10) is

$$P(T_p \in B_1 \cup B_2) = \begin{cases} \frac{\pi(D_{AS}+R_a+R_d)}{2(\pi(R_a+R_d)+a+b)}, & (0 < D_{AS} < R_a + R_d) \\ \frac{\pi(R_a+R_d)}{\pi(R_a+R_d)+a+b}, & (D_{AS} \geq R_a + R_d) \end{cases} \tag{5.61}$$

From (5.7) and (5.61), we can see $D_{AS} \geq R_a + R_d$ is greater than the values of the others, if $C_{AS} > \sqrt{D^2_{AS} - h^2}$ is satisfied. Therefore, the detection probability for detection mode between UAV and USV can reach maximum when $D_{\rm AS} = \min(\sqrt{C_{\rm AS}^2 - h^2}, R_{\rm a} + R_{\rm d})$.

For detection mode between USV and UUV, if $D_{\rm SU} < d_c$ is satisfied, its detection probability is denoted as

$$
\begin{aligned}
& P_{dect}(h_p, D_{\rm SU}, \alpha_{\rm SU}; T_{\rm p} \in \bar{E}_{\rm US}) \\
&= \begin{cases} \frac{\pi \cdot \sqrt{R_d^2 - h_p^2}}{\pi d_c + a + b}, & \text{if } 0 \le h_p < h_u - R_d \text{ and } h_u - R_d > 0 \\ \triangle_3, & \text{if } \max(h_u - R_d, 0) \le h_p \le R_d \\ & \quad \text{and } h_u - R_d < R_d \\ \triangle_2, & \text{if } \max(R_d, h_u - R_d) < h_p \le h_u + R_d \\ 0 & \text{else} \end{cases}
\end{aligned}
\tag{5.62}
$$

where $\triangle_3 = \frac{\pi(D_{\rm SU}\cos(\alpha_{\rm SU}) + r_{ds} + r_{du})}{2(\pi d_c + a + b)}$. With (5.62) and (5.22), the target detection probability for USV-UUV detection reaches maximum if $D_{\rm SU} = \min(C_{\rm US}, 2R_{\rm d})$ and $\alpha_{\rm SU} = \arcsin \frac{h_p}{D_{\rm SU}}$. □

5.4.2 Topology Connection and Convergence Analysis

Theorem 5.2 *Consider a multi-step position prediction strategy under communication constraints in (5.7) and (5.8). If the following conditions are satisfied, the possible positions of UAV, USV and UUV are guaranteed to be connected, i.e.,*

$$\beta_a = k_{ta}/(V_{a,\max}T), 0 < k_{ta} \le \frac{V_{a,\max}T}{2\delta_\#}, \tag{5.63}$$

$$\beta_s = k_{ts}/(V_{su,\max}T), 0 < k_{ts} \le \frac{V_{su,\max}T}{2\delta_\#}, \tag{5.64}$$

$$\beta_u = k_{tu}/(V_{su,\max}T), 0 < k_{tu} \le \frac{V_{su,\max}T}{2\delta_\#}. \tag{5.65}$$

Proof In order to assess the connectivity between UAV and USV, we compute the distance $D_{\rm AS}$ between their possible positions. Referring to (5.31), we have

$$
\begin{aligned}
\left\|\eta_{\rm p,a} - \bar{\mathbf{p}}_{\rm s}\right\| &\le d_{\rm as} + \delta_\#(\left\|\mathbf{F}_{\rm c,1}\right\| + \left\|\mathbf{F}_{\rm t,1}\right\|) \\
&= d_{\rm as} + \delta_\#(k_{\rm ca} + k_{\rm ta}) \\
&= d_{\rm as} + \delta_\#\beta_{\rm a}(d_{\rm as} - R_{\rm sa}) + \delta_\# k_{\rm ta}.
\end{aligned}
\tag{5.66}
$$

Set $\beta_{\rm a} = k_{\rm ta}/(V_{\rm a,max}T)$, thus (5.66) can be rearranged as

$$
\begin{aligned}
\left\|\eta_{\rm p,a} - \bar{\mathbf{p}}_{\rm s}\right\| &\le d_{\rm as} + k_{\rm ta}\delta_\# \left(\frac{d_{\rm as} - R_{\rm sa} + V_{\rm a,max}T}{V_{\rm a,max}T}\right) \\
&\le d_{\rm as} + 2k_{\rm ta}\delta_\#.
\end{aligned}
\tag{5.67}
$$

From (5.67), if $k_{\mathrm{ta}} \leq \frac{V_{\mathrm{a,max}}T}{2\delta_{\#}}$, we can further have

$$\left\|\eta_{\mathrm{p,a}} - \bar{\mathbf{p}}_{\mathrm{s}}\right\| \leq d_{\mathrm{as}} + V_{\mathrm{a,max}}T \leq \sqrt{C_{\mathrm{AS}}^2 - h^2}. \tag{5.68}$$

Therefore, the possible locations $\eta_{\mathrm{p,a}}$ of UAV are within the maximum communication range.

Similar to (5.68), we calculate the distance D_{SU} between the possible positions of USV and UUV. When $k_{\mathrm{ts}} \leq \frac{1}{2\delta_{\#}}V_{\mathrm{su,max}}T$, we can get

$$\begin{aligned}
\left\|\eta_{\mathrm{p,s}} - \bar{\mathbf{p}}_{\mathrm{u}}\right\| &\leq d_{\mathrm{us}} + \delta_{\#}(\left\|\mathbf{F}_{\mathrm{c,2}}\right\| + \left\|\mathbf{F}_{\mathrm{t,2}}\right\|) \\
&= d_{\mathrm{us}} + \delta_{\#}(k_{\mathrm{cs}} + k_{\mathrm{ts}}) \\
&= d_{\mathrm{us}} + k_{\mathrm{ts}}\delta_{\#}(\frac{d_{\mathrm{us}} - R_{\mathrm{su}} + V_{\mathrm{su,max}}T}{V_{\mathrm{su,max}}T}) \\
&\leq d_{\mathrm{us}} + V_{\mathrm{su,max}}T \leq C_{\mathrm{SU}}.
\end{aligned} \tag{5.69}$$

Similarly, if $k_{\mathrm{tu}} \leq \frac{1}{2\delta_{\#}}V_{\mathrm{su,max}}T$, we have

$$\begin{aligned}
&\left\|\eta_{\mathrm{p,u}} - \bar{\mathbf{p}}_{\mathrm{s}}\right\| \leq d_{\mathrm{us}} + \delta_{\#}(\left\|\mathbf{F}_{\mathrm{c,3}}\right\| + \left\|\mathbf{F}_{\mathrm{t,3}}\right\|) \\
&= d_{\mathrm{us}} + \delta_{\#}(k_{\mathrm{cu}} + k_{\mathrm{tu}}) \\
&= d_{\mathrm{us}} + k_{\mathrm{tu}}\delta_{\#}(\frac{d_{\mathrm{us}} - R_{\mathrm{su}} + V_{\mathrm{su,max}}T}{V_{\mathrm{su,max}}T}) \\
&\leq d_{\mathrm{us}} + V_{\mathrm{su,max}}T \leq C_{\mathrm{SU}}.
\end{aligned} \tag{5.70}$$

Therefore, the possible locations $\eta_{\mathrm{p,s}}$ of USV and $\eta_{\mathrm{p,u}}$ of UUV are within the maximum communication range, i.e., the connectivity in (5.7) and (5.8) can be guaranteed. □

Property 5.1 (Lipschitz continuity) *For any real-valued function f with strong convexity, there exists a constant $L_1 > 0$ such that inequality holds for any x and y:*

$$f(y) \leq f(x) + \nabla f(x)(y - x) + \frac{L_1}{2}|y - x|^2, \forall x, y \tag{5.71}$$

Property 5.2 (Strong convexity) *If the function f is strongly convex, then for any two points x and y, we have*

$$f(y) \geq f(x) + \nabla f(x)(y - x) + \frac{\varpi_1}{2}|y - x|^2, \forall x, y \tag{5.72}$$

where ϖ_1 is the strongly convex constant of the function f.

Based on this, we have the following theorem.

Theorem 5.3 *Consider the DDPG-based tracking scheme for UAV, USV and UUV. The convergence of the critic networks can be achieved if the following conditions are satisfied, i.e.,*

$$N \geq \log_{\frac{1}{1-\varpi_1 L_1^{-1}}} \varepsilon^{-1}(Q_\xi(\omega_{\xi,j}) - Q_\xi(\omega_{\xi,0})), \tag{5.73}$$

where $0 < \frac{\varpi_1}{L_1} < 1$ *and* $L_1 > 0$. *The objective function possesses a Lipschitz continuous gradient on* R_n, *and* $Q_\xi(\omega_{\xi,0})$ *is defined as the global optimum function of the target critic network. Additionally, there exists an interval near the global optimum of the function, such that if* f *is convex within this interval, it will lead the objective function of the critic network to converge towards the global optimum of the critic network.*

Proof The objective function satisfies L-Lipschitz continuity. For $L_1 > 0$, the objective function f possesses a gradient that is L-Lipschitz continuous. Let $\delta_\mathrm{j} = \gamma Q_\xi(\omega_{\xi,0}) + c_\xi - Q_\xi(\omega_{\xi,j})$ denote the error of the critic network with respect to the target network after the jth iteration. Correspondingly, after $j+1$ iterations, the error is given by $\delta_{\mathrm{j}+1} = \gamma Q_\xi(\omega_{\xi,0}) + c_\xi - Q_\xi(\omega_{\xi,j+1})$. This pattern can be extended to subsequent iterations. The iteration error between the jth and $(j+1)$th iterations is represented as $\delta_j - \delta_{j+1}$.

By Property 5.1, the following inequality can be derived

$$\begin{aligned} & \delta_j - \delta_{j+1} \\ = \ & Q_\xi(\omega_{\xi,j+1}) - Q_\xi(\omega_{\xi,j}) \\ \leq \ & \nabla_\omega Q_\xi(\omega_{\xi,j+1})(\omega_{\xi,j+1} - \omega_{\xi,j}) + \frac{L_1}{2} \left\| \omega_{\xi,j+1} - \omega_{\xi,j} \right\|^2 \\ = \ & \frac{(\delta_j+1)^2-1}{2L_1} \left\| \nabla_\omega Q_\xi(\omega_{\xi,j}) \right\|^2. \end{aligned} \tag{5.74}$$

Set $\alpha_\omega = \frac{1}{L_1}$, and hence it yields $\delta_j - \delta_{j+1} \leq \frac{(\delta_j+1)^2-1}{2L_1} \left\| \nabla_\omega Q_\xi(\omega_{\xi,j}) \right\|^2$. With Property 5.2, one has

$$\begin{aligned} & Q_\xi(\omega_{\xi,j+1}) - Q_\xi(\omega_{\xi,j}) \\ & \geq \nabla_\omega Q_\xi(\omega_{\xi,j})(\omega_{\xi,j+1} - \omega_{\xi,j}) + \frac{\varpi_1}{2} \left\| \omega_{\xi,j+1} - \omega_{\xi,j} \right\|^2 \\ & \geq (\alpha_\omega \delta_j + \frac{\varpi_1}{2} {\alpha_\omega}^2 {\delta_j}^2) \left\| \nabla_\omega Q_\xi(\omega_{\xi,j}) \right\|^2. \end{aligned} \tag{5.75}$$

From (5.75), one has

$$Q_\xi(\omega_{\xi,0}) - Q_\xi(\omega_{\xi,j}) \leq \frac{(\delta_j+1)^2-1}{2L_1} \left\| \nabla_\omega Q_\xi(\omega_{\xi,j}) \right\|^2. \tag{5.76}$$

From (5.76), we have

$$Q_\xi(\omega_{\xi,0}) - Q_\xi(\omega_{\xi,j}) \geq \frac{(\delta_j+1)^2-1}{2\varpi_1} \left\| \nabla_\omega Q_\xi(\omega_{\xi,j}) \right\|^2. \tag{5.77}$$

From (5.75), one can obtain

$$
\begin{aligned}
& Q_\xi(\omega_{\xi,j+1}) - Q_\xi(\omega_{\xi,0}) \\
\leq\ & Q_\xi(\omega_{\xi,j}) - Q_\xi(\omega_{\xi,0}) + \frac{(\delta_j+1)^2-1}{2L_1}\left\|\nabla_\omega Q_\xi(\omega_{\xi,j})\right\|^2 \\
\leq\ & (1-\frac{\varpi_1}{L_1})(Q_\xi(\omega_{\xi,j}) - Q_\xi(\omega_{\xi,0})). \qquad (5.78)
\end{aligned}
$$

Considering the extended update sequence $(s_i, a_i, s_{i+1}, a_{i+1}, ...)$, one can deduce $Q_\xi(\omega_{\xi,j+1}) - Q_\xi(\omega_{\xi,0}) \leq (1-\frac{\varpi_1}{L_1})^{K+1}(Q_\xi(\omega_{\xi,j-K}) - Q_\xi(\omega_{\xi,0}))$ for $1-\frac{\varpi_1}{L_1} < 1$, i.e., the critic network can converge to the target critic network. □

Corollary 5.1 *After $Q_\xi(\omega_{\xi,j+1})$ converges to the global optimum, actor network can achieve convergence if the following condition is satisfied, i.e., for $0 < \frac{\varpi_1}{L_1} < 1$*

$$
Q_\xi(\theta_{\xi,j+1}) - Q_\xi(\theta_{\xi,0}) \leq (1-\frac{\varpi_1}{L_1})(Q_\xi(\theta_{\xi,j}) - Q_\xi(\theta_{\xi,0})). \qquad (5.79)
$$

where $Q_\xi\left(\theta_{\xi,0}\right)$ is the estimated value after the executing of $\mu_\xi(\cdot|\theta_{\xi,0})$. The objective function possesses a Lipschitz continuous gradient on R_n, and $\mu_\xi(\cdot|\theta_{\xi,0})$ is the global optimum. Additionally, there exists an interval near the global optimum of the function, such that the objective function of the actor network can converge towards the global optimum of the actor network if f is convex within this interval.

Proof One defines $\delta_j^\mu = Q_\xi\left(\theta_{\xi,j+1}\right) - Q_\xi\left(\theta_{\xi,j}\right)$ as the error of the actor network after jth iterations. By Properties 5.1 and 5.2, the following inequality can be derived

$$
Q_\xi\left(\theta_{\xi,j+1}\right) - Q_\xi\left(\theta_{\xi,j}\right) \leq \tfrac{-1}{2L_1}\left\|\nabla_\theta Q_\xi(\theta_{\xi,j})\right\|^2, \qquad (5.80)
$$

$$
Q_\xi\left(\theta_{\xi,j+1}\right) - Q_\xi\left(\theta_{\xi,j}\right) \geq \tfrac{-1}{2\varpi_1}\left\|\nabla_\theta Q_\xi(\theta_{\xi,j})\right\|^2. \qquad (5.81)
$$

One can deduce $Q_\xi(\theta_{\xi,j+1}) - Q_\xi(\theta_{\xi,0}) \leq (1-\frac{\varpi_1}{L_1})^{K+1}(Q_\xi(\theta_{\xi,j-K}) - Q_\xi(\theta_{\xi,0}))$ for $1-\frac{\varpi_1}{L_1} < 1$. From (5.51), the optimal actor network is to minimize the optimal critic network. When $Q_\xi(\theta_{\xi,j+1})$ converges to the global optimum critic network $Q_\xi\left(\theta_{\xi,0}\right)$, the actor network converges to the global optimal actor network. That completes the proof. □

5.5 Technology Practice

In simulation part, we assume that the detection space is limited, e.g., $a = 8000$ m, $b = 12000$ m and $c = 10000$ m. The maximum velocity vectors are $\dot{\mathbf{p}}_{\mathrm{a,max}} = [3, 3, 0]^{\mathrm{T}}$ m/s, $\mathbf{v}_{\mathrm{s,max}} = [2.5, 2.5, \pi/2]^{\mathrm{T}}$ m/s and $\mathbf{v}_{\mathrm{u,max}} = [2, 2, 2, \pi/2]^{\mathrm{T}}$ m/s. The reliable communication distances are $C_{\mathrm{AS}} = 900$ m and $C_{\mathrm{US}} = 420$ m.

In experimental part, the hardware system is mainly composed of five components: (1) *UAV*. The quadcopter drone utilizes an open-source Pixhawk4 software and hardware controller, which is equipped with visual sensor, wireless communication module, image transmission module, and onboard Raspberry Pi computer for mission processing; (2) *USV*. The USV is equipped with NVIDIA/STM32 microcontrollers for task execution, P360 underwater detection sonar for 360° detection, and visual sensors; (3) *UUV*. UUV incorporates NVIDIA microcontrollers, and it is also fitted with a P360 detection sonar; (4) *Communication system*. UAV and USV are equipped with wireless modules for communication for a communication range of over 1 km, and waterborne acoustic communication devices for a communication range of over 500 m; (5) *Localization system*. UUV utilizes Ultra-Short Baseline (USBL) system for underwater localization. In addition, UAV employs a GNSS (global navigation satellite system) for self-localization.

5.5.1 Case Study I: Detection and Tracking Control for a Surface Target

(1) *Detection and tracking solution for a surface target*

In Sect. 5.3.1, the detection mode for a surface target is proposed. The detection radii of UAV and USV are set as $R_a = 500$ m and $R_d = 200$ m, respectively. The initial positions of UAV and USV are randomly distributed within the detection space, while the horizontal distance D_{AS} is set as 600 m. The time step is 1 s and the moving time is 400 s. Accordingly, 10000 times of random detection are conducted by using the linear detection strategy and the curved detection strategy, where the number of the effective detection is shown in Fig. 5.10a. Clearly, the effective detection number with the linear detection strategy is larger than the curved detection strategy, which is coincide with the conclusion in Sect. 5.3.1. After that, we employ the linear detection strategy to detect the surface target. The probabilities of success detection with the variation of D_{AS} are shown in Fig. 5.10b, which are coincide with the theoretical values in (5.13).

After the water surface target is detected, a DDPG-based tracking algorithm is designed for UAV, USV, and UUV in Sect. 5.3.2. To verify its effectiveness, the position vector of surface target is set as $\eta_t = [250, 1380, 0]^T$. Correspondingly, the final desired position vectors are given as follows: $\eta_{D,UAV} = [250, 1380, 150]^T$, $\eta_{D,USV} = [250, 1380, 0]^T$ and $\eta_{D,UUV} = \eta_t$. Based on this, we utilize (5.31), (5.38) and (5.44) to predict the possible locations of the UAV, USV and UUV at time step $k + T$, respectively. We define $T = n\delta$, where n is set to be 10. The aim of the above implementation is to maintain the communication connectivity throughout the tracking process. Based on this, the maximum safety communication distances $R_{sa} = \sqrt{C_{AS}^2 - h^2 - 2V_{a,\max} n\delta\delta_{\#}}$ and $R_{su} = C_{SU} - 2V_{su,\max} n\delta\delta_{\#}$ can be obtained to calculate the parameters β_a, β_s and β_s by (5.63), (5.64) and (5.65). Accordingly, the motion trajectories of UAV, USV and UUV are shown by Fig. 5.11a. The state vec-

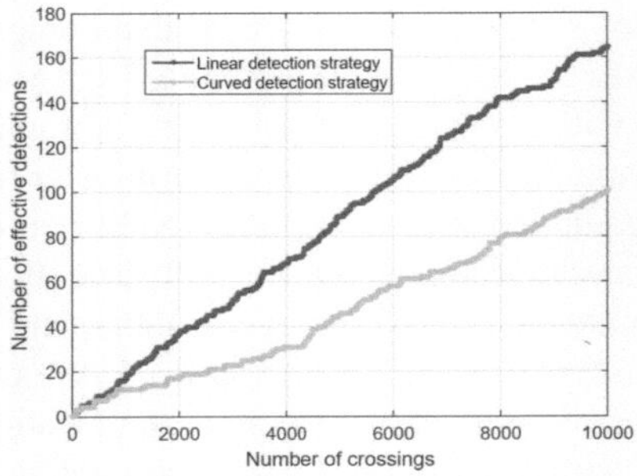

(a) The frequency of random detection of straight and curved lines.

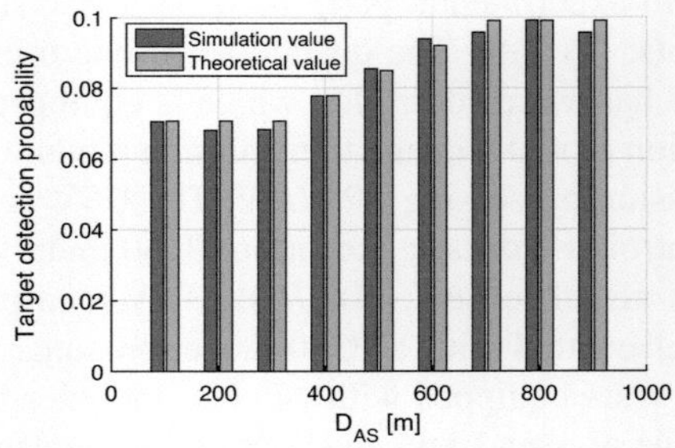

(b) Variation of detection probability with distance D_{AS}.

Fig. 5.10 Detection of surface target with UAV and USV

tors of UAV, USV and UUV are shown in Fig. 5.11b–d, respectively. To show more clearly, the tracking errors d_{ta} of UAV, d_{ts} of USV and d_{tu} of UUV are presented in Fig. 5.11e. Clearly, the tracking errors can eventually converge to zero, which fully validate the feasibility of our tracing algorithm. In addition, the communication distance error between UAV and USV is defined as $e_{as} = \|\bar{\mathbf{p}}_a - \bar{\mathbf{p}}_s\|$, while the communication distance error between USV and UUV is defined as $e_{su} = \|\bar{\mathbf{p}}_s - \bar{\mathbf{p}}_u\|$. Then, the changes of e_{as} and e_{su} are shown in Fig. 5.11f. It is observed that the communication distance errors satisfy the constraints in (5.7) and (5.8), which can verify the effectiveness of the multi-step position prediction strategy.

Besides that, 500 times of iteration are carried out to ensure the convergence of the loss function. As a result, the optimal control strategies τ_a^*, τ_s^* and τ_u^* for UAV, USV and UUV can be shown by Fig. 5.12a–c, respectively. Correspondingly, the convergence processes for the reward functions are presented in Fig. 5.12d. We can find that the values can converge to the optimization.

(2) Comparison with other methods for surface target

In our solution, a multi-step location prediction strategy is incorporated into the tracking procedure, which can guarantee the communication connectivity among UAV, USV and UUV. To verify its merit, we remove the multi-step location prediction strategy from the tracking procedure. Based on this, the trajectories of UAV, USV and UUV are shown in Fig. 5.13a, while the relative communication distances among UAV, USV and UUV are presented in Fig. 5.13b. We can find that the communication connectivity between UAV and USV cannot always be guaranteed since $e_{as} > C_{AS}$ occurs.

Note that the DQN method (e.g., Wei et al., 2023 and Lillicrap et al., 2016) cannot be directly applied to continuous state space since it finds action that maximizes the action value function. In view of this, the DQN method is employed to achieve the tracking task. Accordingly, the tracking errors with DQN method are shown in Fig. 5.13c, and the reward convergence process is depicted by Fig. 5.13d. Besides that, we adopt the Q-learning method to achieve the tracking task, where the state

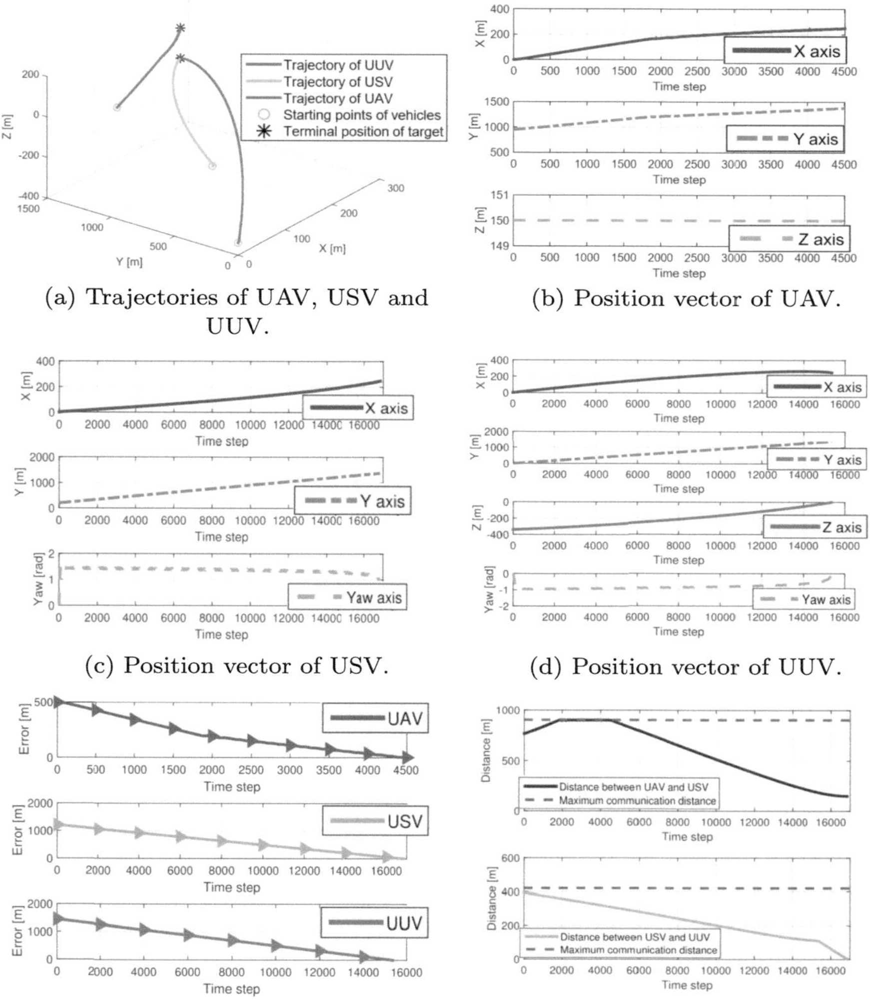

(a) Trajectories of UAV, USV and UUV.

(b) Position vector of UAV.

(c) Position vector of USV.

(d) Position vector of UUV.

(e) Tracking errors.

(f) Communication distances.

Fig. 5.11 Simulation results for the DDPG-based tracking algorithm when a surface target is detected

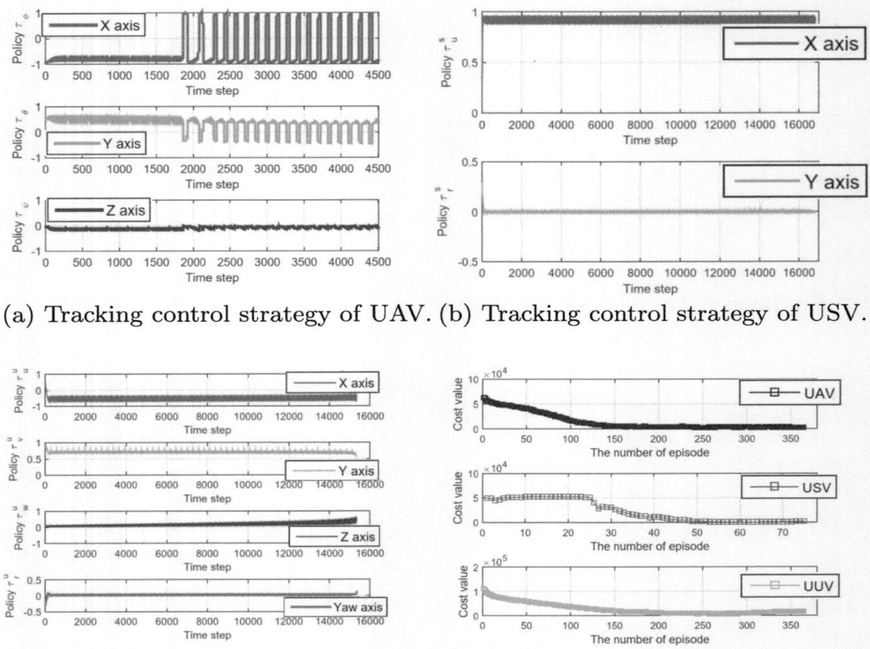

(a) Tracking control strategy of UAV. (b) Tracking control strategy of USV.

(c) Tracking control strategy of UUV. (d) Cost convergence process.

Fig. 5.12 Simulation results for the control strategies and reward convergence process when a surface target is detected

and action spaces are discretized. Then, the tracking errors d_{ta}, d_{ts} and d_{tu} are presented in Fig. 5.13e. We also compare the time consumptions of the UAV, USV and UUV by using the Q-learning, DQN and DDPG tracking algorithms, which can be shown in Fig. 5.13f. From Fig. 5.13a–f, it is evident that the DDPG exhibits superior performance compared to the other algorithms in terms of ensuring communication connectivity and handling continuous states and actions.

(1) Experimental results for tracking a surface target.

In this case, the experimental setup is depicted by Fig. 5.14a, where the pump is considered as the target and it has a large outline shape. Particularly, we use OpenCV to get the outline of the target, and then we turn it into a rectangle. After that, the size of the diagonal for the above rectangular frame can be calculated, and hence, the threshold of the target size is obtained on demand. When the measured size of outline shape is larger than the threshold, the target can be regarded to be successfully detected by UAV. The position vector of target is set as $\eta_t = [22, -24, 0]^{\text{T}}$. The final desired position vectors of UAV, USV and UUV are set as follows: $\eta_{\text{D,UAV}} = [22, -24, 30]^{\text{T}}$, $\eta_{\text{D,USV}} = [22, -24, 0]^{\text{T}}$ and $\eta_{\text{D,UUV}} = \eta_t$. Based on this, the UAV initially operates at an altitude of 30 m by employing visual detection sensor. Meanwhile, the position of the USV is moved to ensure that the horizontal distance D_{AS} remains within a

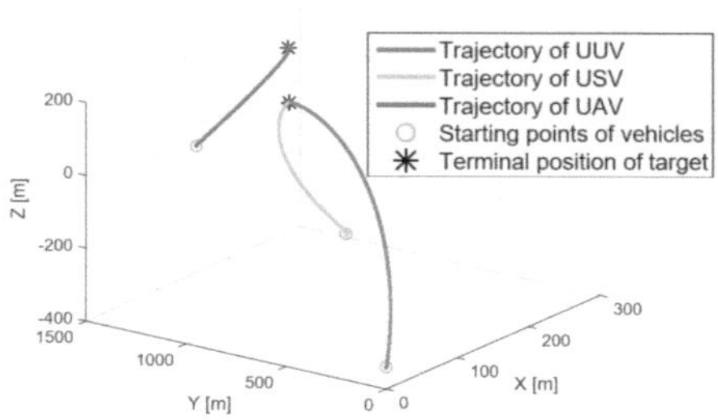

(a) Trajectories of UAV, USV and UUV.

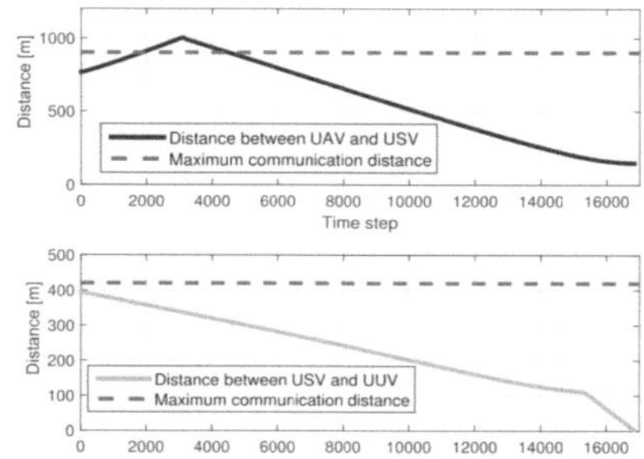

(b) Communication distances.

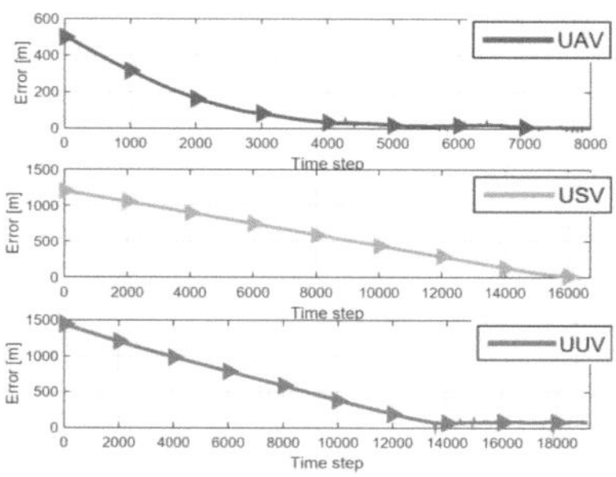

(c) Tracking error based on DQN.

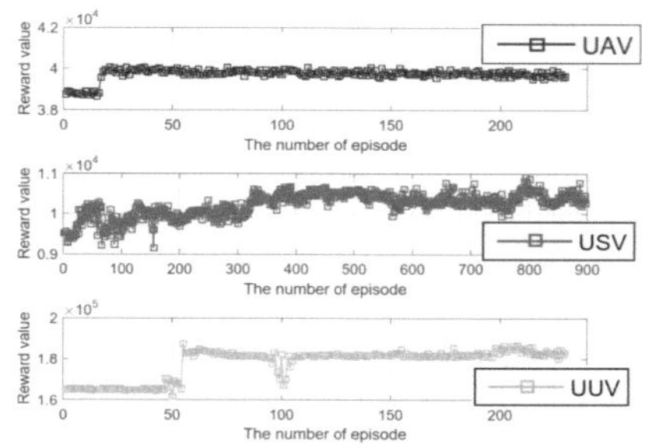

(d) Reward convergence process.

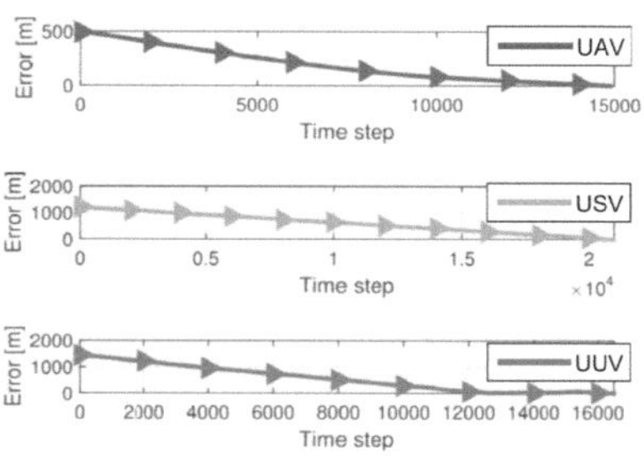

(e) Tracking error via QL.

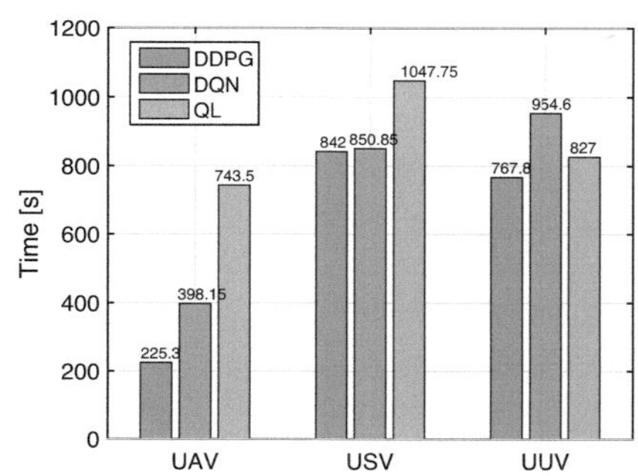

(f) Time comparison.

Fig. 5.13 Comparison with the other algorithms for surface target

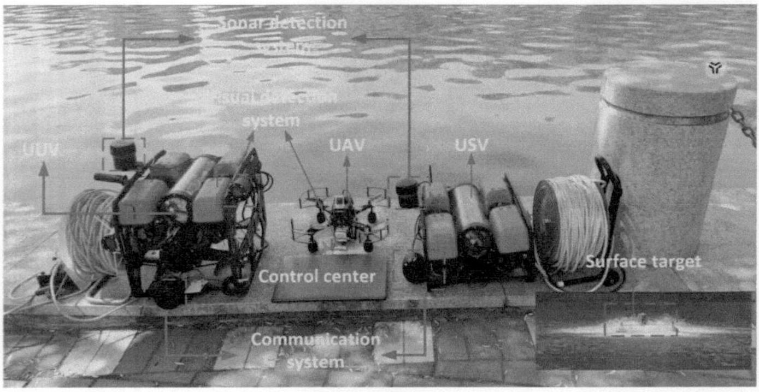

(a) Experimental setup for surface target.

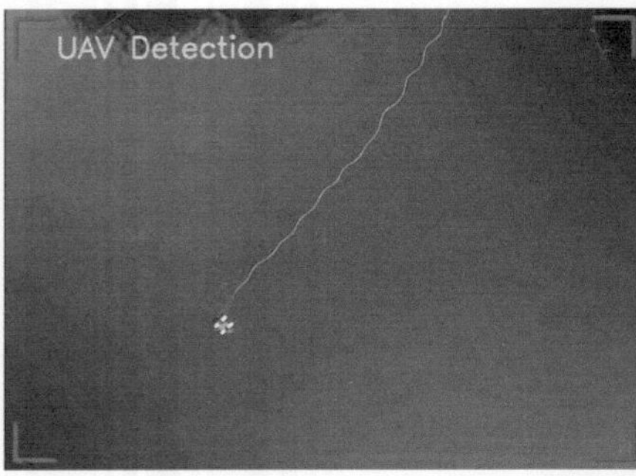

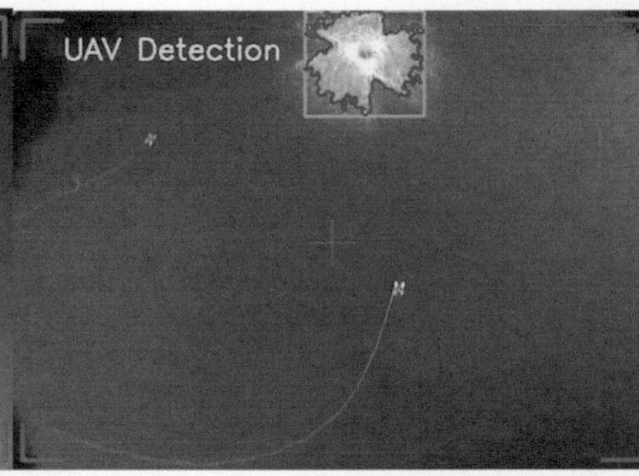

(b) Initial detection image of UAV. (c) Detected visual image by UAV.

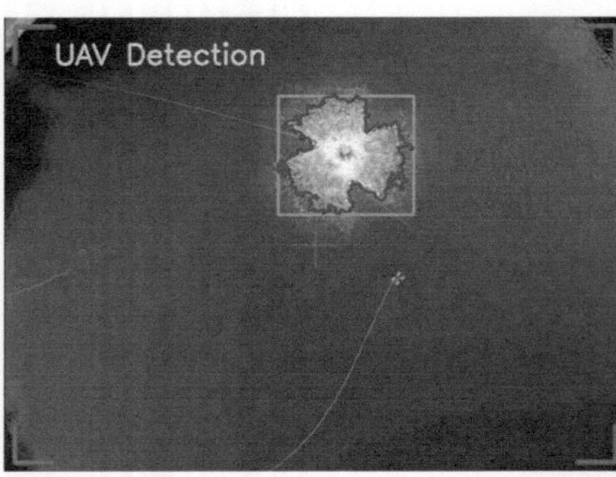

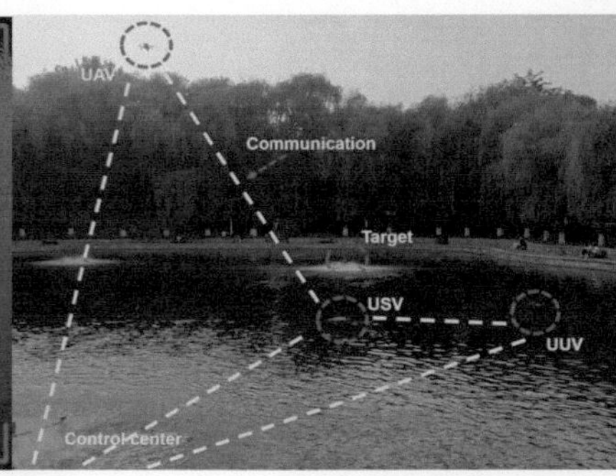

(d) The approaching process of UAV after the target is detected. (e) Cooperation detection process of UAV, USV and UUV.

Fig. 5.14 Experimental setup and results for detecting surface target range of 10 meters with UAV. Subsequently, UUV operates at a depth of 1 m and maintains a distance D_{SU} of 20 m with USV. Furthermore, UAV communicates data with the control center via wireless module, whereas the USV and UUV transmit signals back to the control center through cables. For clearer verification, we adopt a linear motion detection strategy to detect the surface target. At the beginning of the detection, the visual image captured by the UAV is illustrated in Fig. 5.14b. In addition, image processing and recognition units are employed to identify the water surface target. When the surface target is detected by UAV detection of visual, the result of visual detection is shown in Fig. 5.14c. It is clear that the final results of the USV and UUV reach the desired position, as shown by Fig. 5.14d, e, respectively. Meanwhile, the location of the target is transmitted to the USV.

After that, the DDPG-based tracking algorithm is preloaded onto USV and UUV by offline training. Particularly, USV receives the target information and then transmits it to the UUV through underwater communication system. The position information is transferred to the controller to generate motion and the trajectories of UAV, USV and UUV. The related trajectories are shown in Fig. 5.15a, b. Meanwhile, the position vectors of UAV, USV, and UUV on X, Y, and Z axes are shown in Fig. 5.15c–e, respectively. Correspondingly, the tracking errors d_{ta}, d_{ts} and d_{tu} are shown in Fig. 5.15f. The above results demonstrate the effectiveness of DDPG-based algorithm for tracking underwater target in this chapter.

5.5.2 *Case Study II: Detection and Tracking Control for an Underwater Target*

(1) Detection and tracking solution for an underwater target

We consider another general scenario, i.e., an underwater target exists in the monitoring region. Then, the detection mode for underwater target (see Sect. 5.3.1) is adopted to achieve the detection task. Specifically, the initial positions of USV and UUV are randomly distributed, while the horizontal distance D_{SU} is set as 400 m. Additionally, the target is randomly deployed under the water surface within the detection space. Based on this, we conduct five groups of experiment, while each group includes 10000 times of random detection by using the linear detection strategy and the curved detection strategy. Correspondingly, the number of the effective detection can be shown in Fig. 5.16a, where the linear detection strategy has a better detection performance as compared to the curved detection strategy. Furthermore, the formation angle is set as $\alpha_{SU} = \pi/3$, and hence, the variation of detection probability with depth is presented in Fig. 5.16b, which is coincide with the theoretical value in (5.22). In order to verify the influence of formation angle α_{SU} on the detection probability, we conduct experiments by setting different values of α_{SU}, e.g., $\alpha_{SU} = 0$, $\alpha_{SU} = \pi/6$, $\alpha_{SU} = \pi/3$ and $\alpha_{SU} = \pi/2$. By varying the depth h_p, we obtain the corresponding probability values, which are illustrated in Fig. 5.16c. After that, we set the formation angle as $\alpha_{SU} = \pi/6$, and hence, the variation of detec-

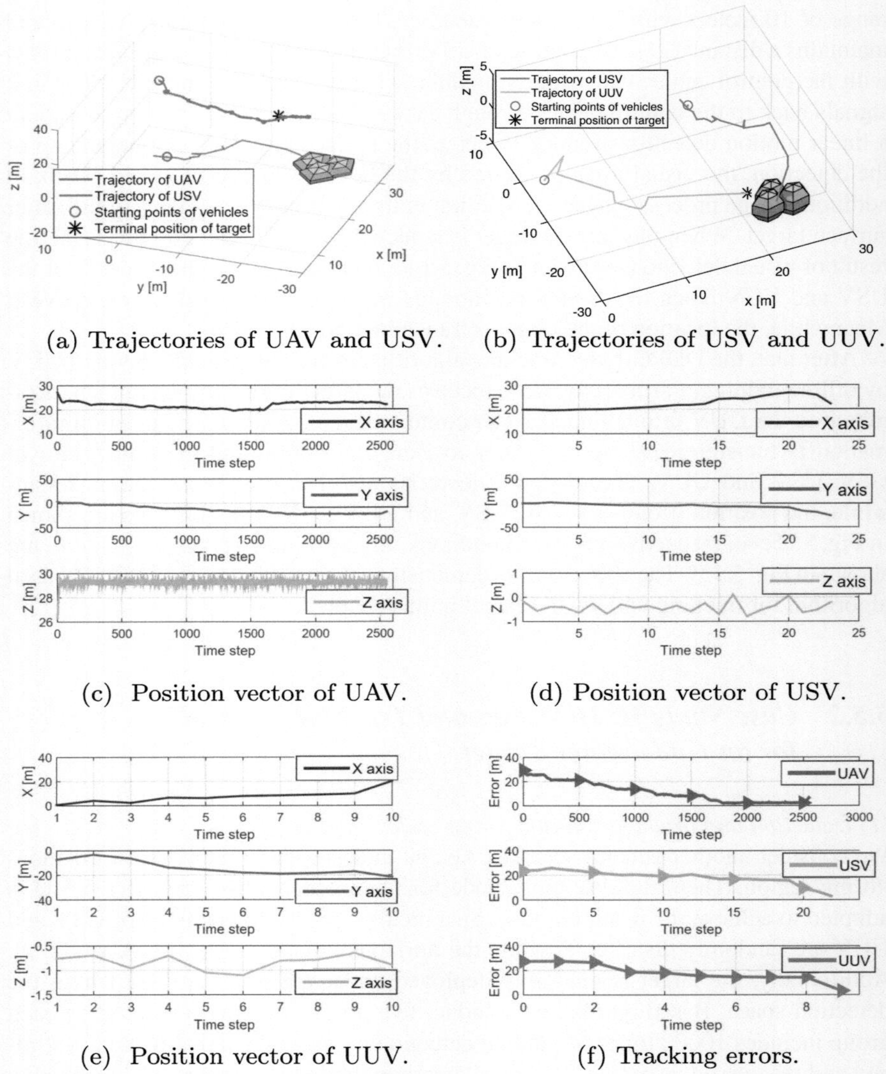

(a) Trajectories of UAV and USV. (b) Trajectories of USV and UUV.

(c) Position vector of UAV. (d) Position vector of USV.

(e) Position vector of UUV. (f) Tracking errors.

Fig. 5.15 Experimental results for tracking surface target

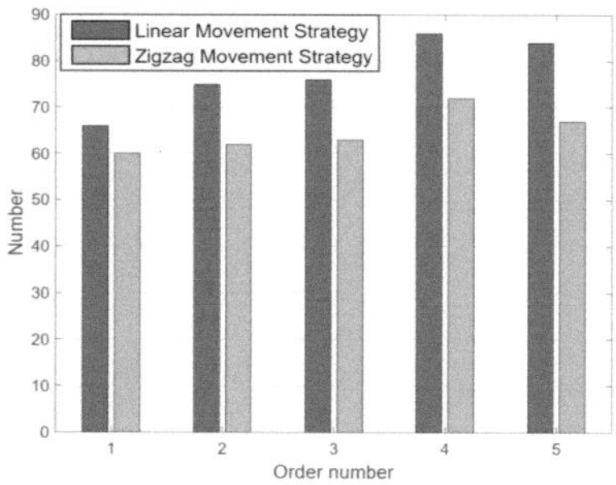

(a) The frequency of random detection of straight and curved lines.

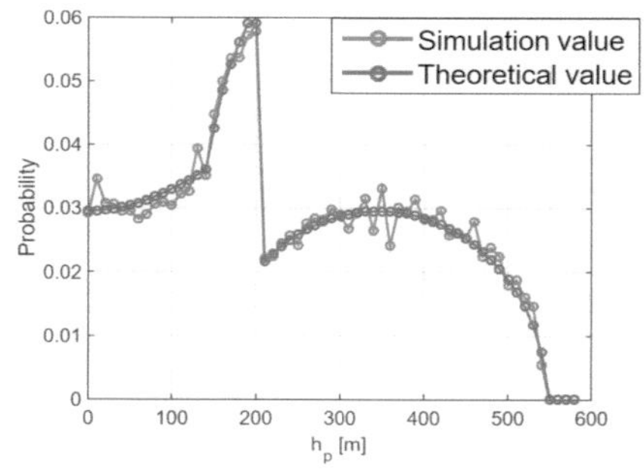

(b) Variation of detection probability with different depth h_p.

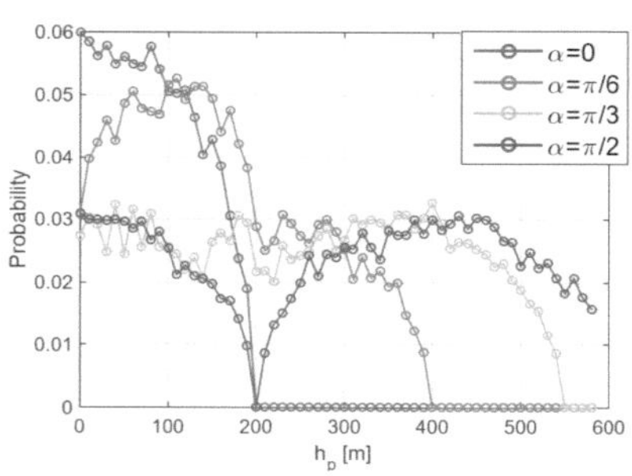

(c) Variation of detection probability with formation angle $\alpha_{\rm SU}$.

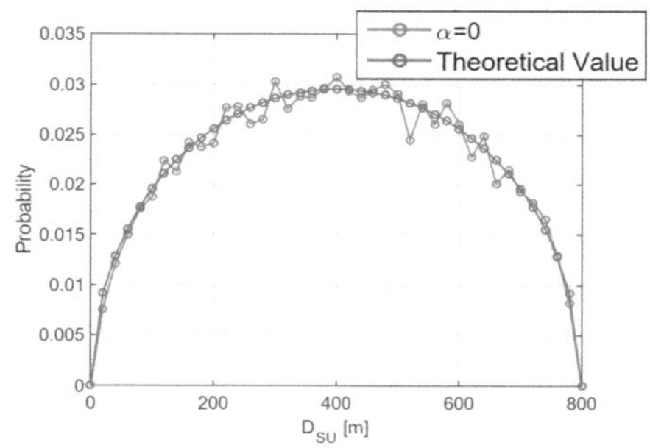

(d) Variation of detection probability with distance $D_{\rm SU}$.

Fig. 5.16 Detection of underwater target with USV and UUV

tion probability with distance $D_{\rm SU}$ is shown in Fig. 5.16d. Clearly, these results are coincide with the theoretical results in Sect. 5.3.1.

To verify the effectiveness of DDPG-based tracking algorithm for underwater target, the position vector of surface target is set as $\eta_{\rm t} = [15, -318, -300]^{\rm T}$. The final desired position vectors are given as follows: $\eta_{\rm D,UAV} = [15, -318, 150]^{\rm T}$, $\eta_{\rm D,USV} = [15, -318, 0]^{\rm T}$ and $\eta_{\rm D,UUV} = \eta_t$. Based on this, we utilize multi-step location prediction strategy to obtain the possible locations of UAV, USV and UUV at time step $k + T$, respectively. To this end, the tracking trajectories of the UAV, USV and UUV are shown in Fig. 5.17a, while their state vectors are shown in Fig. 5.17b–d, respectively. Then, the tracking errors with respect to the target are depicted in Fig. 5.17e. Clearly, the tracking errors can eventually converge to zero. Besides that, the communication distances $e_{\rm as}$ and $e_{\rm su}$ are given in Fig. 5.17f, which can guarantee the communication connectivity. Finally, the optimal control strategies $\tau_{\rm a}^*$, $\tau_{\rm s}^*$ and $\tau_{\rm u}^*$ for UAV, USV and UUV are depicted in Fig. 5.19a–c, respectively. The cost convergence process is depicted by Fig. 5.19d. We find our method can ensure the convergence of loss function by continuous iterations (Fig. 5.18).

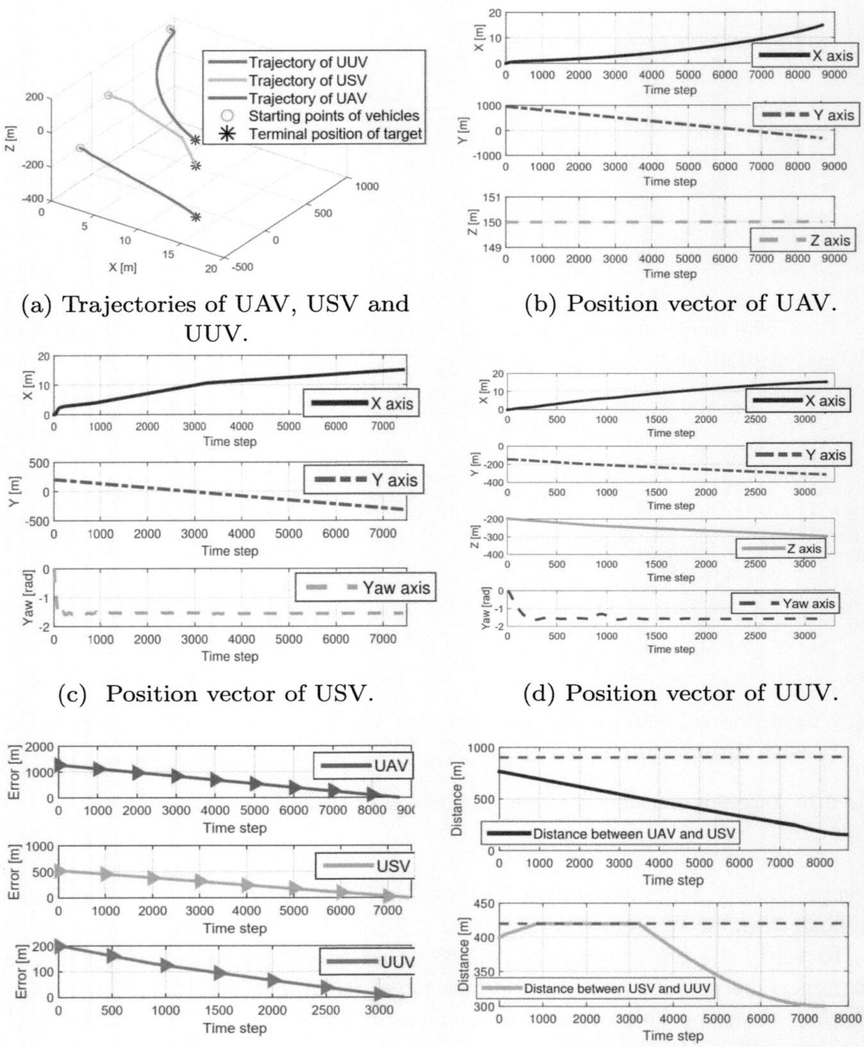

(a) Trajectories of UAV, USV and UUV.
(b) Position vector of UAV.
(c) Position vector of USV.
(d) Position vector of UUV.
(e) Tracking errors.
(f) Communication distances.

Fig. 5.17 Simulation results for the DDPG-based tracking algorithm when an underwater target is detected

(2) Comparison with other methods for underwater target

Similarly, we remove the multi-step location prediction strategy from the tracking procedure. Then, the trajectories of UAV, USV and UUV are presented in Fig. 5.20a. The relative communication distances among UAV, USV and UUV are shown in

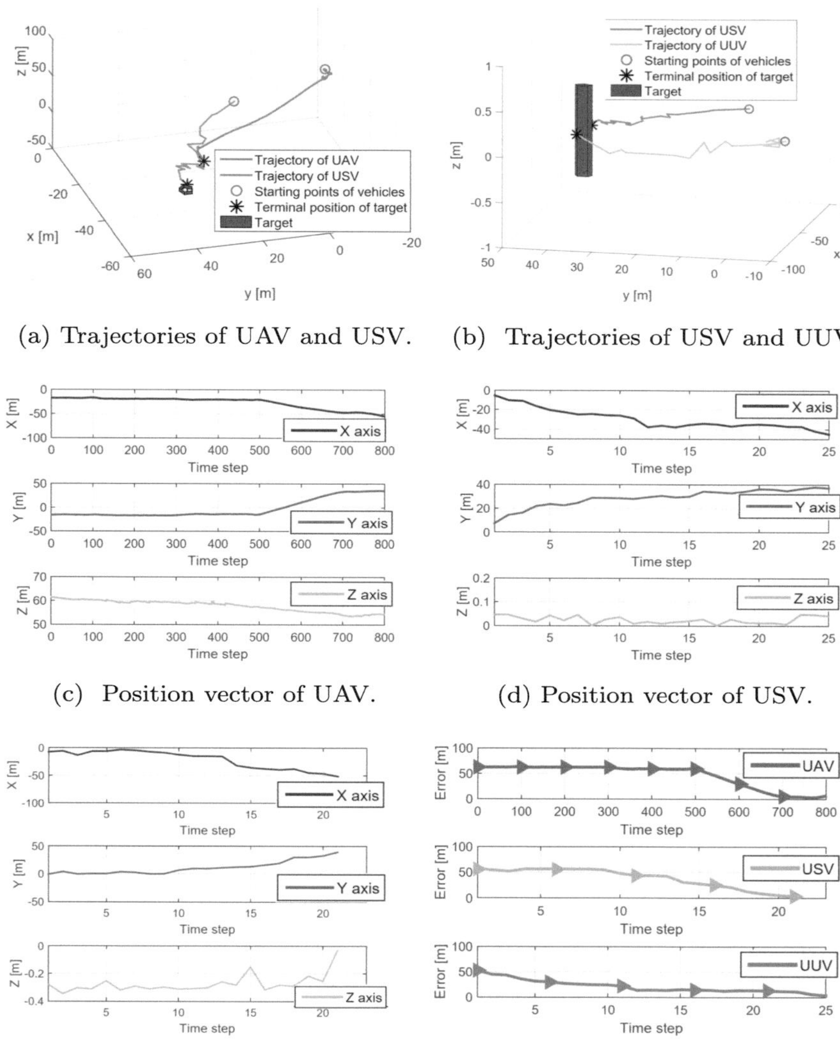

(a) Trajectories of UAV and USV. (b) Trajectories of USV and UUV.

(c) Position vector of UAV. (d) Position vector of USV.

(e) Position vector of UUV. (f) Tracking errors.

Fig. 5.18 Experimental results for tracking underwater target

Fig. 5.20b. It reveals that the value of e_{su} exceeds the maximum communication distance C_{SU}, i.e., the communication connectivity is not guaranteed.

We employ the DQN method (e.g., Wei et al., 2023 and Lillicrap et al., 2016) to achieve the tracking task. Based on this, the tracking errors for UAV, USV and UUV with the DQN method are shown in Fig. 5.20c. In addition, the reward convergence process can be presented by Fig. 5.20d. Along with this, a tracking algorithm based

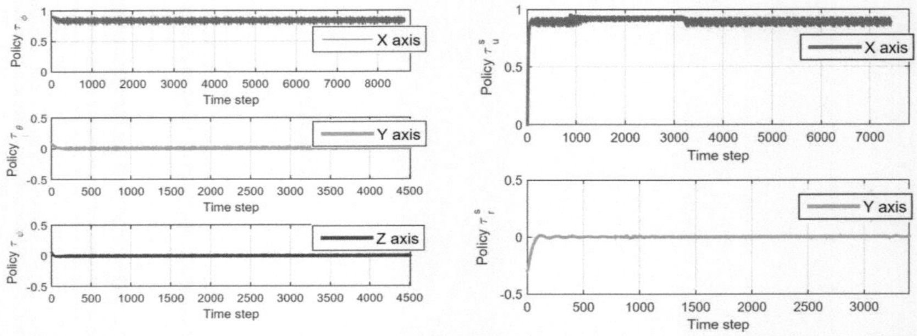

(a) Tracking control strategy of UAV. (b) Tracking control strategy of USV.

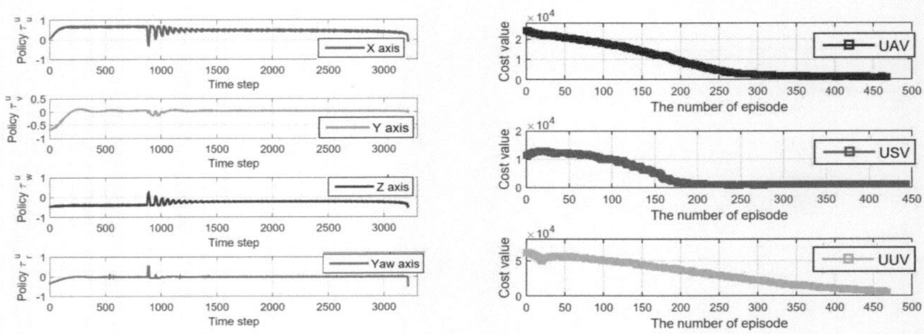

(c) Tracking control strategy of UUV. (d) Cost convergence process.

Fig. 5.19 Simulation results of DDPG-based controller learning for UAV, USV and UUV

on the Q-learning approach is introduced, which utilizes a restricted set of discrete actions for training. Accordingly, the tracking errors d_{ta}, d_{ts} and d_{tu} are shown in Fig. 5.20e. Subsequently, we compare the computational times of the UAV, USV, and UUV by using Q learning, DQN, and DDPG-based tracking algorithms, respectively. The above results are presented in Fig. 5.20f. From Fig. 5.20a–f, we find that DDPG outperforms the other algorithms. Hence, the merit of our underwater tracking algorithm can be demonstrated.

(3) Experimental results for tracking an underwater target.

In this section, we perform experiments to detect and track an underwater target by using a cylindrical object for validation. The experimental setup is depicted by Fig. 5.21a and the position vector of target is $\eta_t = [-48.2, 38.7, -0.5]^{\text{T}}$. Accordingly, the final desired position vectors are shown as follows: $\eta_{\text{D,UAV}} = [-48.2, 38.7, 60]^{\text{T}}$, $\eta_{\text{D,USV}} = [-48.2, 38.7, 0]^{\text{T}}$ and $\eta_{\text{D,UUV}} = \eta_t$. Based on this, UAV performs detection operation at an altitude of 60 meters. The initial position of UUV is within 1 m of depth, while the distance D_{SU} between UUV and USV is set as 20 m. We configure the sonar detection range of USV and UUV to be 10 m. Besides that, a linear motion detection strategy is employed to detect underwater target for UAV, USV and UUV. In view of this, the sonar image (see Fig. 5.21b) represents

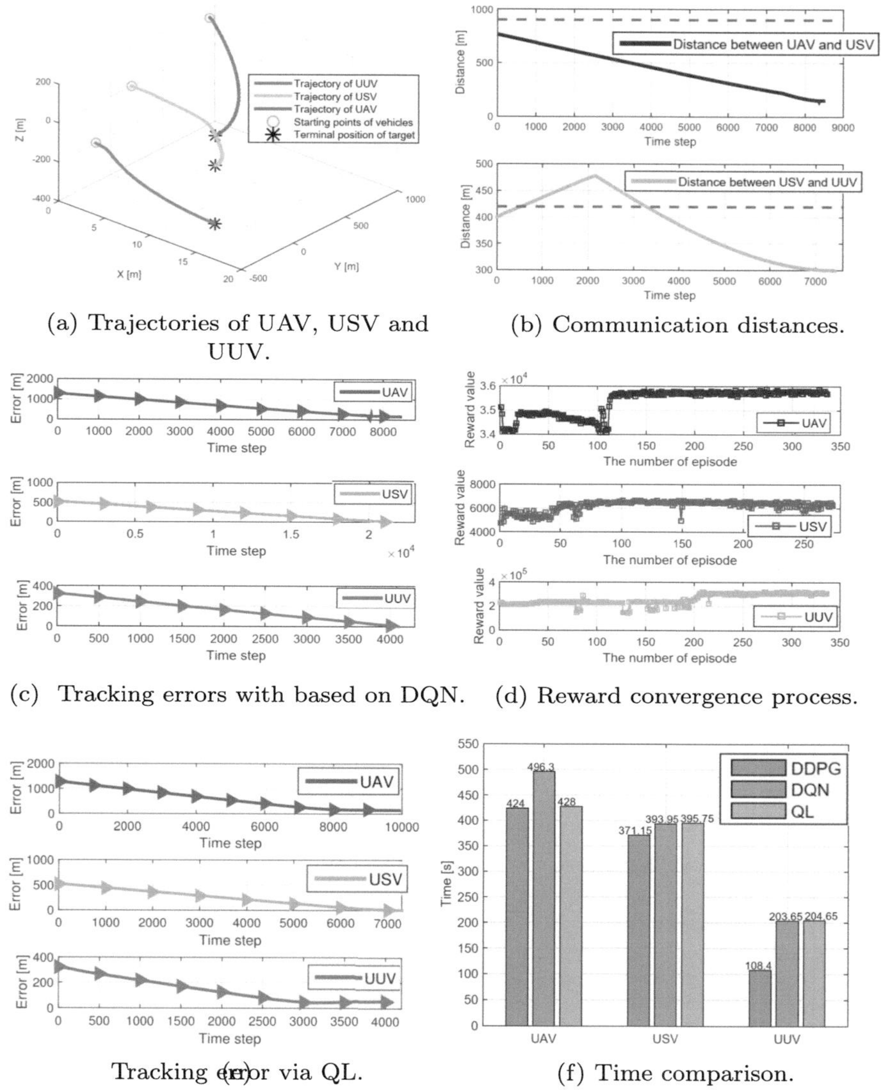

(a) Trajectories of UAV, USV and UUV. (b) Communication distances. (c) Tracking errors with based on DQN. (d) Reward convergence process. (e) Tracking error via QL. (f) Time comparison.

Fig. 5.20 Comparison with the other algorithms for underwater target

the scenario where the target is not detected. In Fig. 5.21c, we find that a strong and concentrated echo is displayed at the detection edge of the sonar image, which means that there is an obvious obstacle. According to the shape and size of the obstacle, it can be considered as the underwater target to be detected. The final results of the USV and UUV reaching the desired position can be easily determined by Fig. 5.21d, e.

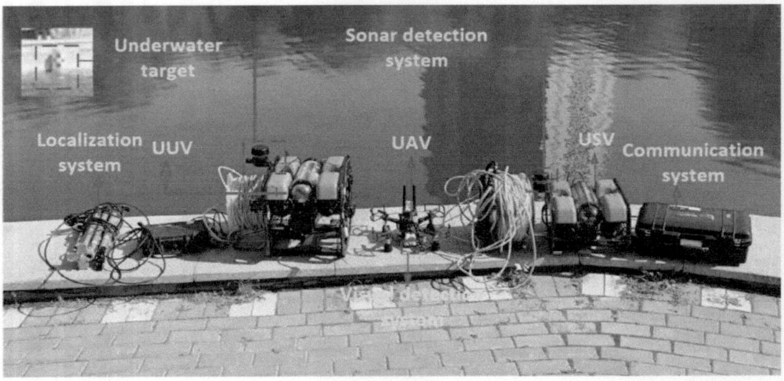

(a) Experimental setup for underwater target.

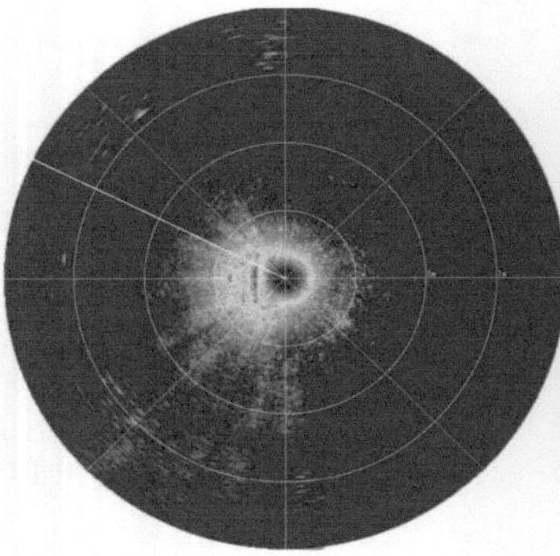

(b) Initial sonar detection of UUV.

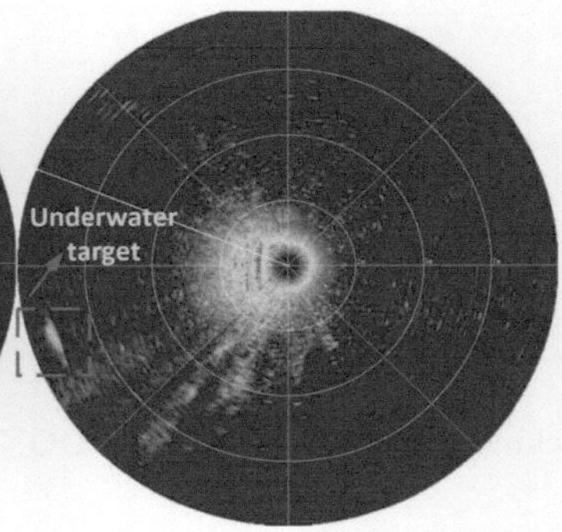

(c) Detected sonar image by UUV.

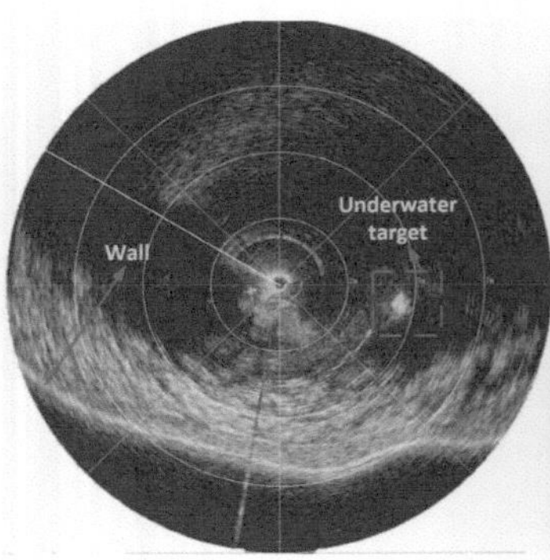

(d) The approaching process of UUV after the target is detected.

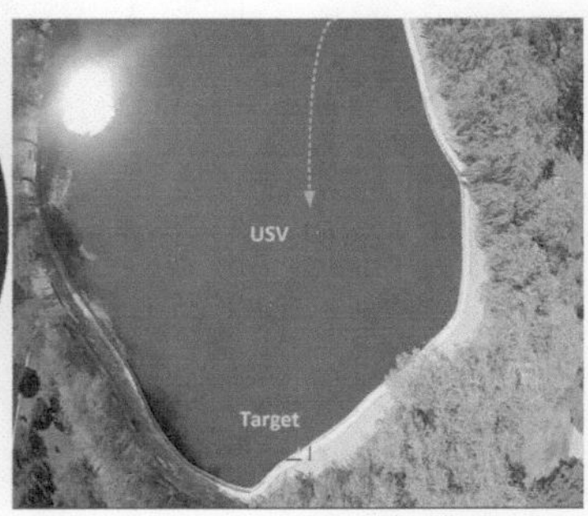

(e) Cooperation detection process of UAV, USV and UUV.

Fig. 5.21 Experimental setup and results for underwater detection

Based on this, the location information of the target is transmitted to the UAV, USV and UUV. The tracking algorithm based on DDPG is used for the USV and UUV separately. Then, the trajectories of USV, and UUV are obtained, as shown in Fig. 5.18a, b, respectively. Particularly, the position vectors of UAV, USV, and UUV on the X, Y, and Z axes are shown in Fig. 5.18c–e, respectively. In addition, the tracking errors d_{ta}, d_{ts} and d_{tu} are shown in Fig. 5.18f. It demonstrates the effectiveness of the DDPG-based tracking algorithm for underwater target.

5.6 Summary

This chapter proposes a solution for UAV, USV and UUV heterogeneous systems to cooperatively detect and track a surface or submerged target. Based on the measure theory, the detection probability for linear motion detection is given. Furthermore, a detection mode between UAV and USV, as well as a detection mode between the USV and UUV, are proposed by adjusting their formation shape to enhance the lower bound of the detection probability. Based on this, a tracking algorithm based on DDPG is designed for UAV, USV and UUV. Along with this, a multi-step location prediction strategy is designed to ensure communication connectivity during the tracking process. Finally, the effectiveness of the proposed method in this chapter is verified by both simulation and experimental results.

References

Basso, E., Schmidt-Didlaukies, H., Pettersen, K., & Sorensen, A. (2023). Global asymptotic tracking for marine vehicles using adaptive hybrid feedback. *IEEE Transactions on Automatic Control, 68*(3), 1584–1599.

Biggs, J., & Holderbaum, W. (July 2009). Optimal kinematic control of an autonomous underwater vehicle. *IEEE Transactions on Automatic Control, 54*(7), 1623–1626.

Cao, W., Yan, J., Yang, X., Luo, X., & Guan, X. (Jan. 2023). Communication-aware formation control of AUVs with model uncertainty and fading channel via integral reinforcement learning. *IEEE/CAA Journal of Automatica Sinica, 10*(1), 1–18.

Caruso, A., Paparella, F., Vieira, L., Erol, M., & Gerla, M. (2008). The meandering current mobility model and its impact on underwater mobile sensor networks. In *IEEE international conference on computer communications* (pp. 771–779).

Cui, S., Wang, Y., Wang, S., Wang, R., Wang, W., & Tan, M. (Apr. 2020). Real-time perception and positioning for creature picking of an underwater vehicle. *IEEE Transactions on Vehicular Technology, 69*(4), 3783–3792.

Enayat, M., & Khorasani, K. (Nov. 2017). Semi-decentralized nonlinear cooperative control strategies for a network of heterogeneous autonomous underwater vehicles. *International Journal of Robust and Nonlinear Control, 27*(16), 2688–2707.

Er, M., Ma, C., Liu, T., & Gong, H. (Apr. 2023). Intelligent motion control of unmanned surface vehicles: A critical review. *Ocean Engineering, 280*(1), 1–15.

Hongwei, Z., Wang, H., Yongsheng, Y., Haiyang, Y., & Haitao, D. (Sep. 2022). Remote passive sonar detection by relative multiscale change entropy. *IEEE Sensors Journal, 22*(18), 18066–18075.

Jia, Z., Lu, H., Li, S., & Zhang, W. (Aug. 2022). Distributed dynamic rendezvous control of the AUV-USV joint system with practical disturbance compensations using model predictive control. *IEEE Journal of Oceanic Engineering, 258*(1), 1–19.
Ke, C., & Chen, H. (2022). Cooperative path planning for airsea heterogeneous unmanned vehicles using search-and-tracking mission. *IEEE Journal of Oceanic Engineering, 262*(1), 1–13.
Ke, C., & Chen, H. (2022). Cooperative path planning for air-sea heterogeneous unmanned vehicles using search-and-tracking mission. *IEEE Journal of Oceanic Engineering, 262*(1), 1–13.
Li, Y., Liu, W., Li, L., Zhang, W., Xu, J., & Jiao, H. (Feb. 2023). Vision-based target detection and positioning approach for underwater robots. *IEEE Photonics Journal, 15*(1), 1–12.
Li, S., Wang, X., Wang, S., & Zhang, Y. (Feb. 2023). Distributed bearing-only formation control for UAV-UWSV heterogeneous system. *Drones, 7*(2), 124–144.
Liang, X., Qu, X., Hou, Y., & Zhang, J. (July 2017). Three-dimensional path following control of underactuated autonomous underwater vehicle based on damping backstepping. *International Journal of Advanced Robotic Systems, 14*(4), 1–9.
Lillicrap, T., Hunt, J., Pritzel, A., Heess, N., Erez, T., Tassa, Y., Silver, D., & Wierstra, D. (2016). Continuous control with deep reinforcement learning. In *Proceeding of International Conference on Learning Representations (ICLR)* (pp. 1–14).
Lindsay, J., Ross, J., Seto, M., Gregson, E., Moore, A., Patel, J., & Bauer, R. (Oct. 2022). Collaboration of heterogeneous marine robots toward multidomain sensing and situational awareness on partially submerged targets. *Ocean Engineering, 47*(4), 880–894.
Luo, H., Wang, J., Bu, F., Ruby, R., Wu, K., & Guo, Z. (May 2022). Recent progress of air/water cross-boundary communications for underwater sensor networks: A review. *IEEE Sensors Journal, 22*(9), 8360–8382.
McCammon, S., Marcon, G., Frantz, M., Welch, T., Best, G., Shearman, R., Nash, J., Barth, J., Adams, J., & Hollinger, A. (Jan. 2021). Ocean front detection and tracking using a team of heterogeneous marine vehicles. *Journal of Field Robotics, 38*(6), 854–881.
McNutt, M. (May 2014). The hunt for MH370. *Science, 344*(6187), 947.
Munaf, A., Canepa, G., & LePage, K. (Oct. 2019). Continuous active sonars for littoral undersea surveillance. *IEEE Journal of Oceanic Engineering, 44*(4), 1198–1212.
Santalo, L. (2004). *Integral geometry and geometric probability*. Cambridge University Press.
Tian, Y., Liu, M., Zhang, S., & Zhou, T. (Mar. 2022). Underwater multi-target passive detection based on transient signals using adaptive empirical mode decomposition. *Applied Acoustics, 190*(1), 1–12.
Wei, W., Wang, J., Fang, Z., Chen, J., Ren, Y., & Dong, Y. (Mar. 2023). 3U: Joint design of UAV-USV-UUV networks for cooperative target hunting. *IEEE Transactions on Vehicular Technology, 72*(3), 4085–4090.
Williams, D. (Jan. 2015). Fast target detection in synthetic aperture sonar imagery: A new algorithm and large-scale performance analysis. *IEEE Journal of Oceanic Engineering, 40*(1), 71–92.
Wu, Y., Low, K. H., & Lv, C. (Jun. 2020). Cooperative path planning for heterogeneous unmanned vehicles in a search-and-track mission aiming at an underwater target. *IEEE Transactions on Vehicular Technology, 69*(6), 6782–6787.
Yan, J., Cao, W., Yang, C.C., & Guan, X. (2022). Communication-efficient and collision-free motion planning of underwater vehicles via integral reinforcement learning. *IEEE Transactions on Neural Networks and Learning Systems*. https://doi.org/10.1109/TNNLS.2022.3226776
Yan, J., Li, X., Yang, X., Luo, X., Hua, C., & Guan, X. (Nov. 2022). Integrated localization and tracking for AUV with model uncertainties via scalable sampling-based reinforcement learning approach. *IEEE Transactions on Systems, Man, and Cybernetics: Systems 52*(11), 6952–6967.
Yan, J., Zhang, Z., Yang, X., Luo, X., & Guan, X. (July 2023). Target detection in underwater sensor networks by fusion of active and passive measurements. *IEEE Transactions on Network Science and Engineering, 10*(4), 2319–2333.
Yang, X., Wang, W., & Huang, P. (Dec. 2020). Distributed optimal consensus with obstacle avoidance algorithm of mixed-order UAVs-USVs-UUVs systems. *ISA Transactions, 107*(12), 270–286.

Zhang, J., Sha, J., Han, G., Liu, J., & Qian, Y. (Mar. 2021). A cooperative-controlbased underwater target escorting mechanism with multiple autonomous underwater vehicles for underwater internet of things. *IEEE Internet Things Journal, 8*(6), 4403–4416.

Zhao, W., Liu, H., Wan, Y., & Lin, Z. (Dec. 2022). Data-driven formation control for multiple heterogeneous vehicles in air-ground coordination. *The IEEE Transactions on Control of Network Systems, 9*(4), 1851–1862.

Chapter 6
Integrated Design of Detection, Communication and Control

6.1 Introduction

Motion planning of an AUV is to determine the collision-free trajectory that enables AUV to arrive at its desired destination. To this end, AUV requires to perform three critical tasks: (1) obstacle detection for effective navigation; (2) communication with surface buoys for accurate self-localization; (3) tracking control for arriving at the destination. We find the detection and communication have overlapping functions due to the reliance on acoustic signals and similar hardware components. However, most existing works (e.g., Khalil et al. 2022; Zhou et al. 2022) separately tackle the detection and communication issues, which results in the competition of spectrum and energy resources. Although the co-design of communication and detection has been developed in radar system (e.g., Liu et al. 2017), the small aperture of underwater sonar and the narrow bandwidth of acoustic channel make it hard to deploy the radar-suited integration system to AUV. Meanwhile, the existing control methods (e.g., Yu et al. 2024; Wu et al. 2025) always assume the detection and communication are perfectly achieved, but harsh underwater environment makes it difficult to hold this assumption.

Apart from that, we note that some learning-based obstacle avoidance methods have been developed for AUV to adapt the complex and dynamic underwater environments, e.g., the deep Q network (DQN) in Yang et al. (2022) and the deep deterministic policy gradient (DDPG) in Yan et al. (2023). However, the above methods suffer from the drawback of local optimum due to the insufficient exploration. Under the integrated framework of detection, communication and control, how to design a learning-based motion planning solution for AUV that can both avoid local optimum and obstacle collision is still unsolved.

However, when the tasks to be performed contain complex activities (e.g., grasp in a very cluttered and unexplore region), the full autonomy is still so far to accomplish the tasks. In this context, underwater teleoperation system can be utilized to extend a human operator's intelligence and ability to perform complex tasks in remote site. As

J. Yan and X. Guan, *Maritime Unmanned Systems*, Cognitive Intelligence and Robotics 1, https://doi.org/10.1007/978-981-95-6907-6_6

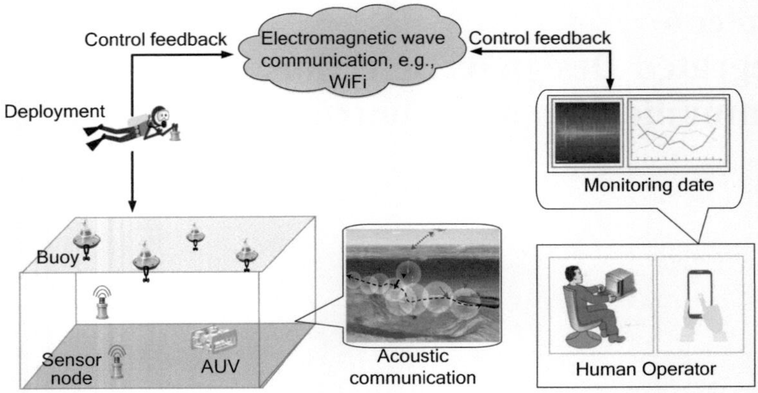

Fig. 6.1 Underwater teleoperation system toward HOTL

shown in Fig. 6.1, AUV continuously transmits or reflects back its measurements to human operator through the acoustic and WiFi links. With the reverse communication links, human operator exerts a force on the master manipulator, such that the same force can be performed on AUV in remote underwater environment.

From the viewpoint of cybernetics, the main objectives of teleoperation include two parts: (1) *Stability.* Keep stability of the closed-loop control system under the behaviors of environment and human operator; (2) *Transparency*. Supply the human operator with an ideal sense of direct action, which is regarded as the transparency between human operator and underwater environment. To this end, many bilateral teleoperation systems have been developed for terrestrial vehicles or robots (Hokayem and Spong 2006; Moniruzzaman et al. 2022). Nevertheless, the unique characteristics of underwater environment make it challenging to construct the underwater teleoperation system, which includes as follows

(1) *Unavailable position information*. The radio signals rapidly attenuate in water, and hence, the global positioning system (GPS) is not available for AUV. Although the inertial navigation system can be employed to estimate the position, it suffers from accumulative errors due to ambient (e.g., current and tides) and manmade (e.g., pumps and drilling) noises.

(2) *Weak communication*. Up to now, acoustic wave is the only available medium for underwater communication more than a hundred meters (Sun et al. 2022). However, the bandwidth of acoustic communication is low, which results in low data rate. Meanwhile, its propagation delay is five orders of magnitude higher than in radio frequency channels. Besides, multi-path and high-energy consumption cannot be ignored.

(3) *Unknown model parameter of AUV.* Ocean exists unknown current and tide, which are regarded as external disturbances. The relationships between AUV and manipulator are complicated and uncertain. So, it is difficult if not impossible to

acquire the accurate dynamic models of AUV, which leads to the failure of model-based controllers for terrestrial vehicles.

In light of the above challenges and opportunities, this chapter investigates a unified design philosophy that integrates detection, communication, and control within underwater intelligent systems. Departing from the conventional separated design paradigm, it emphasizes the tight coupling among sensing, networking, and decision-making processes under the constraints of acoustic bandwidth, energy, and latency. Specifically, the chapter develops an integrated signal and control co-design framework that enables underwater vehicles to jointly perceive, communicate, and act in dynamic environments. It further extends this framework from fully autonomous motion planning to human-on-the-loop teleoperation, such that complex marine missions can be accomplished. Through theoretical formulation, algorithmic design, and experimental validation, the chapter demonstrates how integrated co-design can substantially enhance the efficiency, adaptability, and robustness of underwater detection, communication, and control systems.

6.2 Co-design of Detection, Communication and Control for Underwater Target Tracking

6.2.1 Model Construction and Problem Formulation

AUV is driven by the propulsion systems, and its position information is acquired by communicating with surface buoys. The surface buoys are installed with Global Positioning System (GPS) to provide self-localization reference for AUV. Since AUV's depth (i.e., z) can be measured by depth gauge, three buoys are used in our scenario. With regard to this, one defines $(x_{B,i};\, y_{B,i})$ as the position vector of buoy i for $\forall i \in \{1, 2, 3\}$. At time $\hat{t}_{A,i}$, AUV sends out a message to buoy i, and the timestamp of receiving the reply is $\check{t}_{i,A}$. Thus, the relative distance between buoy j and AUV is given by $\check{d}_{A,i} = \frac{1}{2}\mathbf{c}(\check{t}_{i,A} - \hat{t}_{A,i})$, where $\mathbf{c} = 1500$ m/s is the acoustic speed. Thereby, the horizontal positions of AUV (i.e., x and y) are acquired by solving the following equation, i.e., for $i = 1, 2, 3$

$$\sqrt{(x - x_{B,i})^2 + (y - y_{B,i})^2 + z^2} = \check{d}_{A,i}. \tag{6.1}$$

Meanwhile, AUV requires to send out a detection wave to sense the existence of obstacles. The mth obstacle can be covered by a column O_m, and the detected obstacle set at time t is denoted by

$$\Omega(t) \subset \{O_1(\sigma_{o_1}, \varepsilon_{o_1}, h_1), \ldots, O_m(\sigma_{o_m}, \varepsilon_{o_m}, h_m)\}, \tag{6.2}$$

where σ_m, ε_m and h_m are the center, radius and height, respectively.

The position and velocity vectors of AUV are $\boldsymbol{\eta} = [x, y, z, \psi]^{\mathrm{T}}$ and $\mathbf{v} = [u, v, w, r]^{\mathrm{T}}$, respectively. ψ is the yaw angle, while u, v and w are the linear velocities on surge, sway and heave, respectively. r is the yaw angle velocity. The control input vector is $\boldsymbol{\tau} = [\tau_u, \tau_v, \tau_w, \tau_r]^{\mathrm{T}}$. Thus, the dynamic model of AUV is

$$\dot{\boldsymbol{\eta}} = \mathbf{J}(\boldsymbol{\eta})\mathbf{v}, \ \ \mathbf{M}\dot{\mathbf{v}} + \mathbf{C}(\mathbf{v})\mathbf{v} + \mathbf{D}(\mathbf{v})\mathbf{v} + \mathbf{G}(\boldsymbol{\eta}) = \boldsymbol{\tau}, \tag{6.3}$$

where $\mathbf{M}$ is the inertia matrix, $\mathbf{C}(\mathbf{v})$ is the Coriolis and centrifugal matrix, $\mathbf{D}(\mathbf{v})$ is the hydrodynamic damping matrix, $\mathbf{G}(\boldsymbol{\eta})$ is the hydrostatic force vector, and $\mathbf{J}$ is the transformation matrix.

An integrated communication and detection wave is designed to avoid the competition of spectrum and energy resources, as reproduced from Guo et al. (2025). After that, AUV requires to safely reach the destination vector (x_t, y_t, z_t). Accordingly, the main purpose of this chapter is given as follows.

- *Localization-detection*: Get (x, y) and Ω by an integrated wave;
- *Collision-free motion control*: Drive $(x, y, z) \rightarrow (x_t, y_t, z_t)$.

6.2.2 Integrated Waveform Design

The main procedure of our solution is described by Fig. 6.2. Specifically, we first adopt the orthogonal chirp division multiplexing (OCDM) to co-design the communication and detection waves. It can demonstrate superior performance in mitigating multipath fading and Doppler effects compared to conventional orthogonal frequency division multiplexing (OFDM) based solution (Zhou and Wang 2014). With the integrated wave, the self-localization and detection tasks can be simultaneously achieved. After that, a collision-free tracking controller is developed to drive

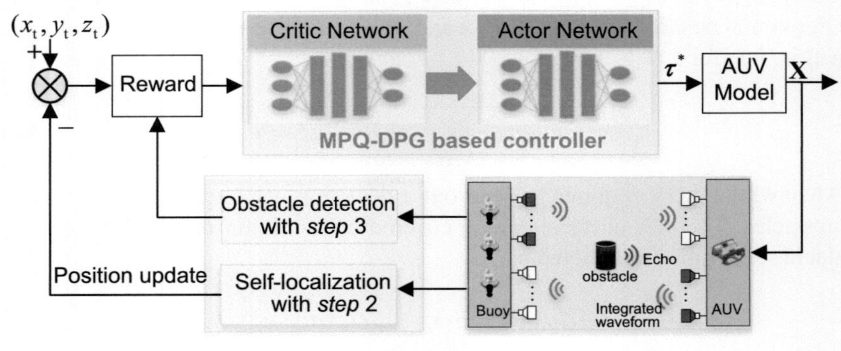

Fig. 6.2 Main procedure of the motion planning via the integrated design

AUV to the destination. Accordingly, the above procedure can be detailed as the following steps.

Step 1. Integrated Waveform Design

The detection-based chirp signal is regarded as the fundamental waveform, through which the communication symbols are modulated on the above waveform. To achieve multi-carrier modulation, each fundamental waveform should satisfy mutual orthogonality, and hence, the nth chirp waveform can be expressed as

$$\zeta_n(t) = e^{j\frac{\pi}{4}} e^{-j\pi\frac{N}{T^2}(t-n\frac{T}{N})^2}, \ t \in [0, T], \tag{6.4}$$

where N is the number of chirp waveform, and T is the period.

Denote the communication data symbol as $\mathbf{x}$, and the nth element is $\mathbf{x}_n$. With the nth chirp waveform in (6.4), we map $\mathbf{x}_n$ on the integrated multi-carrier waveform $S(t)$, and it can be defined as

$$S(t) = \sum_{n=0}^{N-1} \mathbf{x}_n \zeta_n(t), \tag{6.5}$$

where zero padding is utilized to suppress inter-symbol interference.

Step 2. Self-Localization with the Integrated Waveform

AUV transmits the integrated waveform $S(t)$ to surface buoys, which contains the AUV's status information, environmental data collected by multiple sensors, and an identification code for distinguishing the target buoy. After that, the buoy first verifies whether it is the intended recipient by matching the identification code. If the match is successful, the buoy sends the demodulated data to the control center while transmitting reply signal $S_i(t)$ to AUV for self-localization references. Due to the amplitude of echo is significantly lower than that of the communication wave, we set a voltage threshold to distinguish the echo from the communication wave, i.e., the voltage of the extracted waveform is considered as a communication signal when it is greater than the voltage threshold, otherwise it is echo.

The received waveform for buoy $i = 1, 2, 3$ can be expressed by

$$r_i(t) = S(t) \star \check{h}(t, \beta_i) + n_0(t), \tag{6.6}$$

where $\star$ is the convolution operation, $\check{h}(t, \beta_i) = \sum_{\check{p}=1}^{\check{P}} A_{\check{p}} \check{\delta}(\beta_i - \beta_{\check{p},i}) e^{j2\pi f_{d,\check{p}} t}$ is underwater communication channel model, $\check{P}$ is the total number of paths, $A_{\check{p}}$ is the amplitude of the $\check{p}$th path, $\beta_{\check{p},i}$ is the delay of the $\check{p}$th path, $f_{d,\check{p}}$ is the doppler shift of the $\check{p}$th path and $n_0(t)$ is white Gaussian noise.

The signal orthogonality is adopted to demodulate the received waveform, and the demodulation process for buoy $i = 1, 2, 3$ is

$$\widetilde{\mathbf{x}}_{n,i} = \int_0^T r_i(t) \zeta_n^* dt, \tag{6.7}$$

where ζ_n^* represents the complex conjugate of ζ_n, and its role is to ensure orthogonality in the inner product calculation.

After that, each buoy relays a message to AUV, i.e., $S_i(t)$ for $i = 1, 2, 3$. Then, AUV receives and demodulates the message, through which the relative distance information $\check{d}_{A,i}$ can be calculated. That means the self-localization task can be accomplished by (6.1).

Step 3 Obstacle Detection with the Echo

When the integrated waveform $S(t)$ interacts with an obstacle, an echo is generated, and hence, it is received by the AUV. Specifically, the received echo signal $\tilde{\vartheta}_m(t)$ for obstacle m is expressed as

$$\tilde{\vartheta}_m(t) = S(t) \star \widehat{h}(t, \chi_m) + n_0(t), \tag{6.8}$$

where $\widehat{h}(t, \chi_m) = \sum_{\widehat{p}=1}^{\widehat{P}} A_{\widehat{p}} \widehat{\delta}(\chi_m - \chi_{m,\widehat{p}}) e^{j2\pi f_{d,\widehat{p}} t}$ is underwater detection channel model, $\widehat{P}$ is the total number of reflection paths, $\chi_{m,\widehat{p}}$ represents the time difference between the transmission and reception of the integrated signal, $A_{\widehat{p}}$ is the amplitude decay of the $\widehat{p}$th path, and $f_{d,\widehat{p}}$ is the doppler shift of the $\widehat{p}$th path.

From (6.8), AUV converts $\tilde{\vartheta}_m(t)$ into discrete-time data, such that $\chi_{m,\widehat{p}}$ and $f_{d,\widehat{p}}$ can be extracted by echo detector and spectral analysis, respectively. Finally, the position of obstacle m can be estimated.

6.2.3 Design of Tracking Controller

For the convenience of design, model (6.3) is rearranged as

$$\boldsymbol{\xi}(k+1) = f(\boldsymbol{\xi}(k)) + \mathbf{h}(\boldsymbol{\xi}(k))\boldsymbol{\tau}(k), \tag{6.9}$$

where $f(\boldsymbol{\xi}) = [\boldsymbol{\eta} + \delta \mathbf{J}(\psi)\mathbf{v}; \mathbf{v} - \delta \mathbf{M}^{-1}(\mathbf{C}(\mathbf{v})\mathbf{v} + \mathbf{D}(\mathbf{v})\mathbf{v} + \mathbf{G}(\boldsymbol{\eta}))]$, $\mathbf{h}(\boldsymbol{\xi}) = [\mathbf{0}_{4\times4}; \delta \mathbf{M}^{-1}]$, $\boldsymbol{\xi} = [\boldsymbol{\eta}; \mathbf{v}]$, and δ is sampling interval.

With the location information in Step 2 and the detection result in Step 3, the reward function $R(k)$ for AUV at time step k is given as

$$R(k) = -\alpha_1 \|\mathbf{X}(k) - \mathbf{X}_t\| - \alpha_2 \sum_{l=1}^{m} \frac{1}{\| \|\mathbf{X}(k) - \sigma_l\| - \varepsilon_l \|}, \tag{6.10}$$

where $\mathbf{X}(k) = [x_k, y_k, z_k]^{\mathrm{T}}$ is the position vector of the AUV, and $\mathbf{X}_t = [x_t, y_t, z_t]^{\mathrm{T}}$ is the destination vector. α_1 and α_2 are positive weights. σ_l and ε_l are the center and radius of the lth obstacle. It can be seen that the aim of this reward function is to drive the AUV to the destination while avoiding the collision with the obstacle.

Based on (6.10), the value function of AUV is set as

$$Q(\boldsymbol{\xi}(k), \boldsymbol{\tau}(k)) = E(R(k+1) + \gamma \arg\max_{\boldsymbol{\tau}} Q(\boldsymbol{\xi}(k+1), \boldsymbol{\tau}(k+1))), \tag{6.11}$$

where $E[\cdot]$ is the expectation, and $\gamma \in (0, 1]$ is the discount factor.

The optimal input $\boldsymbol{\tau}^*(k)$ is selected by the following procedure

$$\boldsymbol{\tau}^*(k) = \arg\max_{\boldsymbol{\tau}} Q(\boldsymbol{\xi}(k), \boldsymbol{\tau}(k)). \tag{6.12}$$

To this end, AUV interacts with the environment to collect experience $\{\boldsymbol{\xi}(k), \boldsymbol{\tau}(k), R(k+1), \boldsymbol{\xi}(k+1)\}$, and stores it in experience pool $\mathbf{R}_{buf}$. Then, the exploration action can be denoted by

$$\boldsymbol{\tau}(k) = (1/p) \sum_{\bar{p}=1}^{p} \mu(\boldsymbol{\xi}(k)|\lambda_k^{\bar{p}}) + \mathcal{N}^*, \tag{6.13}$$

where p denotes the number of actor networks, μ is the subgreedy policy, $\lambda_k^{\bar{p}}$ is weight of the $\bar{p}$th actor network, and $\mathcal{N}^*$ is the noise.

With (6.13), a random minibatch of size κ is sampled from $\mathbf{R}_{buf}$, and the Bellman error is calculated. Denote q as the number of critic networks, and then the Bellman error of the $\bar{q}$th critic network is

$$B_{\bar{q}} = (1/\kappa) \sum |Q(\boldsymbol{\xi}(k), \boldsymbol{\tau}(k)|\omega_k^{\bar{q}}) - Y(k+1)|, \ \ \bar{q} \in q, \tag{6.14}$$

where Y is its target value, and $\omega_k^{\bar{q}}$ denotes its weight.

Define the number of the worst critic network as $c = \arg\max B_{\bar{q}}$, and its weight ω_k^c is optimized by minimizing the loss function

$$L(\omega_k^c) = (1/\kappa) \sum (Q(\boldsymbol{\xi}(k), \boldsymbol{\tau}(k)|\omega_k^c) - Y^c(k+1))^2, \tag{6.15}$$

where $Y^c(k+1) = R(k+1) + \frac{\gamma}{q-1} \sum_{\bar{q}=1, \bar{q} \neq c}^{q} Q(\boldsymbol{\xi}(k+1), \boldsymbol{\tau}(k+1)|\omega_k^{\bar{q}})$.

With regard to (6.15), ω_k^c is updated by gradient descent, i.e.,

$$\nabla_{\omega^c} L(\omega_k^c) = \nabla_{\omega^c} E(Q(\boldsymbol{\xi}(k), \boldsymbol{\tau}(k)|\omega_k^c) - Y^c(k+1))^2, \tag{6.16}$$

and hence, the worst critic network weight ω_k^c is updated by

$$\omega_{k+1}^c = \omega_k^c - \gamma_1 \nabla_{\omega^c} L(\omega_k^c), \tag{6.17}$$

where $\gamma_1 \in (0, 1)$ is the learning rate for updating critic weight.

For an arbitrary actor network $\bar{p}$, the actor network weight $\lambda_k^{\bar{p}}$ can be updated by maximizing the following loss function $L(\lambda_k^{\bar{p}})$, i.e.,

$$L(\lambda_k^{\bar{p}}) = E(Q(\boldsymbol{\xi}(k), \mu(\boldsymbol{\xi}(k))|\lambda_k^{\bar{p}})). \tag{6.18}$$

Similar to ω_k^c, the $\lambda_k^{\bar{p}}$ can be updated by gradient ascent, i.e.,

$$\nabla_{\lambda^{\bar{p}}} L(\lambda_k^{\bar{\mathbf{p}}}) = \nabla_{\lambda^{\bar{p}}} E(Q(\boldsymbol{\xi}(k), \mu(\boldsymbol{\xi}(k)|\lambda_k^{\bar{p}}))), \tag{6.19}$$

and its update procedure is given as

$$\lambda_{k+1}^{\bar{p}} = \lambda_k^{\bar{p}} + \gamma_2 \nabla_{\lambda^{\bar{p}}} L(\lambda_k^{\bar{p}}), \tag{6.20}$$

where $\gamma_2 \in (0, 1)$ is the learning rate for updating actor weight.

Repeat the above iteration procedure until $\|\lambda_k^{\bar{p}} - \lambda^{\bar{p}*}\| \leq \epsilon_k$ and $\|\omega_k^{\bar{q}} - \omega^{\bar{q}*}\| \leq \varrho_k$, where ϱ_k and ϵ_k are the positive decimals. Accordingly, the optimal weights $\omega^{\bar{q}*}$ and $\lambda^{\bar{p}*}$ can be obtained. Substituting $\omega^{\bar{q}*}$ and $\lambda^{\bar{p}*}$ into exploration action network (6.13). Thus, the optimal action of AUV at time step k can be acquired, i.e.,

$$\boldsymbol{\tau}^*(k) = (1/p) \sum_{\bar{p}=1}^{p} \mu(\boldsymbol{\xi}(k)|\lambda_k^{\bar{p}*}). \tag{6.21}$$

From (6.11) and (6.14), we find that the controller in (6.21) is model-free since the model parameters of AUV are not required. Compared with the DDPG-based solution (Yan et al. 2023), the multiple Q-function estimators in (6.11) and the different state-action region exploration in (6.13) can prevent premature convergence to suboptimal policies, encourage thorough exploration, and reduce overestimation bias.

6.3 Co-design of Detection, Communication and Control for Underwater Teleoperation

6.3.1 Model Construction and Problem Formulation

Denote $\boldsymbol{\eta}_{\mathrm{m}} = [x_{\mathrm{m}}, y_{\mathrm{m}}, z_{\mathrm{m}}, \psi_{\mathrm{m}}]^{\mathrm{T}}$ and $\mathbf{v}_{\mathrm{m}} = [u_{\mathrm{m}}, v_{\mathrm{m}}, w_{\mathrm{m}}, r_{\mathrm{m}}]^{\mathrm{T}}$ as the state and velocity vectors of master human operator, respectively. Similarly, denote $\boldsymbol{\eta}_{\mathrm{s}} = [x_{\mathrm{s}}, y_{\mathrm{s}}, z_{\mathrm{s}}, \psi_{\mathrm{s}}]^{\mathrm{T}}$ and $\mathbf{v}_{\mathrm{s}} = [u_{\mathrm{s}}, v_{\mathrm{s}}, w_{\mathrm{s}}, r_{\mathrm{s}}]^{\mathrm{T}}$ as the state and velocity vectors of slave AUV, respectively. The elements in the above vectors represent the states in surge, sway, heave and yaw, respectively. Hence, the dynamic model is given by

$$\begin{aligned}
&\mathbf{M}_{\mathrm{m}}\dot{\mathbf{v}}_{\mathrm{m}} + \mathbf{C}_{\mathrm{m}}(\mathbf{v}_{\mathrm{m}})\mathbf{v}_{\mathrm{m}} + \mathbf{G}_{\mathrm{m}}(\boldsymbol{\eta}_{\mathrm{m}}) = \boldsymbol{\tau}_{\mathrm{m}} + \mathbf{F}_{\mathrm{h}},\\
&\mathbf{M}_{\mathrm{s}}\dot{\mathbf{v}}_{\mathrm{s}} + \mathbf{C}_{\mathrm{s}}(\mathbf{v}_{\mathrm{s}})\mathbf{v}_{\mathrm{s}} + \mathbf{D}_{\mathrm{s}}(\mathbf{v}_{\mathrm{s}})\mathbf{v}_{\mathrm{s}} + \mathbf{G}_{\mathrm{s}}(\boldsymbol{\eta}_{\mathrm{s}}) = \boldsymbol{\tau}_{\mathrm{s}},
\end{aligned} \tag{6.22}$$

where $\mathbf{M}_k$ is inertia and mass matrix, $\mathbf{C}_k$ is the coriolis matrix, $\mathbf{G}_k$ is the gravity and buoyancy matrix, $\boldsymbol{\tau}_k$ is the input torque for $k \in \{\mathrm{m,s}\}$. In addition, $\mathbf{J}_{\mathrm{h}}$ and $\mathbf{F}_{\mathrm{h}}$ denote the transformation matrix and input force of human operator, respectively. $\mathbf{D}_{\mathrm{s}}$ denotes the hydrodynamic damping matrix. Transform the motion of body-fixed reference frame into inertial reference frame, the system kinematic model is denoted as

$$\dot{\boldsymbol{\eta}}_{\mathrm{m}} = \mathbf{J}_{\mathrm{m}}\mathbf{v}_{\mathrm{m}}, \ \dot{\boldsymbol{\eta}}_{\mathrm{s}} = \mathbf{J}_{\mathrm{s}}\mathbf{v}_{\mathrm{s}}, \tag{6.23}$$

where $\mathbf{J}_{\mathrm{k}}$ represent the transformation matrix.

After that, it is necessary to acquire the time delay between master human operator and slave AUV. To this end, the one-way time delay is denoted as

$$d = \frac{1}{c_1} \left\| \mathbf{x}_{\mathrm{m}} - \mathbf{x}_{\mathrm{buoy}} \right\| + \frac{1}{c_2} \left\| \mathbf{x}_{\mathrm{buoy}} - \mathbf{x}_{\mathrm{s}} \right\| + d_1, \tag{6.24}$$

where $c_1 \approx 3 \times 10^8$ m/s and $c_2 \approx 1500$ m/s denote the speed of electromagnetic wave and acoustic wave, respectively. $\mathbf{x}_{\mathrm{m}}$, $\mathbf{x}_{\mathrm{buoy}}$ and $\mathbf{x}_{\mathrm{s}}$ are the position vectors of master human operator, buoy and slave AUV, respectively. d_1 denotes the fixed processing delay.

Accordingly, our aim is to design the hardware of underwater teleoperation system, through which the state information between the master site and slave site can be exchanged. Based on this, we develop a model-free controller to enable the slave AUV move according to the intention of human operator.

6.3.2 *Construction of Underwater Teleoperation Solution*

Hardware Design

The hardware design of underwater teleoperation system mainly consists of AUV, surface buoy, human operator and cross-medium communication network.

(1) *AUV.* As shown in Fig. 6.3, the BlueROV2, which is one of the most affordable high-performance underwater vehicles, acts as the major structure of AUV in our system. It also includes buoyancy shells, search lamps, thrusters, a manipulator, and binocular cameras. To ensure AUV can move freely, the gravity of AUV is set to be equal to its buoyancy. Two vertical and four horizontal thrusters are actuated on the

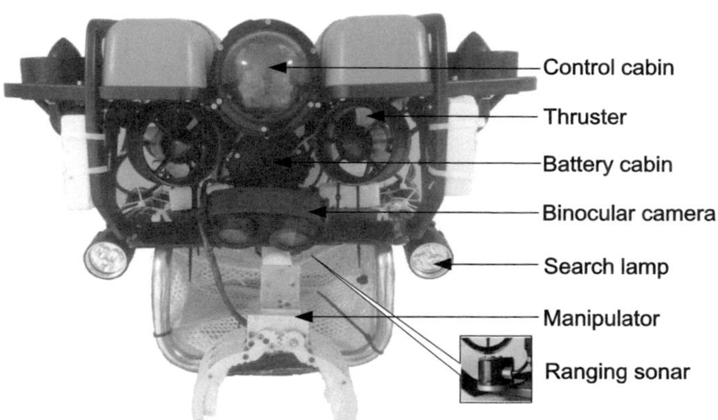

Fig. 6.3 AUV used in the underwater teleoperation system

AUV frame, where each thruster can generate a maximum 5 Kgf propulsive force, enabling AUV to reach a maximum speed of 3 knots. Besides that, a manipulator equipped with force sensor and encoder is installed at the bottom of the AUV frame, which can drive the end-effector by the servo to perform a given task. The binocular camera is used to monitor the underwater target closely to provide visual guidance. What is more, the power for the entire system is supplied by a 24 V battery.

In the control cabin, STM32 processor is employed to be the mircocontroller, whose aim is to receive the data from image processor, buoy and on-board sensors. The image processor receives video streams from binocular cameras, such that the short-distance guidance function can be achieved. Meanwhile, the buoy provides long-distance self-localization reference for AUV. In addition, the on-board sensors, including inertial measurement unit (IMU), bathymeter and sonar, measure the real-time state of AUV.

(2) *Buoy.* The main role of buoy can be summarized as follows: (1) Self-localization reference for AUV; (2) Relay the data between human operator and AUV. To this end, a SBL system developed by Waterlink is adopted to locate the position of AUV, as shown by Fig. 6.4a. The SBL system is composed of four transducers and an acoustic modem, which are mounted on buoy and AUV, respectively. Each transducer receives echo from the acoustic modem. Then the time difference of arrival (TDOA) is embedded into the processor to estimate the position information. As have been pointed out in SBL Homepage By Waterlinked (2022), underwater localization accuracy of SBL system can reach sub-meter level within a few hundred meters.

To guarantee effective data relay, buoy is installed with radio transmission equipment for WiFi communication with human operator, and OFDM (Orthogonal Fre-

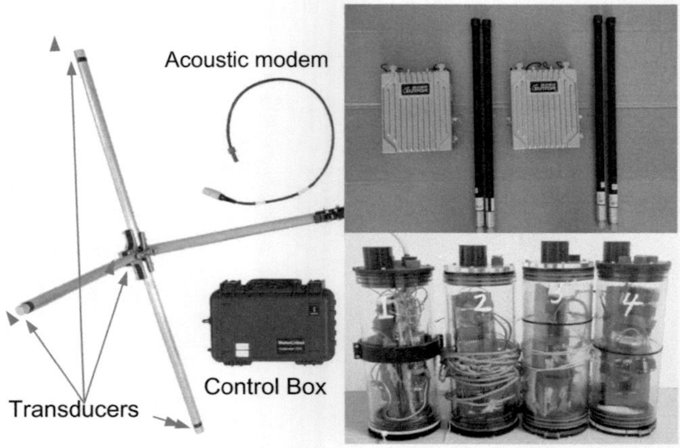

(a) SBL system. (b) WiFi-acoustic communication.

Fig. 6.4 Localization and communication equipments on buoy quency Division Multiplexing) transmission equipment for acoustic communication with AUV, as shown in Fig. 6.4b. Besides that, the umbilical connects buoy with AUV, whose aim is to receive and transmit the video streams between AUV and buoy.

(3) *Human operator*. The data and video stream relayed by buoy are first decoded, and then be displayed on the host computer developed by PyQt5. Accordingly, human operator analyzes these information to make real-time decisions. Particularly, human interacts with underwater remote environment through a master manipulator, such that the transparency can reflect how and how fast the human operator. Of note, the control signal is generated by the master manipulator, and further transmitted to the AUV in the opposite direction over the cross-medium communication channel, through which the underwater teleoperation can be realized.

(4) *Communication network*. The following three types of transmission channels are utilized: (1) Data transmission between buoy and AUV by the acoustic communication; (2) Video transmission between buoy and AUV by the umbilical communication. (3) Data and video transmissions between buoy and human operator by the WiFi communication.

The work frequency band of the acoustic communication system (see the bottom of Fig. 6.4b) is within 21 kHz to 27 kHz, and it adopts the OFDM mode. In order to overcome the multipath and Doppler effects, the cyclic prefix is extended and the frequency interval is enlarged. Due to this, it has a stable communication rate of 300 bps and a maximum communication rate of 2000 bps. The design of system bandwidth is related to transducer technology, frequency response characteristics and water environment. On the other hand, the umbilical is made of Kevlar fiber, whose two ends are connected with power carrier modules. The communication protocol used by umbilical is IEEE 1901.1 with frequency band of 30 MHz, and its maximum transmission rate can reach 500 Mbps.

The WiFi equipment produced by SUNTOR (see the top of Fig. 6.4b) can transmit the data and video stream to somewhere 10 km away through a 6 dB omnidirectional antenna under unobstructed condition. The fixed processing delay is less than 20 ms, and the allocated frequency range is $1240 \sim 1300$ MHz with the variable bandwidth $5 \sim 40$ MHz. The transmitting power can be adjusted among $100 \sim 2000$ MW according to the system requirement.

Software Design

The software design of underwater teleoperation system includes the self-localization of AUV, visual guidance scheme and the teleoperation controller, as shown in Fig. 6.5.

(1) *Self-localization of AUV*. The TDOA algorithm is employed to estimate the position of AUV. Four transducers are fixed on the T-shaped baseline, which is carried by the buoy. We define $t_{\rm t}$ as the timestamp when AUV broadcasts its localization request, and $\mathbf{x}_{\rm s}$ is its unknown position. Similarly, define the timestamp when each transducer receives the data as t_i, where the known position of transducer

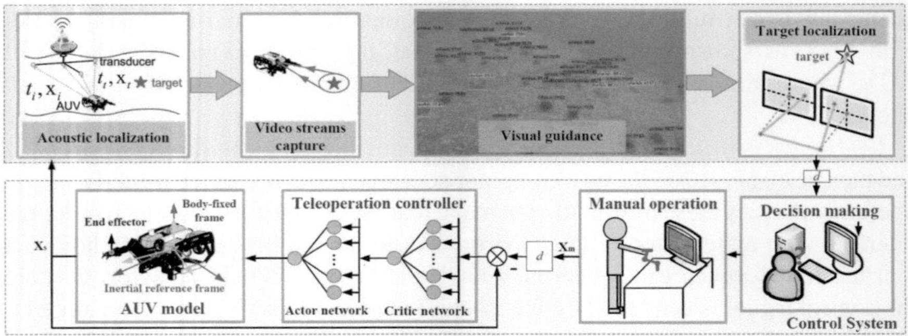

Fig. 6.5 Block diagram of our solution. With the data and video streams of AUV relayed by buoy, human operator makes real-time decision and controls the master manipulator, such that the transparency between human operator and underwater environment is guaranteed

$i \in \{1, 2, 3, 4\}$ is denoted by $\mathbf{x}_i$. Following the principle of TDOA, the time difference between transducers i and j is

$$\begin{aligned} t_{i,j} &= (t_i - t_t) - (t_j - t_t), \\ &= \frac{1}{c_1}(\|\mathbf{x}_i - \mathbf{x}_s\| - \|\mathbf{x}_j - \mathbf{x}_s\|) + n_{i,j}, i \neq j, \end{aligned} \tag{6.25}$$

where $c_1 \approx 1500$ m/s is acoustic speed. $n_{i,j}$ is measurement noise, obeying the zero-mean Gaussian distribution.

Based on (6.25), the time difference measurements among different transducers can be stacked in

$$\underbrace{\begin{bmatrix} \tilde{t}_{1,2} \\ \tilde{t}_{1,3} \\ \tilde{t}_{1,4} \\ \tilde{t}_{2,3} \\ \tilde{t}_{2,4} \\ \tilde{t}_{3,4} \end{bmatrix}}_{:= \tilde{t}} = \frac{1}{c_1} \underbrace{\begin{bmatrix} \|\mathbf{x}_1\text{-}\mathbf{x}_s\| - \|\mathbf{x}_2\text{-}\mathbf{x}_s\| \\ \|\mathbf{x}_1\text{-}\mathbf{x}_s\| - \|\mathbf{x}_3\text{-}\mathbf{x}_s\| \\ \|\mathbf{x}_1\text{-}\mathbf{x}_s\| - \|\mathbf{x}_4\text{-}\mathbf{x}_s\| \\ \|\mathbf{x}_2\text{-}\mathbf{x}_s\| - \|\mathbf{x}_3\text{-}\mathbf{x}_s\| \\ \|\mathbf{x}_2\text{-}\mathbf{x}_s\| - \|\mathbf{x}_4\text{-}\mathbf{x}_s\| \\ \|\mathbf{x}_3\text{-}\mathbf{x}_s\| - \|\mathbf{x}_4\text{-}\mathbf{x}_s\| \end{bmatrix}}_{:= \varphi(\mathbf{x}_s)} + \underbrace{\begin{bmatrix} n_{1,2} \\ n_{1,3} \\ n_{1,4} \\ n_{2,3} \\ n_{2,4} \\ n_{3,4} \end{bmatrix}}_{:= \bar{\mathbf{w}}} \tag{6.26}$$

where $\tilde{t}$ denotes the measurement vector, $\varphi(\mathbf{x}_s)$ is a nonlinear function related with the position of AUV, and $\bar{\mathbf{w}}$ denotes the noise vector with covariance matrix $\bar{\mathbf{S}}$. Along with this, the self-localization of AUV can be achieved by solving the following optimization problem, i.e., $\hat{\boldsymbol{x}}_s^* = \text{argmin}_{\hat{\boldsymbol{x}}_s}\{[\tilde{t} - \varphi(\hat{\boldsymbol{x}}_s)]^T\bar{\mathbf{S}}^{-1}[\tilde{t} - \varphi(\hat{\boldsymbol{x}}_s)]\}$, where $\hat{\boldsymbol{x}}_s^*$ and $\hat{\boldsymbol{x}}_s$ are the optimal and estimated positions of AUV, respectively. The closed solution of (6.24) is difficult to be obtained due to the nonlinear function $\varphi(\mathbf{x}_t)$. Readers can refer to our previous works Yan et al. (2021, 2022) for further details about the underwater localization.

(2) *Visual guidance scheme*. The aim of visual guidance scheme is to assist human operator to make decision-making in short-range region, including target recognition and localization. Particularly, the target recognition function relies on the YOLOv4-tiny network (Guan et al. 2022). Specifically, the recognition process is divided into offline training and online recognition. In offline training procedure, the preprocessed and labeled dataset are employed to train the YOLOv4-tiny network. After that, the online recognition of input video stream can be achieved by the YOLOv4-tiny network embedded in AUV. Of note, the frame rate of target recognition in our system can reach 17 frames per second (FPS) in the image processor. For the target localization function, the binocular vision residual method is adopted. Firstly, the binocular camera is calibrated to obtain its owner focal length and optical center distance. Based on this, the horizontal distance difference between the left and right cameras is calculated. According to the principle of similar triangles, we can further get the target position in the camera frame, through which the global position of target can be obtained by coordinate transformation. With the detected category and the estimated position of underwater target, the motion guidance of AUV can be eventually accomplished.

(3) *Integral reinforcement learning based controller*. With the communication delay in (6.24), an IRL-based controller is designed. Particularly, the main idea of IRL is to establish a cause-and-effect relationship between the rewards and the actions, which can obtain the optimal control policy by solving the Hamilton-Jacobi-Bellman (HJB) equation in real-time (Lewis and Vrabie 2009; Song et al. 2016). Specifically, the human operator controls the master manipulator according to the feedback states and video streams. Meanwhile, AUV is driven by the slave controller, whose aim is to bilaterally track the movement of the master manipulator. To this end, the cost functions for master manipulator and slave manipulator are designed as

$$g_{\rm m} = \mathbf{E}_{\rm m}^{\rm T} Q_{\rm m} \mathbf{E}_{\rm m} + \boldsymbol{\tau}_{\rm m}^{\rm T} R_{\rm m} \boldsymbol{\tau}_{\rm m}, \ g_{\rm s} = \mathbf{E}_{\rm s}^{\rm T} Q_{\rm s} \mathbf{E}_{\rm s} + \boldsymbol{\tau}_{\rm s}^{\rm T} R_{\rm s} \boldsymbol{\tau}_{\rm s}, \tag{6.27}$$

where $\mathbf{E}_{\rm m} = \mathbf{X}_{\rm m} - \mathbf{X}_{\rm s}(t-d)$ and $\mathbf{E}_{\rm s} = \mathbf{X}_{\rm s} - \mathbf{X}_{\rm m}(t-d)$. Specifically, $\mathbf{X}_{\rm m} = [\mathbf{x}_{\rm m,d}; \psi_{\rm m,d}; \mathbf{q}_{\rm m}; \mathbf{v}_{\rm m}]$ and $\mathbf{X}_{\rm s} = [\hat{\boldsymbol{x}}_{\rm s}^*; \psi_{\rm s}; \mathbf{q}_{\rm s}; \mathbf{v}_{\rm s}]$ denote the states of master and slave manipulators, respectively. $\mathbf{x}_{\rm m,d}$ is the desired position vector of slave manipulator in inertial reference frame, within the grasping feasible region. $\psi_{\rm m,d}$ and $\psi_{\rm s}$ are the desired and actual yaw angles of AUV, respectively. $\mathbf{q}_{\rm m}$ and $\mathbf{q}_{\rm s}$ are the joint angle vectors of master and slave manipulators, respectively. $\mathbf{v}_{\rm m}$ and $\mathbf{v}_{\rm s}$ are the velocity vectors of master and slave manipulators, respectively. $Q_{\rm m}$, $R_{\rm m}$, $Q_{\rm s}$ and $R_{\rm s}$ are positive definite matrices. $\boldsymbol{\tau}_{\rm m}$ and $\boldsymbol{\tau}_{\rm s}$ are the control torques of master and slave manipulators, respectively. In this way, the cost function $g_{\rm m}$ can ensure master site track the motion of slave site, and $g_{\rm s}$ can guarantee the slave site move autonomously according to the intention of human operator. For more clear depiction, the control framework of slave site is shown by Fig. 6.6.

Along with this, the value functions of master and slave manipulators are defined by $V(\mathbf{X}_{\rm m}(t))$ and $V(\mathbf{X}_{\rm s}(t))$, respectively. The optimal problem is formulated as finding optimal control policies $\boldsymbol{\tau}_{\rm m}^*$ and $\boldsymbol{\tau}_{\rm s}^*$ to minimize the value functions. To this end, the policy iteration is adopted to seek the optimal policies, which includes the fol-

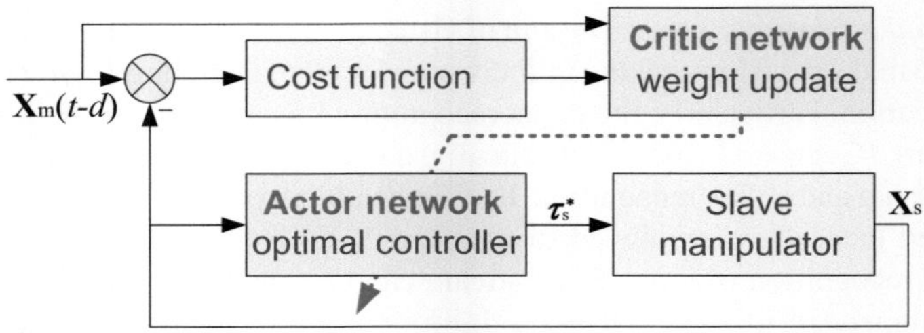

Fig. 6.6 Control framework of the slave manipulator

lowing phases. 1) *Initialization:* Choose the admissible policies $\tilde{\boldsymbol{\tau}}_{\mathrm{m}}$ and $\tilde{\boldsymbol{\tau}}_{\mathrm{s}}$, while setting the initial value functions $V(\mathbf{X}_{\mathrm{m}}(0))$ and $V(\mathbf{X}_{\mathrm{s}}(0))$ as 0; 2) *Policy Evaluation:* Determine the values of the current policies by value functions $V_\sigma(\mathbf{X}_{\mathrm{m}}(t))$ and $V_\sigma(\mathbf{X}_{\mathrm{s}}(t))$ at iteration time σ. 3) *Policy Improvement:* Update the control policies $\boldsymbol{\tau}_{\mathrm{m},\sigma+1}$ and $\boldsymbol{\tau}_{\mathrm{s},\sigma+1}$ for master and slave manipulators, respectively.

Note that the $V(\mathbf{X}_{\mathrm{m}}(t))$, $V(\mathbf{X}_{\mathrm{s}}(t))$, $\boldsymbol{\tau}_{\mathrm{m},\sigma+1}$ and $\boldsymbol{\tau}_{\mathrm{s},\sigma+1}$ are unknown to master and slave manipulators. In view of this, the approximate dynamic programming is adopted to determine the optimal control solution by measuring the states of the system forward in time. The theory is based on the neural network approximation and integral temporal difference error. Particularly, we use critic and actor networks to approximate the above unknown functions, which can be denoted as $\hat{V}_{\mathrm{m},\sigma}$, $\hat{V}_{\mathrm{s},\sigma}$, $\hat{\boldsymbol{\tau}}_{\mathrm{m},\sigma+1}$ and $\hat{\boldsymbol{\tau}}_{\mathrm{s},\sigma+1}$. The weights of the above networks can be adaptively updated by using the recursive least-squares or gradient descent methods, until the integral temporal difference errors of master and slave converge manipulators converge to 0.

6.4 Technology Practice

6.4.1 Case Study I: Integrated Design for Target Tracking Control

We first verify our solution in the indoor pool, as shown in Fig. 6.7a. Particularly, the obstacle (i.e., the yellow bucket) position is set as follows: $O = ((3\mathrm{m}, 2.5\mathrm{m}, -1\mathrm{m}), 0.3\mathrm{m}, 1.2\mathrm{m})$. The initial parameters of network are $p = q = 4$, $\alpha_1 = 0.59$, $\alpha_2 = 0.02, \gamma = 0.99, \gamma_1 = 10^{-4}$ and $\gamma_2 = 10^{-3}$. The communication-related parameters are set as follows: $N = 64$, $\Upsilon = 2/3$, $T = 5\mathrm{ms}$, and $T_{zp} = T/4$. Accordingly, the motion trajectory of AUV is shown by Fig. 6.7b. We can find AUV can arrive at its desired destination while avoiding collision with obstacle. To highlight superiority, the bandwidth utilization comparison between integrated design in this chapter and the separate design in Khalil et al. (2022), Zhou et al. (2022) is shown by Fig. 6.7c. It can be seen that the integrated design of detection and communication signals in this chapter can improve spectrum efficiency. Besides that, the integrated waveform

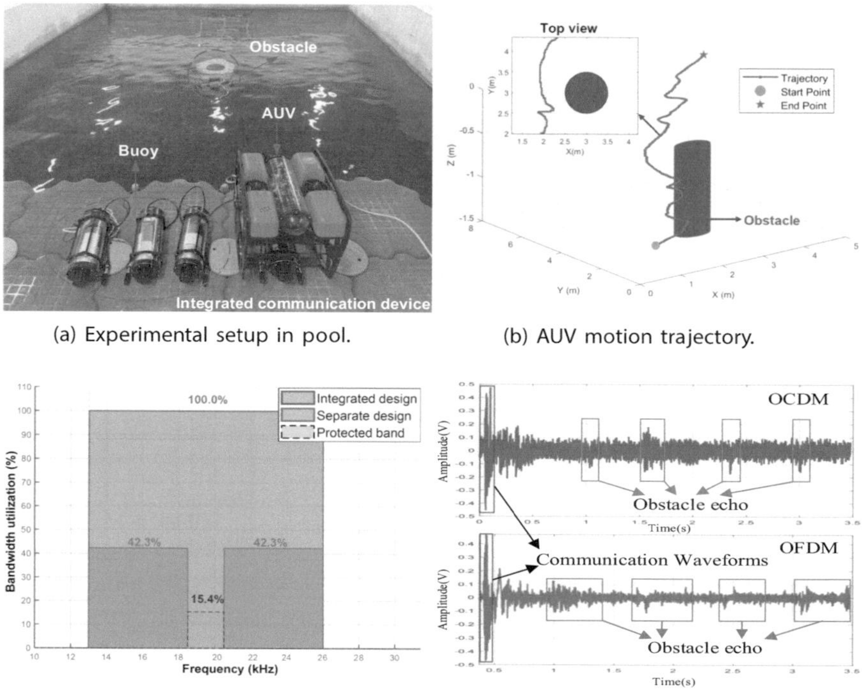

(a) Experimental setup in pool.

(b) AUV motion trajectory.

(c) Bandwidth utilization comparison.

(d) Received wave comparison.

Fig. 6.7 Experimental results in the indoor pool

comparison between OCDM in this chapter and the OFDM in Yan et al. (2024) is shown by Fig. 6.7d. Compared with the OFDM mode, the OCDM mode in this chapter has the following advantages: (1) it can improve the detection accuracy due to the concentration of echo energy; (2) the multipath superposition is weak due to the time-frequency characteristics, i.e., OCDM waveforms mitigate multipath effects and improve the detection performance.

Next, we give the experimental results in lake, as shown in Fig. 6.8a. The motion trajectory of AUV with MPQ learning-based algorithm in this chapter is provided by Fig. 6.8b. Meanwhile, the motion trajectory with DDPG-based algorithm (e.g., Yan et al. 2023) is also presented in Fig. 6.8b. To show more clearly, the relative distances from AUV to obstacle are depicted in Fig. 6.8c. These results reflect that both methods are capable of guiding AUV to its destination. However, DDPG-based controller occasionally leads to collisions with obstacles due to its tendency to local optimum. Finally, the received wave comparison on AUV is provided in Fig. 6.8d. We can find that the OCDM waveform yields clearer obstacle echoes than OFDM, further verifying its enhanced detection capability. Through extensive experiments, we find that the proposed integrated solution not only achieves collision-free motion planning

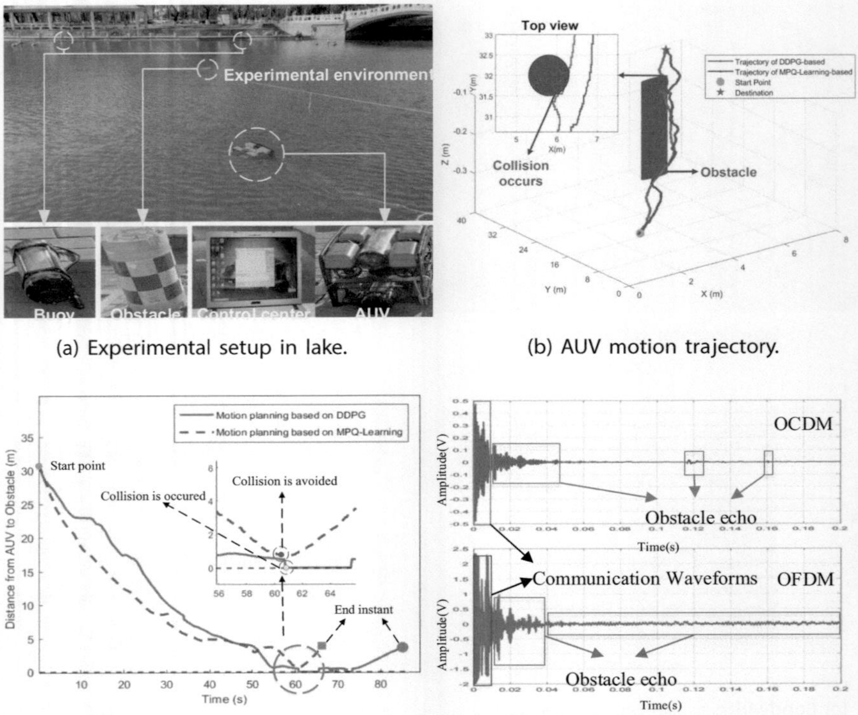

(a) Experimental setup in lake. (b) AUV motion trajectory.

(c) Distance from AUV to obstacle. (d) Received wave comparison.

Fig. 6.8 Experimental results in the lake

but also simultaneously improves detection performance and spectrum utilization efficiency compared to existing approaches.

6.4.2 Case Study II: Integrated Design for Teleoperation Control

In order to verify our approach, we first construct the teleoperation system at Yanming pool in Yanshan University. The deployment map for such system is shown in Fig. 6.9.

Specifically, human operator is located on the bridge, whose aim is to remotely drive AUV to reach the target. Buoy provides self-localization reference and data replay services for AUV. In addition, AUV patrols the underwater monitoring region, and then reaches the target (e.g., sea cucumber in our experiment). According to the AUV state and target position relayed by buoy, human operator drives AUV to the target point successfully. The screenshots for the tracking control of AUV in differ-

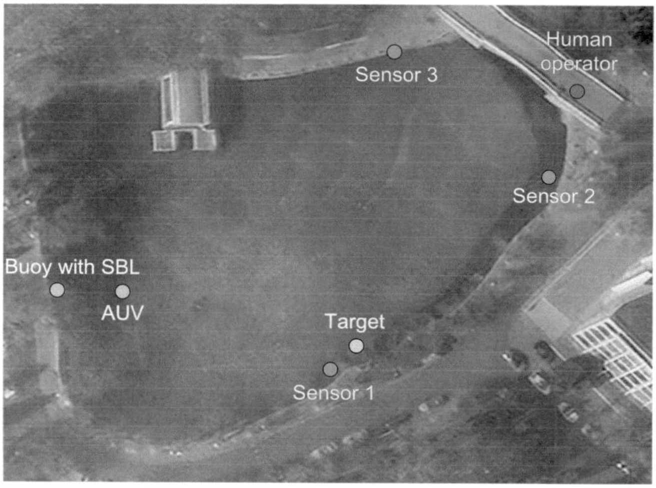

Fig. 6.9 Deployment map for the underwater teleoperation system

ent time instants are shown in Fig. 6.10, while the experiment setup is depicted by Fig. 6.11. Note that the target is hung at the bottom of another AUV at a depth of 1.26 m, as shown by Fig. 6.10. Define buoy as the origin point, while $\mathbf{x}_{t,x}$, $\mathbf{x}_{t,y}$ and $\mathbf{x}_{t,z}$ are the positions of AUV on X, Y and Z axes, respectively. Correspondingly, $\mathbf{e}_{t,x}$, $\mathbf{e}_{t,y}$ and $\mathbf{e}_{t,z}$ are the tracking errors between AUV and target in X, Y and Z axes, respectively. Then, the trajectory of AUV during the teleoperation procedure is shown in Fig. 6.12, while the tracking errors are shown in Fig. 6.13. Clearly, AUV can reach the target point, and the tracking errors can converge to 0. That means the teleoperation in our system is achieved.

In this section, we construct the teleoperation system in the coastal area of the Bohai Sea, which is a shallow inland sea with an average water depth of 18 m in China. Specifically, one human operator is deployed on the beach to remotely control the slave AUV through the master unit, whose geographic coordinate is set as E 119°27'32.23", N 39°48'32.46". Meanwhile, AUV in the sea receives the decision from master site and then carries out the tracking task for target.

To this end, we set $\mathbf{x}_{t,d} = [7.147\text{m}; 21.883\text{m}; -4.853\text{m}]$ as the desired position vector of AUV. The data interaction between human operator and AUV spans 2 Km with the signal-noise radio of 36 dB. Through the wireless communication links, the bidirectional transmission between master and slave sites can be constructed. Besides that, the SBL localization system can provide real-time position of AUV, such that the tracking errors can be measured for the closed-loop control system. Combining the bidirectional transmission into the closed-loop control system, the human operator remotely controls the AUV. Based on this, the marine experiment

Fig. 6.10 Screenshots for the tracking of AUV in different time instants

Fig. 6.11 Experiment setup, where the operator remotely drives AUV to reach the target

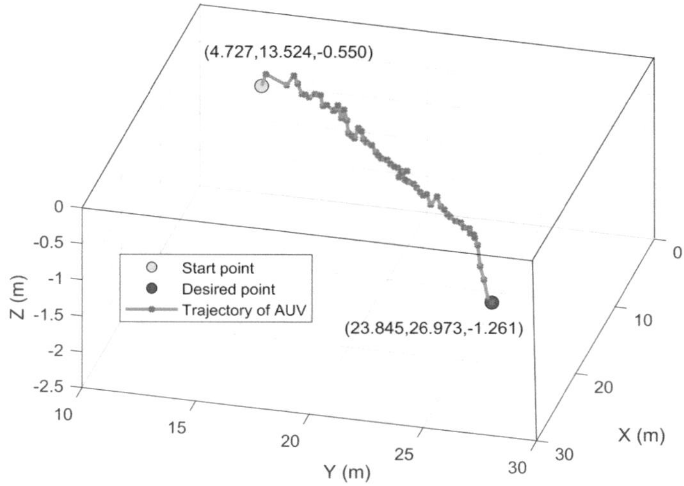

Fig. 6.12 Trajectory of AUV during the teleoperation control

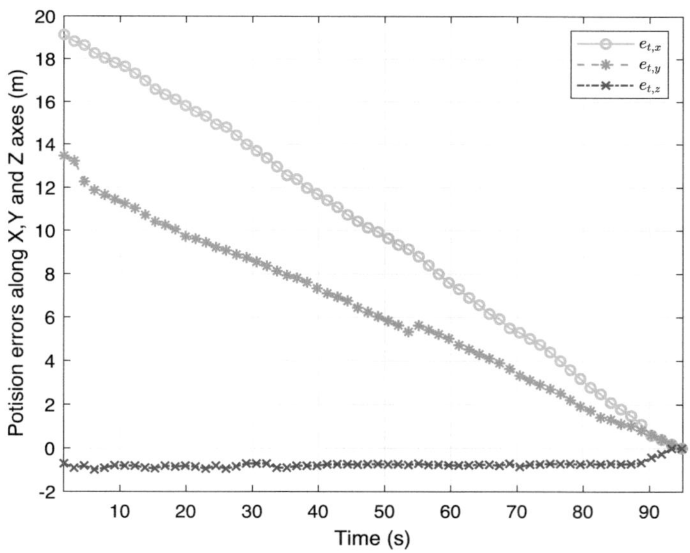

Fig. 6.13 Position errors between AUV and target

Fig. 6.14 Setup of the marine experiment in the coastal area of the Bohai Sea

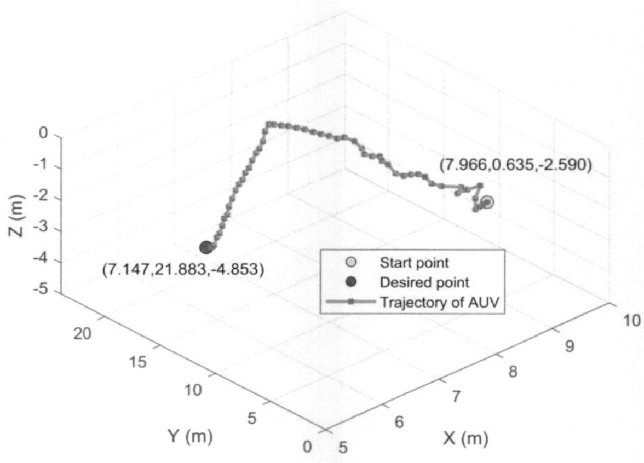

Fig. 6.15 Trajectories of AUV in marine environment

setup is shown in Fig. 6.14, where the persons on board only need to deploy the AUV in a specific area. By using the learning-based controller, the position trajectories and the tracking errors of AUV are shown in Figs. 6.15 and 6.16, respectively. Clearly, AUV is steered to the desired position. Meanwhile, the position errors on X, Y and Z axes can converge to 0. These results demonstrate the effectiveness of the proposed underwater teleoperation system in real marine environment.

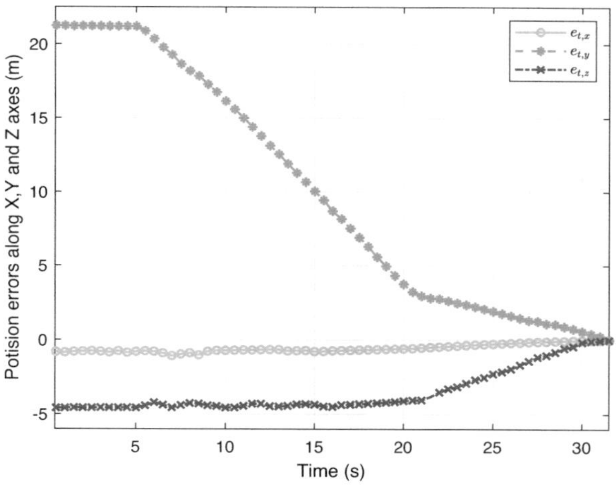

Fig. 6.16 Tracking errors of AUV

6.5 Summary

This chapter has explored an integrated design paradigm that unifies detection, communication, and control for intelligent underwater systems. By merging the physical-layer signal design with high-level decision and control strategies, the proposed framework enables underwater vehicles to perceive, exchange, and act in a tightly coupled and energy-efficient manner.

The integration of OCDM-based communication detection waveforms with learning-driven control algorithms allows autonomous underwater vehicles to simultaneously achieve self-localization, environmental awareness, and adaptive motion planning, thereby overcoming the traditional separation between sensing and control. Extending this principle to human-centered operations, the human-on-the-loop teleoperation architecture further demonstrates how bidirectional acoustic radio links, visual perception, and integral reinforcement learning can collaboratively ensure transparency and stability under long communication delays and uncertain dynamics.

Experimental validations in both controlled pools and real marine environments confirm that the integrated co-design not only enhances spectral and energy efficiency but also significantly improves control accuracy, robustness, and responsiveness.

References

Guan, L., Jia, L., Xie, Z., & Yin, C. (2022). A lightweight framework for obstacle detection in the railway image based on fast region proposal and improved YOLO-Tiny network. *IEEE Transactions on Instrumentation and Measurement, 71*(1), 1–16.

Guo, T., Yan, J., Yang, X., Zhang, T., Chen, C., & Guan, X. (2025). Motion planning of an autonomous underwater vehicle via the integrated design of detection, communication and control. *IEEE/CAA Journal of Automatica Sinica*. https://doi.org/10.1109/JAS.2025.125543

Hokayem, P., & Spong, M. (2006). Bilateral teleoperation: An historical survey. *Automatica, 42*(12), 2035–2057.

Khalil, R., Saeed, N., Babar, M., Jan, T., & Din, S. (2022). Bayesian multidimensional scaling for location awareness in hybrid-internet of underwater things. *IEEE/CAA Journal of Automatica Sinica, 9*(3), 496–509.

Lewis, F., & Vrabie, D. (2009). Reinforcement learning and adaptive dynamic programming for feedback control. *IEEE Transactions on Systems, Man, and Cybernetics: Systems, 9*(3), 32–50.

Liu, Y., Liao, G., Xu, J., Yang, Z., & Zhang, Y. (2017). Adaptive OFDM integrated radar and communications waveform design based on information theory. *IEEE Communications Letters, 21*(10), 2174–2177.

Moniruzzaman, M., Rassau, A., Chai, D., & Islam, S. (2022). Teleoperation methods and enhancement techniques for mobile robots: A comprehensive survey. *Robotics and Autonomous Systems, 150*, 1–31.

SBL homepage by Water Linked. Accessed: Nov. 2022. [Online]. Available: https://waterlinked.com/underwater-gps

Song, R., Lewis, F., Wei, Q., & Zhang, H. (May2016). Off-policy actor-critic structure for optimal control of known systems with disturbances. *IEEE Transactions on Cybernetics,46*(5), 1041–1050.

Sun, Z., Guo, H., & Akyildiz, I. (2022). High-data-rate long-range underwater communications via acoustic reconfigurable intelligent surfaces. *IEEE Communications Magazine, 60*(10), 96–102.

Wu, H., Zhang, K., Mei, X., et al. (2025). Heuristic strategy-empowered real-time path following for autonomous surface vessel with adaptive line-of-sight guidance. *IEEE Journal of Oceanic Engineering, 50*(1), 307–323.

Yan, J., Li, X., Yang, X., Luo, X., Hua, C., & Guan, X. (2022). Integrated localization and tracking for AUV with model uncertainties via scalable sampling-based reinforcement learning approach. *IEEE Transactions on Systems, Man, and Cybernetics: Systems*, *52*(11), 6952-6967.

Yan, J., Zhao, H., Meng, Y., & Guan, X. (2021). Localization in underwater sensor networks. Springer.

Yan, J., Cui, L., Yang, X., Chen, C., & Guan, X. (2024). Design of an embedded system for integrated underwater communication and detection. *IEEE Embedded Systems Letters, 17*(2), 83–86.

Yang, X., Shi, Y., Liu, W., Ye, H., Zhong, W., & Xiang, Z. (2022). Global path planning algorithm based on double DQN for multi-tasks amphibious unmanned surface vehicle. *Ocean Engineering,266*(1), 1–14.

Yan, J., Zhang, L., Yang, X., Chen, C., & Guan, X. (2023). Communication-aware motion planning of AUV in obstacle-dense environment: A binocular vision-based deep learning method. *IEEE Transactions on Intelligent Transportation Systems, 24*(12), 14927–14943.

Yu, L., Qiao, L., & Shen, C. (2024). High-speed obstacle avoidance of a large-scale underactuated autonomous underwater vehicle under a finite field of view. *IEEE Transactions on Automation Science and Engineering, 22*, 1996–2005.

Zhou, Z., Liu, J., & Yu, J. (2022). A survey of underwater multi-robot systems. *IEEE/CAA Journal of Automatica Sinica, 9*(1), 1–18.

Zhou, S., & Wang, Z. (2014). *OFDM for underwater acoustic communications*. NJ, USA: Wiley.

Chapter 7
Demonstration Application in Port Monitoring System

7.1 Introduction

Port activities, such as increased shipping traffic, intensive fishing operations, and offshore energy extraction, coupled with climate change, have placed unprecedented pressure on marine ecosystems Zhuang et al. (2025), Fitzpatrick et al. (2023). For example, there were at least 2400 incidents in maritime ports as reported by the data from 2022 New Data Reveals Risk of Maritime Incidents in Ports and Terminals (2023). For that reason, the effective monitoring of port activities is essential to address these challenges, ensuring the sustainable use of ocean resources and the protection of marine biodiversity. Specifically, the key areas of port activity monitoring include ship tracking, port operations and security, shipwreck detection and recovery, search and rescue operations for drowning persons, and related port activities.

Traditionally, the port monitoring has been carried out through human visual inspections or radar Tran et al. (2024). However, human visual inspections depend on the experience, skill, and judgment of the inspector, which may lead to potential variability in the results. Besides that, human vision can only detect surface-level problems, and it cannot provide long-term monitoring especially in harsh underwater environment. On the other hand, the radar-based system cannot monitor subsurface conditions, which is expensive to install, maintain and operate. For the aforementioned reasons, the unmanned aerial vehicles (UAVs) are increasingly applied for port monitoring. Particularly, UAVs are remotely controlled or fly autonomously based on pre-programmed flight plans or real-time data, and they can provide more cost-effective solution for monitoring large areas compared to traditional human inspection methods. In view of this, an UAV-driven solution for assisting structural health monitoring solution was developed in Ref. Tsaimou et al. (2023), where the structural health monitoring practices were developed to asset management framework. In Peng et al. (2021), UAVs were employed to monitor the ship exhaust emissions, where the collected data on UAVs can be transmitted wirelessly to the ground station. In order to extend battery life, ships can be adopted to act the mobile supply bases of

J. Yan and X. Guan, *Maritime Unmanned Systems*, Cognitive Intelligence and Robotics 1, https://doi.org/10.1007/978-981-95-6907-6_7

UAVs, and hence, the joint routing and scheduling solution was developed to monitor the pollution from vesselsShen et al. (2022). A LabVIEW-based online monitoring and safety evaluation system for UAVs was designed in Ref. Zhang et al. (2024), and its aim was to offer a flight guarantee for autonomous flying. Nonetheless, the UAV-based monitoring systems suffer from the limited battery life and payload capacity Khalid et al. (2023), Adil et al. (2023), which are not sufficient for large-scale or continuous monitoring of vast port areas. In addition, UAVs are sensitive to weather conditions. For example, the strong winds, heavy rain, or poor visibility (such as fog) limit their performance, resulting in missed monitoring opportunities and compromised data quality Xia et al. (2023). These properties limit the application fields of UAVs in the port monitoring activities.

Based on the above considerations, some researchers attempt to employ autonomous underwater vehicles (AUVs) to monitor the port activities. Compared with the UAV-based monitoring system, AUVs can inspect the hulls of ships to identify corrosion, damage, or biofouling (e.g., the accumulation of marine organisms on the hull). This helps reduce the need for expensive dry-docking or manual inspections, providing more frequent and detailed assessments. Meanwhile, AUVs can monitor the condition of port region, including checking for sedimentation, pollution levels, and overall water quality. Inspired by this, an AUV-assisted sequence algorithm for towed force estimation was developed Choi et al. (2005). In McNelly et al. (2023), an AUV prototype was designed to support the protection challenges. Followed by this, a swarm of AUVs were employed to measure the submesoscale dynamics Jaffe et al. (2017), which can measure the underlying horizontal flows within the swarm. Based on the idea of mean-shift exploration, a swarm strategy for terrestrial vehicle was developed to assemble highly complex shapes Sun et al. (2023). However, the proposed solutions in Refs. Choi et al. (2005), McNelly et al. (2023), Jaffe et al. (2017), Sun et al. (2023) focused on the motion control of vehicles, and they ignored the sensing, communication and control abilities of vehicles during the monitoring procedure. As far as we know, the port environment is harsh, which makes it hard to accurately acquire the sensing information of marine targets. On the other hand, surface and subsurface wireless communications are often limited, which make real-time data transmission challenging. Although the seafloor telecom cables can be deployed to sense the underwater activities Sladen et al. (2019), Lin et al. (2024), the telecom cables have poor mobility ability, which may significantly reduce the application domain for dynamic environment. With consideration of the harsh underwater environment, how to systematically design the AUV-assisted system for port monitoring is largely unsolved.

Here, we propose a ubiquitous monitoring solution for port activities by using the cooperation of AUVs, as depicted by Fig. 7.1a. The main idea is summarized as follows: when AUVs are on the surface, they recognize the surface target (e.g., ship or buoy) with the onshore cameras, such that the target attitudes can be estimated by using deep learning technology; when AUVs are under the water surface, they detect the underwater target (e.g., shipwreck or drowning person) with the onshore sonars. Meanwhile, the radio and acoustic waves are both adopted to ensure the bilateral communication between human operator and AUVs. Finally, the experimental results

(a) Description of the monitoring system.

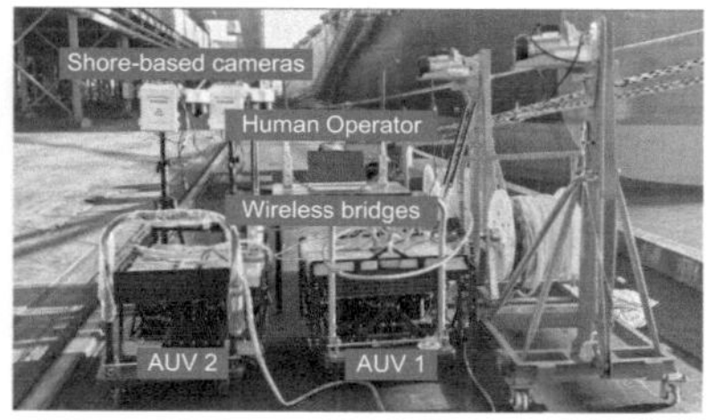

(b) Experimental setup, including AUVs, human operator, onshore cameras, and communication network.

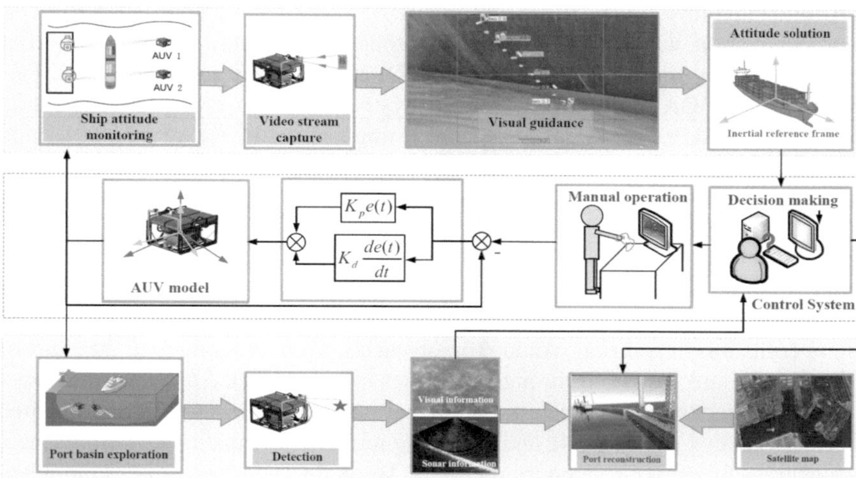

(c) Block diagram of our monitoring, where human operator makes remote operations based on the collected measurements.

Fig. 7.1 Ubiquitous monitoring of port activities

verified the effectiveness of our solution. Compared with existing port monitoring methods, the main contributions of our solution are as follows:

(1) **Systematical design of the AUV-assisted monitoring system.** Key components include two AUVs, a human operator, two onshore cameras, and a wireless communication network. The proposed system can recognize the ship waterline, and more importantly, it can detect the port basin with the assistance of human in a boat/vessel on water offshore;
(2) **On-line ship waterline recognition algorithm**. With regard the the dynamic ocean environment, we establish the target dataset to offline train the YOLOv8-obb network, such that an on-line recognition algorithm is developed to estimate the state of ship;

(3) **Teleoperation control of AUVs**. A remote teleoperation controller is developed to drive AUVs to the neighboring areas of ship and port basin with the help of human on the loop, making it an essential tool for complex underwater missions.

7.2 Integration Methods for Port Monitoring System

7.2.1 Self-localization and Motion Control

As mentioned above, AUVs use GPS for surface localization and UGPS for underwater localization, which allow AUVs to determine their position information. After that, they employ proportional derivative (PD) controller for motion control. Particularly, the TDOA algorithm is utilized by the UGPS to estimate the position information of AUVs. To simplify without losing generality, we take AUV 1 as an example, through which the principle for self-localization of AUV 1 via UGPS can be provided as follows. When performing 3D path planning, the influence of the surrounding port environment is explicitly considered. The water surface defines the upper operational boundary of the AUV to prevent it from accidentally surfacing except when GPS localization is required. The seafloor defines the lower limit of navigation depth, which is obtained from depth sensors or bathymetric data to avoid ground collision. In addition, volumetric obstacles, such as submerged facilities or moored ships, are identified through sonar sensing and treated as restricted zones during route generation. These environmental boundaries ensure that the planned trajectory remains within a safe and navigable region of the underwater workspace.

Four transducers are carried by the buoy. We define t_h as the time stamp when AUV 1 broadcasts its localization request, and let the AUV position vector be $\mathbf{p} \in \mathbb{R}^3$. Similarly, we denote the time stamp when each transducer receives the signal as t_i, where the known position of transducer $i \in \{1, 2, 3, 4\}$ is $\mathbf{r}_i \in \mathbb{R}^3$. By the principle of TDOA, the time difference between transducers i and j is

$$
\begin{aligned}
t_{i,j} &= (t_i - t_h) - (t_j - t_h) \\
&= \frac{1}{c_w}(||x_i - x_h|| - |||x_j - x_h||) + n_{i,j}, i \neq j
\end{aligned}
\tag{7.1}
$$

where $c_w \approx 1500$m/s denotes the average underwater acoustic speed, and $n_{i,j}$ denotes the measurement noise. Of note, $n_{i,j}$ follows a zero-mean Gaussian distribution.

Using (7.1), the time-difference measurements are stacked as

$$\underbrace{\begin{bmatrix} \widetilde{t}_{1,2} \\ \widetilde{t}_{1,3} \\ \widetilde{t}_{1,4} \\ \widetilde{t}_{2,3} \\ \widetilde{t}_{2,4} \\ \widetilde{t}_{3,4} \end{bmatrix}}_{:=\widetilde{\mathbf{t}}} = \frac{1}{c_{\mathrm{w}}} \underbrace{\begin{bmatrix} \|\mathbf{r}_1 - \mathbf{p}\| - \|\mathbf{r}_2 - \mathbf{p}\| \\ \|\mathbf{r}_1 - \mathbf{p}\| - \|\mathbf{r}_3 - \mathbf{p}\| \\ \|\mathbf{r}_1 - \mathbf{p}\| - \|\mathbf{r}_4 - \mathbf{p}\| \\ \|\mathbf{r}_2 - \mathbf{p}\| - \|\mathbf{r}_3 - \mathbf{p}\| \\ \|\mathbf{r}_2 - \mathbf{p}\| - \|\mathbf{r}_4 - \mathbf{p}\| \\ \|\mathbf{r}_3 - \mathbf{p}\| - \|\mathbf{r}_4 - \mathbf{p}\| \end{bmatrix}}_{:=\phi(\mathbf{p})} + \underbrace{\begin{bmatrix} n_{1,2} \\ n_{1,3} \\ n_{1,4} \\ n_{2,3} \\ n_{2,4} \\ n_{3,4} \end{bmatrix}}_{:=\mathbf{w}}, \tag{7.2}$$

where $\widetilde{\mathbf{t}}$ is the measurement vector, $\phi(\mathbf{p})$ is a *vector-valued nonlinear* function of the AUV position, and $\mathbf{w}$ is the noise vector with covariance matrix $\mathbf{S}$.

With the estimated position information, the current position vector of AUV 1 can be obtained. Then, AUV 1 is required to arrive at the neighboring areas of the ship. The PD controller calculates the error between the current position and the target position, through which the control input is adjusted to move the AUV towards the desired state. Specifically, the motion control of the AUV can be divided into position control and heading control. The goal of position control is to move AUV 1 from its current position to the target position. To this end, the error $\mathbf{p}_{\mathrm{t}}$ is given by

$$\mathbf{e}(t) = \mathbf{p}_{\mathrm{t}} - \mathbf{p}(t). \tag{7.3}$$

The PD controller output for *position* control is

$$\mathbf{u}(t) = K_p\, \mathbf{e}(t) + K_d\, \dot{\mathbf{e}}(t), \tag{7.4}$$

where K_p and K_d are scalar gains, and $\dot{\mathbf{e}}(t)$ is the time derivative of the position error vector.

The goal of heading control is to drive the AUV to the desired heading θ_{t}. Let the heading error be

$$\Delta\theta(t) = \theta_{\mathrm{t}} - \theta(t). \tag{7.5}$$

Then the PD controller output for *heading* control is

$$u_\theta(t) = K_{p\theta}\, \Delta\theta(t) + K_{d\theta}\, \dot{\Delta}\theta(t), \tag{7.6}$$

where $K_{p\theta}$ and $K_{d\theta}$ are scalar gains for heading control.

Accordingly, the STM32 processor computes the pulse-width modulation (PWM) signals to drive the thrusters. Since the thruster power is proportional to the applied voltage, the PWM duty cycle is adjusted to change the thruster voltage. When necessary (e.g., in port emergencies), a human operator can directly modify the control signal. Through the above procedure, the control strategy can be applied by AUVs in the actual environment.

7.2.2 Ship Attitude Monitoring

Ship attitude monitoring includes baseline identification and attitude estimation. Specifically, the baseline recognition process consists of two stages: offline training and online recognition. In the offline training stage, a network model is trained to meet the target recognition requirements. In the online recognition stage, the real-time baseline detection is performed by using the trained model.

Before performing the offline training, we give the evaluation index of YOLOv8-obb network. It consists of four components: localization loss function L_{loc}, object loss function L_{obj}, regularization loss function L_{reg} and classification loss function L_{cls}. The loss function is expressed as follows

$$L_{OSS} = \lambda_{cls} L_{cls} + \lambda_{loc} L_{loc} + \lambda_{obj} L_{obj} + \lambda_{angle} L_{angle} + \lambda_{reg} L_{reg}, \tag{7.7}$$

where λ_{cls}, λ_{loc}, λ_{obj} and λ_{reg} denote the weight coefficients for each loss component. Specifically, λ_{cls} controls the contribution of the classification accuracy, λ_{loc} adjusts the influence of the localization precision, λ_{obj} determines the sensitivity to the presence of target objects, and λ_{reg} regularizes the network parameters.

In (7.7), the classification loss function L_{cls} is defined as

$$L_{cls} = \sum_{i=1}^{S^2} \sum_{j=1}^{C} l_{ij}^{obj} [\hat{P}_i^j(C) \log P_i^j(C) + (1 - \hat{P}_i^j(C)) \log(1 - P_i^j(C))], \tag{7.8}$$

where C is the class of the detected object. P_i^j and $\hat{P}_i^j$ are the predicted and actual probabilities, respectively.

In (7.7), the classification loss function L_{loc} is defined as

$$L_{loc} = 1 - IOU(A, B) + \frac{\rho^2(A_c, B_c)}{m} + \alpha v, \tag{7.9}$$

with

$$\alpha = \frac{v}{[1 - IOU(A, B)] + v}, \tag{7.10}$$

$$v = \frac{4}{\pi^2} \left(\arctan \frac{w^{gt}}{h^{gt}} - \arctan \frac{w}{h} \right)^2, \tag{7.11}$$

where $IOU(A, B)$ is the intersection over union between the predicted bounding box and the ground truth bounding box. $\rho^2(A_c, B_c)$ is the squared Euclidean distance between the center points of the predicted and ground truth bounding boxes. m is the diagonal distance of the minimal enclosing region that can contain both the predicted and ground truth bounding boxes. w^{gt} and h^{gt} are the width and height of ground truth bounding box, respectively. w and h are the width and height of the predicted bounding box, respectively.

In (7.7), the classification loss function L_{obj} is defined as

$$L_{obj} = -[y_{obj} \log(\hat{p}_{obj}) + (1 - y_{obj}) \log(1 - \hat{p}_{obj})], \tag{7.12}$$

where y_{obj} is the object existence label, and $\hat{p}_{obj}$ is the predicted object score.

Meanwhile, the classification loss function L_{angle} is defined as

$$L_{angle} = \text{SmoothL1}(\sin(\theta) - \sin(\theta^{gt}) + \text{SmoothL1}(\cos(\theta) - \cos(\theta^{gt}))), \tag{7.13}$$

with

$$\text{SmoothL1}(x) = \begin{cases} 0.5x^2 & \text{if } |x| < 1, \\ |x| - 0.5 & \text{otherwise}, \end{cases} \tag{7.14}$$

where θ is the predicted angle, and θ^{gt} is the ground truth angle.

The classification loss function L_{reg} is defined as

$$L_{reg} = \sum\nolimits_{w \in \Xi} w^2, \tag{7.15}$$

where Ξ is the parameter set in the neural network model, and it need to be optimized.

The above index can improve the detection accuracy and robustness of the model in dealing with target baseline. Meanwhile, it can enhance the generalization ability of the model by regularization to prevent overfitting. The whole procedure of ship attitude monitoring is provided in detail.

Step 1. Construction of the dataset
We choose different weather, sea conditions, and targets for data collection. Specifically, the dataset in our system consists of 500 images of the target's baseline at the bow (shoreward and offshore) and stern (shoreward and offshore). After data cleaning, the dataset is split into 400 training images and 100 test images. Through the above operation, different parts of the images are labeled. To maintain timeliness under dynamically changing port conditions, an incremental dataset updating mechanism is planned. The system periodically collects new samples from AUV and onshore cameras during missions, automatically evaluating data novelty through distribution drift detection. Newly acquired samples are annotated and merged into the existing dataset using an active learning pipeline, ensuring that the YOLOv8-obb model continues to generalize across varying water clarity, lighting, and seasonal variations.

Step 2. Model training
The training images are then fed into the YOLOv8-obb network model, with the input image size adjusted to 640 × 640 to maintain consistency in input dimensions. For the model after offline training, the precision-confidence and recall-confidence curves are jointly adopted to evaluate the performance. Note that the precision measures the proportion of true positive predictions among all positive predictions, and

recall measures the proportion of true positive predictions among all actual positive instances. Thereby, the precision-confidence probability $P_{\text{precision}}$ can be defined as

$$P_{\text{precision}} = \frac{TP}{TP + FP}, \tag{7.16}$$

and the recall-confidence probability P_{recall} is defined as

$$P_{\text{recall}} = \frac{TP}{TP + FN}, \tag{7.17}$$

where TP denotes the number of correctly predicted positive instances, FP denotes the number of incorrectly predicted positive instances, FN denotes the number of actual positive instances missed by the model, and TN denotes the number of correctly predicted negative instances.

Based on (7.16) and (7.17), the model parameters are adjusted, so that it can learn patterns, relationships, and structures inherent in the data.

Step 3. Online target baseline recognition

The trained coding and decoding network is used to segment the position of the target's baseline in the image. To balance accuracy and real-time performance, the recognition module is designed with an adaptive processing strategy. In actual operation, when the computational load increases, the module will give priority to real-time response and use a lightweight model to process data at the same time. When the computing resources are stable, it is restored to a full-precision model. After that, the YOLOv8-obb network is used to identify the baseline. Based on it, the draft value of the target is calculated via the mapping between unit pixel and actual distance.

Step 4. Ship attitude estimation

Based on Step 3, we take the ship attitude estimation as an example. Particularly, the draft data (i.e., the data D_{BP} for the shoreward end of the bow, the data D_{SP} for the shoreward end of the stern, the data D_{BS} for the offshore end of the bow, and the data D_{SS} for the offshore end of the stern) can be received. After that, we conduct the target attitude estimation. Specifically, the *pitch estimation* refers to the difference in draft between the bow and stern of a target, indicating whether the target is leaning forward or backward. Then, it can be calculated by

$$H_{\text{pitch}}(T) = \left(\frac{D_{BP} + D_{BS}}{2}\right) - \left(\frac{D_{SP} + D_{SS}}{2}\right), \tag{7.18}$$

and meanwhile, the *roll estimation* refers to the difference in draft between the port (near shore) and starboard (away from shore) sides, indicating whether the ship is tilted to port or starboard. Based on this, it is calculated by

$$H_{\text{roll}}(T) = \left(\frac{D_{BP} + D_{SP}}{2}\right) - \left(\frac{D_{BS} + D_{SS}}{2}\right). \tag{7.19}$$

By using the draft measurement of the hull, the trim and balance of the target can be estimated effectively, which provides data support for the safety of the target.

We used Google Satellite Offline Maps and images taken on-site to reconstruct the 3D scene of the underwater environment using Unity 3D. The Google Satellite Offline Maps provide geographic coordinates and terrain data for constructing the 3D scene of the underwater environment. The on-site images provide visual information for the 3D scene, helping to fill in the details that may be missing from the offline maps. The on-site images include important features such as the docks, ships, port boundaries, buildings, and the water surface. The terrain is created using Unity's terrain tools, and external models (such as docks, ships, and gantry cranes) are imported to complete the underwater environment. Based on this, images or sonar data collected by AUVs will be displayed in the virtual environment. This data can provide a foundation for subsequent structural analysis and 3D modeling of the underwater environment.

7.2.3 Management Platform Design

It uses a combination of AUVs and shore-based cameras to monitor the attitude of surface/underwater target. AUVs are equipped with sonar for detecting underwater features and cameras for capturing images of the target's baseline. The sonar provides images of underwater features, while the cameras capture the baseline of the target, further supplementing the sonar data with visual information. The AUVs are deployed to monitor the baseline on the offshore side of the target, while shore-based cameras are positioned on the dockside to monitor the visible baseline of the target. Data from both AUVs and shore-based cameras are transmitted to the host computer via wireless bridges and wired connections. This setup allows us to obtain images of the baseline from different directions. By applying baseline recognition algorithms and target attitude estimation algorithms, we can determine the real-time attitude of the target. We develop a user-friendly interface using Qt, which integrates data from the AUVs and shore-based cameras, providing a centralized monitoring platform for human operators. The platform displays real-time sensor data, the target's attitude, and collected image information. The platform supports both automatic and manual control of the AUVs, allowing operators to manage the monitoring system with flexibility. Additionally, it includes real-time data logging and alert mechanisms to ensure the safety of the targets.

7.3 Construction of Port Monitoring System

7.3.1 Hardware Design

As shown in Fig. 7.1b, the hardware design of the proposed monitoring system mainly consists of AUVs, human operator, onshore cameras, and wireless communication networks.

(1) *AUV*. Note that the collective motion of biology can create adaptable, efficient, and robust swarms capable of performing diverse tasks in various environments Slade et al. (2024), Xiao et al. (2024). Particularly, each biological system (such as bird, fish and insect) interacts with its neighboring individuals to perform complex and intelligent behavior, which enable the transition from individual intelligence to collective intelligence. Inspired by this, we adopt the cooperation of AUVs to perform the monitoring task. Specifically, the size of AUV in our system is 75 cm×57 cm×55 cm, and its weight is 90 kg. It includes buoyancy modules, searchlight, thrusters, control cabin, camera and detection sonar. Specifically, the buoyancy modules allow AUV to achieve neutral buoyancy, maintain depth control, optimize energy usage, and remain stable throughout its mission. The searchlight can provide a maximum brightness of 2200 lumens, and its role is to enhance visibility during AUV operations. Six thrusters (i.e., two vertical thrusters and four horizontal thrusters) are deployed on each AUV, which enable AUV to move through the water, control its speed, direction, and orientation, and carry out complex monitoring tasks. Each thruster is driven via a PWM control signal generates by the STM32 processor, allowing fine adjustment of thrust magnitude. Of note, each thruster provides 8 kgf of thrust, and it allows the AUV to reach a maximum speed of 4 knots. The roles of camera and detection sonar are to detect the target (i.e., the surface and underwater targets) by optic and acoustic measurements, respectively. Note that the sonar is produced by AOHI Detection sonar data sheet (2024), which has a maximum operating frequency of 1.2MHz, a maximum number of beams of 512, a horizontal beam angle of 130°/130°, and a vertical beam angle of 20°/20°. In addition, the entire AUV is powered by a power supply with an output of 24V.
In the control cabin, the Jetson Nano development board functions as the high-level control unit of the AUV, responsible for real-time image processing, sensor fusion. The STM32 processor serves as the low-level controller, executing thruster control, sensor data acquisition, and safety monitoring. A master Cslave architecture is established between the Jetson Nano and STM32 via serial communication, enabling real-time command execution and feedback. This architecture ensures deterministic response in time-critical tasks while allowing the Jetson Nano to handle computationally intensive operations. It provides the necessary computational power for real-time data processing and computer vision tasks in an energy-efficiency way. In order to drive the thrusters and receive the measurement data, the STM32 processor is also installed on AUV. Additionally,

the GPS, IMU, depth sensor and ranging sonar, are jointly employed to capture the real-time motion of AUV.

(2) *Human Operator*. The measured data and video stream are first decoded, and then displayed by the host computer in a real-time way. Of note, the human operator analyzes this information to make real-time decision. Specifically, the human operator interacts with the underwater remote environment through the manipulator, which allows the precise control in hazardous port environment. Notably, the control commands are initiated by the human operator, and then be transmitted back to the AUV by wireless bridges. Accordingly, the underwater teleoperation can be accomplished.

(3) *Onshore Camera*. As shown by Fig. 7.1a, onshore cameras are deployed to monitor the coastal surface target. The cameras are produced by Hikvision Onshore camera data manual (2024), with 400w pixels and 360-degree rotation with internal servos. The onshore cameras collect images of the target's baseline (e.g., the ship waterline) and transmit them to host computer, through which host computer can estimate target's attitude via the embedded deep learning model.

(4) *Telecommunication Network*. The following three transmission channels are adopted in our system: (a) the wired transmission is employed to achieve data transmission between onshore cameras and host computer; (b) the umbilical communication is employed to achieve data transmission between AUV and cable drum trolley/cart; (c) the wireless bridges are employed to achieve data transmission between cable drum trolley/cart and host computer. For the wired transmission, it adopts IEEE standard 802.3 protocol IEEE Std 802.3-2008, IEEE Standard for Ethernet (2008), whose maximum transmission rate can reach 1000 Mb/s. For the umbilical communication, it adopts IEEE standard 1901.1 protocolIEEE Std 1901.1.1-2020 (2020), with a frequency band of 30 MHz and a maximum transmission rate of up to 500 Mb/s. The bridge equipment uses a 6db omnidirectional antenna to transmit data. Based on this, a self-organizing network configuration can be constructed, whose broadband peak frequency is 120 Mbps. This hybrid communication method of wired and wireless to ensure communication reliability. When communication interference or temporary disconnection occurs, a redundant failover mechanism switches the active channel based on signal quality monitoring.

7.3.2 Software Design

As shown in Fig. 7.1c, the software design includes the self-localization of AUVs, the target attitude monitoring, the monitoring environment three dimensional (3D) reconstruction, the construction of management platform, and the integration with DeepSeek model.

(1) *Self-Localization of AUVs*. The role of the localization unit is to provide the real-time position information of AUVs. When AUVs are on the water surface,

GPS is adopted to acquire the position information, and its accuracy is within 0.1 m. When AUVs are under the water surface, underwater GPS produced by WaterLinked is adopted. Of note, the Time Difference of Arrival (TDOA) algorithm is conducted by underwater GPS, where four receivers are installed on the baseline of T-shaped structure to receive localization signals from the AUVs. Besides that, IMU serves as a supplementary localization unit, especially when the signals from GPS or underwater GPS are not well received due to the environmental noises. It is noted that the Kalman filtering algorithm is conducted by the IMU.

(2) *Target Attitude Monitoring*. The procedure of target attitude monitoring includes the following three steps: (a) Construct the preprocessed and labeled target baseline datasets, through which the offline models of YOLOv8-obb network and decoder coding network can be constructed; (b) Collect the target baseline images by the cameras mounted on AUVs and shore; (c) Acquire the positions of baseline within the images and identificate them by using the YOLOv8-obb model, such that the target's draft values are obtained; (d) With the above draft values, one estimates the attitude of target by difference calculation.

(3) *Monitoring Environment 3D Reconstruction*. The role of the 3D reconstruction is to create a detailed, accurate, and interactive 3D model of the monitoring environment. This allows us to accurately document potential defects and visualize the environment in a highly intuitive and interactive way. To this end, the Unity 3D (U3D) software is employed to reconstruct the monitoring environment, and its procedure consists of two main parts: a) *Offline environment setup*. A 3D model of the monitoring environment is constructed by using satellite map and on-site photographs; b) *Online recognition and recording*. The optical and acoustic data collected by AUVs are used to identify and annotate potential defects, (e.g.., cracks and exposed reinforcement bars). Clearly, the above implementation makes it convenient to monitor long-term structural integrity and detect any defects in port monitoring environment.

(4) *Management Platform*. The role of the management platform is to control the entire system and realize monitoring of the port activities. Its functions include data visualization/recording, AUV motion control, safety precaution, and database storage. To this end, the QT software IEEE Std 802.3-2008, IEEE Standard for Ethernet (2008) is employed to set the management platform, and its procedure consists of four main parts: (a) *Software interface design*. Using Qt Designer, one uses the layout manager to design interface (e.g., video streams display windows, control buttons and data windows), and then, the User Interface Compiler(UIC) tool is used to convert the interface into C++ code for convenience of debug; (b) *Data processing and analysis*. The data from onshore cameras and AUVs are displayed on the interface, while the human operator analyses the data and sends out the control comments to AUVs; (c) *Data logging and security alerts*. A record is embedded into the management platform, such that the data from draft, sensors, sonar and cameras can be displayed; (d) *Monitoring environment database*. In order to bridge the gap between the envi-

ronment and the 3D reconstruction, the MySQL software is employed to set up the monitoring environment database.

(5) *Integration with the DeepSeek Model.* It integrates DeepSeek to augment operator capabilities in critical judgment, situational understanding, and operational reasoning. Specifically, DeepSeek supports mission planning by generating adaptive patrol routes based on environmental conditions and historical detection data. Its framework comprises four key components:

   (a) *Comprehensive data integration.* According to the maritime monitoring regulations, one builds a unified database that synthesizes AUV information and the monitoring environment;
   (b) *Intelligent data inter-connectivity.* One deploys a standardized MCP (Model Context Protocol) interfaces to facilitate seamless, high-bandwidth data exchange between the central control system and DeepSeek's AI engine, such that the instantaneous processing of monitoring data and dynamic terminal status information can be ensured;
   (c) *AI-driven decision support.* The host system generates intelligent operational recommendations through DeepSeek-processed analytics, and hence, it is presented via an intuitive visual interface that enables effective human-AI collaboration for critical decision cycles;
   (d) *Continuous optimization.* One implements an adaptive closed-loop system that progressively refines monitoring protocols, ensuring optimal coordination between AUVs while maintaining operational resilience. This synergistic integration of advanced AI with port monitoring infrastructure can deliver transformative improvements in both operational intelligence and dynamic environmental responsiveness across all maritime domains.

7.4 Technology Practice

7.4.1 Case Study I: Effectiveness of Detection, Communication and Control

See Figs. 7.2, 7.3 and 7.4.

7.4.2 3D Reconstruction of Port Basin

$->$ **Recognition of the ship waterlines**. Taking the bulk carrier as an example (see Fig. 7.2a), we establish the dataset of ship waterlines, where some selected images are shown in Fig. 7.2b. Specifically, the collected data includes video streams, 20 h

(a) Waterline shape for the bulk carrier.

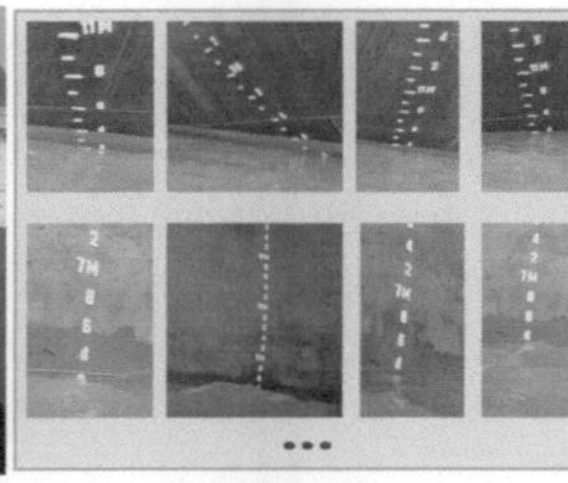

(b) Establishment of the ship waterline dataset.

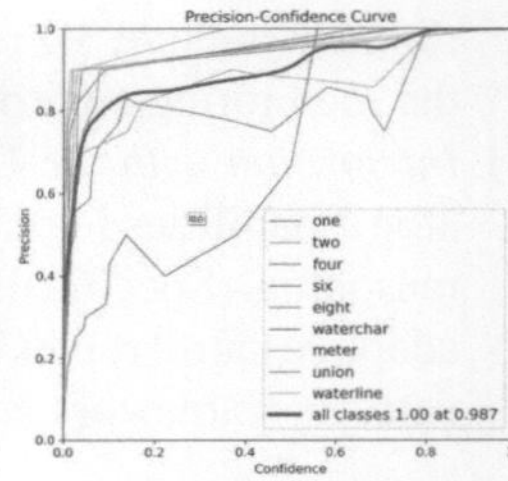

(c) Precision-confidence curve for the classification model.

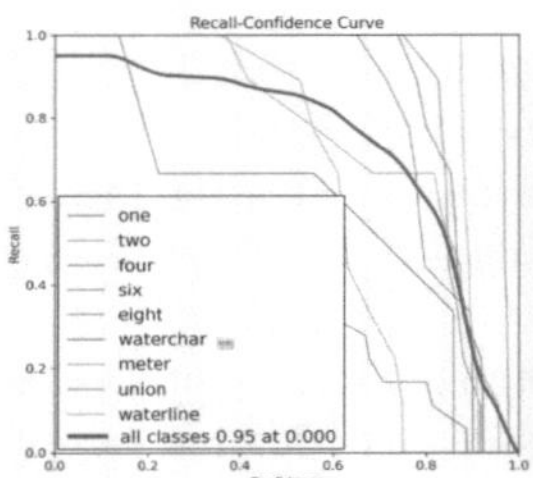

(d) Recall-confidence curve for the classification model.

(e) Segmentated image.

(f) The recognized waterline regions of ship

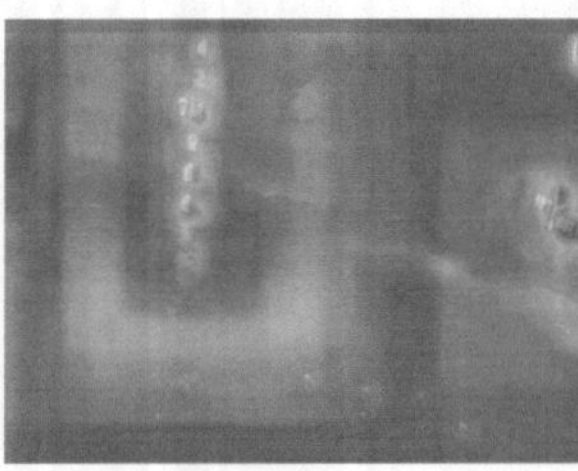

(g) The recognized waterline regions of ship

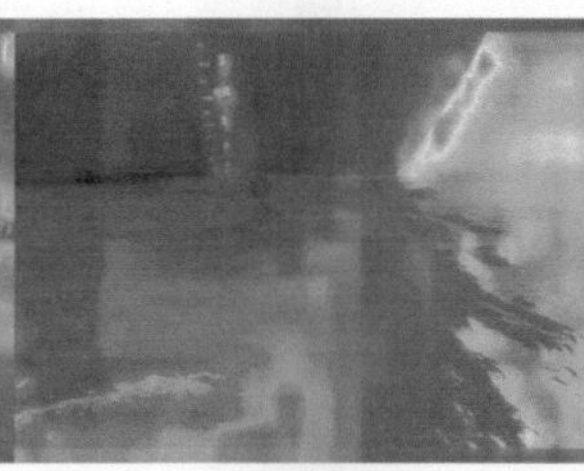

(h) The recognized waterline regions of ship

(i) Recognized ship waterlines

Fig. 7.2 Recognition of the ship waterlines

of continuous video clips and image frames from AUV, and 5000 annotated frames with visible waterlines. Based on this, the offine training is conducted, where the evaluation indexes such as Precision-Confidence and Recall-Confidence curves are shown in Fig. 7.2c–d, respectively. From Fig. 7.2c, one knows the trade-off between recognition precision and confidence can be balanced by the offline model. Meanwhile, the trade-off between recall and confidence can also be balanced, as depicted in Fig. 7.2d. Before implementing the online recognition, the collected image is segmentated, as shown in Fig. 7.2e. After that, the online identification operation is implemented by the cameras. The waterline regions at different positions are recognized by Fig. 7.2f–h. The recognized ship waterline is shown in Fig. 7.2i.The average accuracy of the recognition model in identifying ship water gauges reaches 92.4%, and the frame processing rate remains at 28 FPS. Clearly, the recognition algorithm can effectively recognize the ship waterlines. That means the ship attitudes can be indirectly estimated, i.e., the ship has a draught of 8.5 m.

− > **Effectiveness of the communication network**. The experiment setup is conducted in Tangshan Port, which is an important deep-water port for the northern region of China. The port water depth ranges from 12–15 m, and wave heights below 0.5 m, indicating a relatively calm surface state. The visibility in air was approximately 2 km, while the underwater optical visibility was estimated at 2-3 m, suitable for camera-based recognition tasks. The deployment of the communication equipments is shown in Fig. 7.3a, where the initial distance between AUV and host computer is 100 m. Based on this, the data transmission rate from AUV to the control center is shown in Fig. 7.3b. We can find that the communication rate is larger than 9.8 Mbp/s. Besides that, we also test the data transmission integrity, which is an important index to evaluate the network quality and data integrity. From Fig. 7.3c, the sequence numbers are presented to track the ordered transmission and reliability of data, where the packet loss rate is very small since the curves of receiver windows

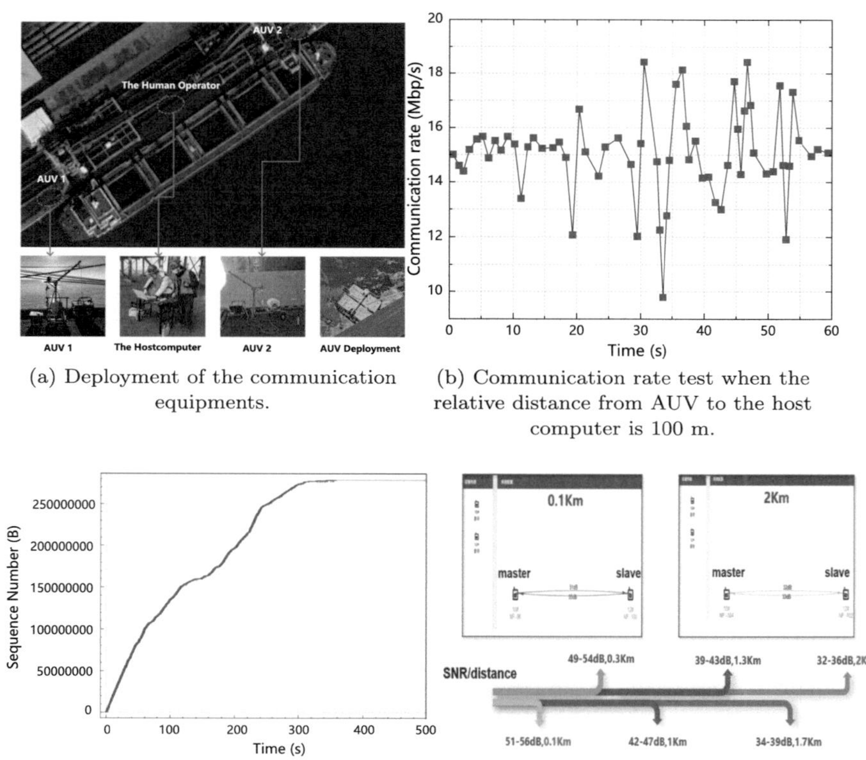

(a) Deployment of the communication equipments.

(b) Communication rate test when the relative distance from AUV to the host computer is 100 m.

(c) Data transmission integrity test.

(d) Communication qualities when the relative distance is from 100 m to 2000 m.

Fig. 7.3 Communication effectiveness at Tangshan Port in China size, the sending data and the acknowledged sequence number are almost coincident. That means the communication quality can be ensured. In order to further verify the reliability of telecommunication, we adjust the communication distance from 100 to 2000 m at Bohai Sea Qinhuangdao area. Correspondingly, the communication qualities (i.e., SNRs) at different distances are shown in Fig. 7.3d. Clearly, the quality of the communication network can be guaranteed in the port area, since the maximum port berth in Tangshan Port is 1.5 km. The above results reflect the communication effectiveness of our monitoring system.

$-$ > **Effectiveness of the motion controller for AUV**. After receiving the monitoring request, AUVs require to reach the desired monitoring area by the motion controller. To test its effectiveness, the target point are shown in Fig. 7.4a. Without loss of generality, the average wind is 3 beaufort scale. An obstacle (i.e., the ship in Fig. 7.4a is set in our scenario. Accordingly, the trajectories for one AUV are shown in Fig. 7.4b. The detection data for the obstacle with the sonar are shown in Fig. 7.4c. From Fig. 7.4a–c, we can find the tracking control can be accomplished. The obstacle can be effectively detected, and more importantly, the obstacle collision can be

(a) Experimental setup for the motion control at the waterway.

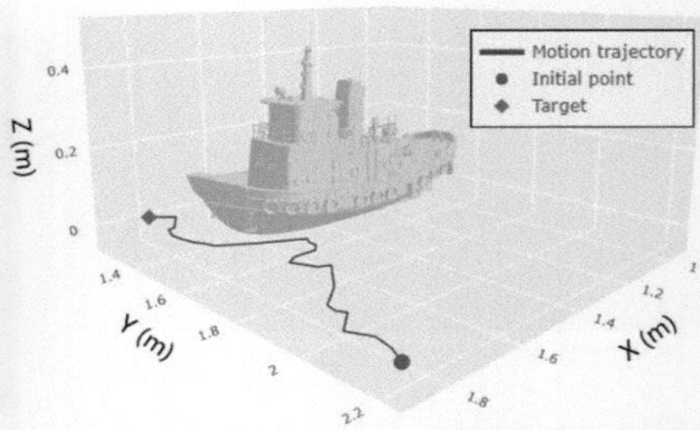

(b) Position trajectory of the AUV towards the target point.

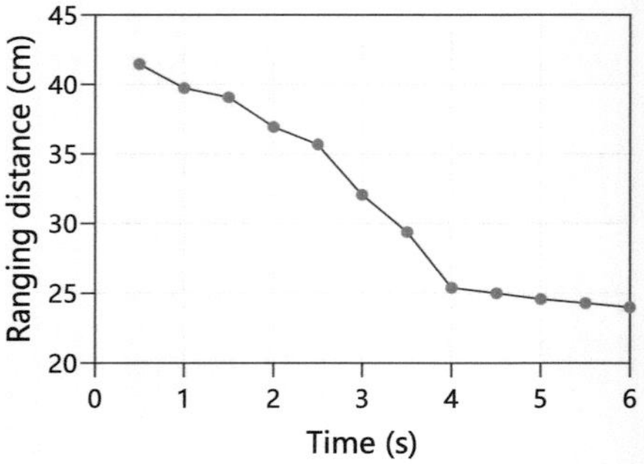

(c) Detection data for the obstacle with the sonar.

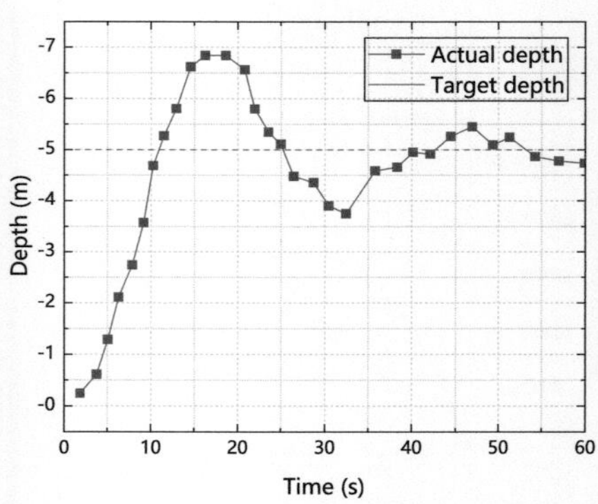

(d) Position trajectory of the AUV for depth control, where the desired depth is 5 m.

Fig. 7.4 AUV motion control

avoided since all the relative distances are larger than zero meter. Besides that, we set the desired depth is -5 m, and hence, the depth control for the AUV is verified by Fig. 7.4d. Clearly, AUV has the diving ability to monitor the port basin at a specific depth.

7.4.3 Case Study II: Construction of Management Platform

$->$ **Effectiveness of port basin 3D reconstruction**. The 3D reconstruction of port basin plays an important role in reducing the risks and enhancing the overall functionality of port facilities. For considering this, we acquire the offline Google map of Tangshan Port, as shown in Fig. 7.5a. Meanwhile, some on-site images are also employed, as provided by Fig. 7.5b–c. In order to capture features such as exposed rebar or cracks in the port basin, we simulate a scenario with exposed rebar in the water tank (see Fig. 7.5d), and then, the feature-related dataset (see Fig. 7.5e) can be established. Correspondingly, the recognized deficiency by using sonar in the water tank can be shown in Fig. 7.5f. After that, AUV is employed to collect visual and acoustic data from the port basin, as shown in Fig. 7.5g–h. Finally, the U3D software is adopted to generate the 3D model of port basin, as shown in Fig. 7.5i.

$->$ **Construction of the management platform**. To enable real-time visualization and decision support, we construct a management platform for the ubiquitous monitoring of ship and port basin. Specifically, it displays sensor data, video streams from AUVs and onshore cameras, as shown in Fig. 7.6. By integrating real-time data, automation, and analytic, the constructed management platform can enhance the operational performance, reduce the costs, and provide better services to human operator.

$->$ **Effectiveness of the DeepSeek-based motion control**. Experimental setup is implemented in the pool, as shown in Fig. 7.7a–b. DeepSeek runs on the shore-based host, assisting operators with mission planning and task prioritization by analyzing real-time data from multiple AUVs. Chosen for its lightweight, modular design and high inference speed, it s fine-tuned on marine data (sonar imagery, navigation logs, operational records) to improve domain awareness and terminology handling. The model integrates with the AUV control system via standardized APIs; latency, stability, and resource use are continuously monitored, with automatic fallback to manual control on anomalies. Communications are protected by multi-layer security, and the model executes in an isolated environment to mitigate cyber risks. Overall, DeepSeek boosts autonomy and efficiency enabling smoother multi-AUV coordination, faster task reallocation, and reduced operator workload without increasing onboard compute. The experiment setup is conducted in Yanmin Pool at Yanshan University. By using DeepSeek model, one enters into the text of "Emergency alert! Dispatch patrol units immediately!". After that, the DeepSeek model replies the text of "Received, patrol has been initiated!", as shown in Fig. 7.7c. Specifically, AUVs construct a maintainable contextual dialogue list, which includes the monitoring tasks, the sens-

(a) offline Google map of Tangshan Port. (b) On-site image such as the portal cranes. (c) On-site image such as the scheduling center office.

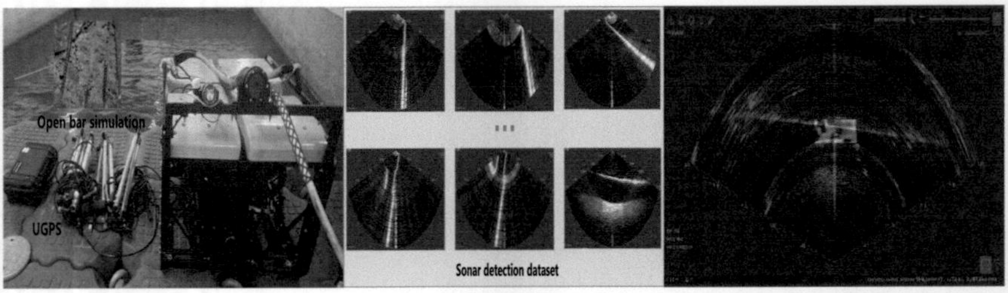

(d) Exposed rebar in the water tank. (e) Feature-related dataset, including cracks, corrosion, and exposed rebars in underwater infrastructures. (f) Recognized deficiency by using sonar in the water tank.

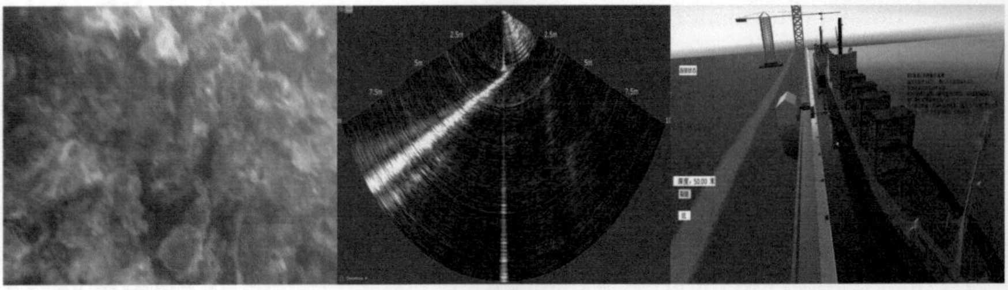

(g) Visual data from the port basin. (h) Visual data from the port basin. (i) 3D model of port basin with the U3D software.

Fig. 7.5 Effectiveness of port basin 3D reconstruction

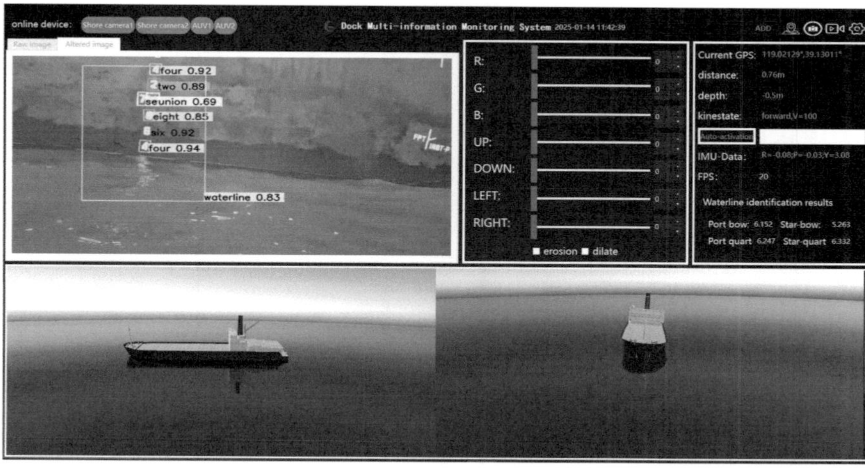

Fig. 7.6 Construction of the management platform. The upper-left section of the platform is used to display the raw and altered image from AUVs and onshore cameras; the upper-middle section is the image processing operation area; the upper-right section displays AUV parameters data, autonomous movement target point settings, and ship waterline recognition results; the lower section shows the 3D attitude of the ship

ing data, and the decisions for each dialogue round. Along with this, AUVs integrate the DeepSeek decisions into control commands, through which AUVs are driven to detect the existence of target, as shown in Fig. 7.7d.

$->$ **Effectiveness of the sonar-based detection capacity**. Following the pre-programmed search patterns illustrated in Fig. 7.7d, AUV conducts a systematic sonar sweep of the designated monitoring zone. The onboard sonar successfully identifies the target object at T+03:55 (mission elapsed time), with the acquisition confirmed by visual returns as depicted in Fig. 7.7e). These results confirm the effectiveness of our monitoring system for underwater target.

7.5 Summary

AUVs are employed in this work to achieve ubiquitous monitoring of port activities. Different from the simple motion control of AUVs, we systematically design the AUV-assisted monitoring system for port activities toward human-in-the-loop, such that the human intelligence and abilities can be brought to bear on complex underwater tasks. Our system is inspired by the basis idea in Pendleton et al. (2017), but it is fundamentally different from the solution in Pendleton et al. (2017). Particularly, the main differences can be summarized as follows: (1) *Sensing unit*. The radar is adopted in the above literature to obtain the point cloud data, however the radar can-

(a) Experimental setup in the pool.

(b) Experimental setup in the pool.

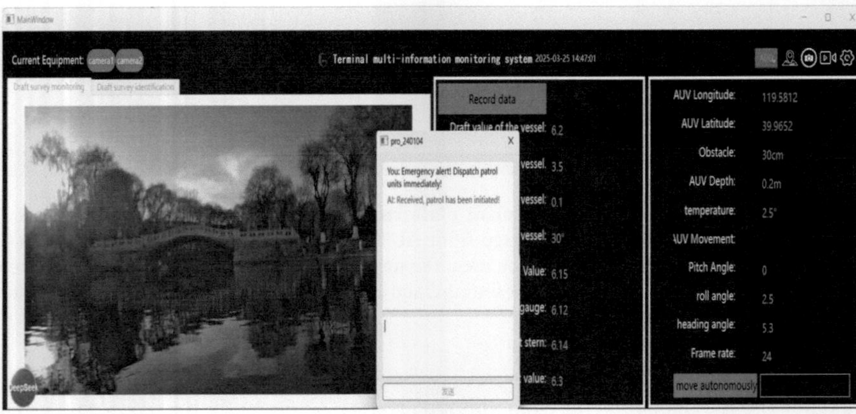

(c) Integration with the DeepSeek model.

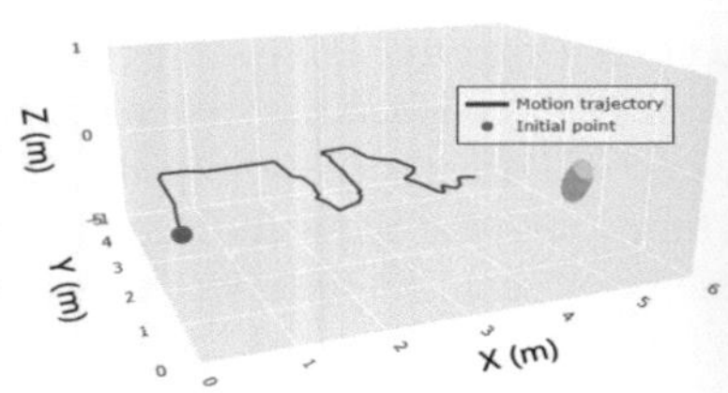

(d) Position trajectories of one AUV during the detection of underwater target.

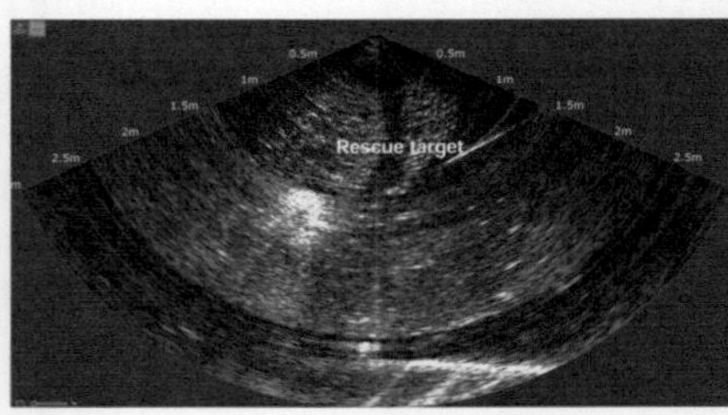

(e) Detection data for the target with the sonar.

Fig. 7.7 Experimental results for the underwater target

not be applied to underwater environment due to light scattering and absorption. For that reason, the sonar is used in our system to detect the underwater target; (2) *Communication unit*. The electromagnetic wave communication is adopted in Pendleton et al. (2017) to guarantee the communication transmission, however, it cannot be used in underwater system, and hence, the acoustic communication is employed in our system to guarantee the communication transmission; (3) *Localization unit*. GPS is used to achieve vehicle localization in Pendleton et al. (2017), but GPS signals are entirely blocked in underwater environment. For that reason, the UGPS is adopted in our system to achieve precise underwater localization by TDOA algorithms; 4) *Control unit*. Autonomous vehicles only need to consider two-dimensional planar motion, while AUVs need to consider three-dimensional spatial motion. Therefore, the three-dimensional motion planning is considered in our system, which is ignored in Pendleton et al. (2017). Meanwhile, our system not only boosts the capability to handle unforeseen or dynamic situations but also optimizes human resources, ensuring effective and intelligent oversight of AUV operations. Therefore, our system is fundamentally different from Pendleton et al. (2017).

Our system can also be applied to the other application scenarios, such as port monitoring Ratnarajah et al. (2023), off-shore explorationLee et al. (2019), submarine rescue Shlesinger (2006), cooperation navigation Yang et al. (2024), Zhang et al. (2019), pipeline patrolZhao et al. (2024) and tsunami warning Selva et al. (2021). Nevertheless, there are still some unsolved issues for the ubiquitous monitoring of port activities. For example, our system lacks the air space monitoring ability, and hence, it cannot monitor the surface element of a target, e.g., the deck of a ship. Thereby, it is necessary to employ UAVs, USVs, and AUVs to cooperatively achieve the monitoring task, as depicted by Fig. 7.8. Based on the above consideration, we discuss some important yet unresolved issues in the cross-domain swarm collaboration monitoring process, as well as directions worthy of further exploration in the future.

- *Development of Cross Domain Swarm Standards*. The cross-domain swarm monitoring of port activities can comprehensively utilize various technologies and resources. It can overcome the limitations of a single field, achieve efficient col-

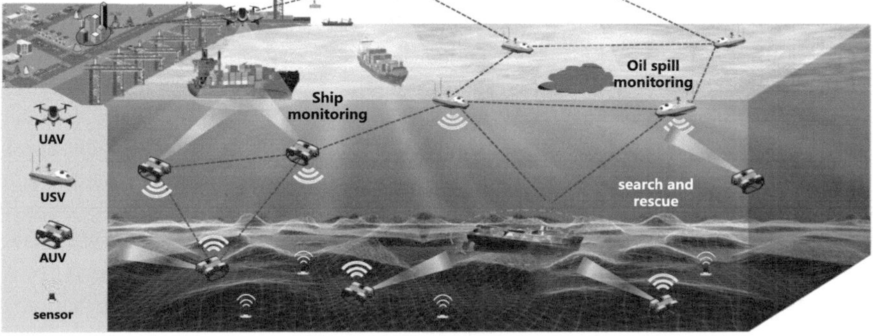

Fig. 7.8 Air-surface-underwater integration cross-domain monitoring system. UAVs perform aerial monitoring and provide wide surface coverage, while they also act as the relay nodes for data to the shore-based center. USVs conduct surface monitoring, while they relay the aerial and underwater data. AUVs perform underwater tasks such as harbor basin exploration through swarm collaboration. Accordingly, the control center integrates global information to optimize task planning and remote control

laboration of multiple tasks, and bring unprecedented opportunities for port monitoring. However, different fields may adopt different technical standards and specifications, while each field may have unique operational processes and regulations. At the same time, different countries or organizations have different management standards and legal provisions for cross domain swarm. Therefore, it is necessary to develop a unified framework and standardized protocols for cross domain cooperation. By developing unified technical interfaces, the integration/collaboration between different fields can be promoted. As a result, the overall efficiency and availability of cross domain swarm systems can be improved.

- *Integrated Architecture of Sensing, Communication and Control.* At present, the sensing, communication, and control functions of cross-domain swarm are designed independently. That means the related devices compete with each other in energy consumption and frequency band. Due to their size limitations, it is hard to improve the perception and communication interaction functions. Note that the sensing, communication and control have similarities in working principles, system architecture, signal processing, and operating frequency bands. The above property brings possibilities for integrated design of sensing, communication control. MealeyMealey (1963) first proposed the integration idea of radar detection and communication in the 1960s. With consideration of the port characteristics, how to design an integrated exploration and control architecture for cross domain swarm remains a challenge to be solved in recent years.
- *Cross-Media Communication Technology.* In order to ensure the communication, it is necessary to interact with various types of manned/unmanned platforms, such as UAVs, USVs and AUVs. Therefore, the various communication methods such as underwater acoustics, optics, electromagnetism, and others can be integrated to enhance the reliability and transmission speed of cross medium communication. On the other hand, one can develop the relay-based communication technology, where the effective information communication with other unmanned platforms can be achieved. The implementation of the above functions requires to overcome multiple technologies, such as optoelectronic conversion technology, radio frequency conversion technology, protocol conversion and relay technology.
- *Cross Domain Swarm Driven by Artificial Intelligence.* Driven by the artificial intelligence, cross domain swarm can achieve more efficient and intelligent applications. For example, various platforms can make intelligent decisions through machine learning and inference algorithms. Meanwhile, it can enhance the coordination ability with other platforms to achieve better overall results. In addition, the artificial intelligence technology can enable cross domain swarm platforms to have intelligent perception and data analysis capabilities, thereby extracting valuable information and helping them make more accurate and timely decisions. However, it is not possible to mechanically transplant artificial intelligence technology into cross domain swarm. Instead, we should fully consider the characteristics and requirements of cross domain swarm, through the cross-domain swarm in combination with the advantages and limitations of artificial intelligence technology can be effectively guaranteed.

- *Application Platform in Real Ocean Environment*. Till now, most of the results are verified through simulation software. While the simulations are incredibly valuable, the real ocean testing is still required to verify the performance of ship and port basin monitoring in dynamic environments. For example, the search and rescue missions should be verified in real lake or ocean environment Serra et al. (2020), rather than on the simulation software. This helps uncover problems that may not have been anticipated during simulation (e.g., the sensor calibration issues, the network interference, or the regulatory constraints). However, it is hard or costly to perform the real-world testing. To this end, researchers should bridge the gap between simulation results and practical deployment. Meanwhile, the collaboration and data sharing with maritime organizations could also be a way to carry out these tests more effectively.

References

Adil, M., Jan, M. A., Liu, Y., Abulkasim, H., Farouk, A., & Song, H. (2023). A systematic survey: Security threats to UAV-aided IoT applications, taxonomy, current challenges and requirements with future research directions. *IEEE Transactions on Intelligent Transportation Systems, 24*, 1437–1455.

Choi, J.-K., Sakai, H., & Tanaka, T. (2005). Autonomous towed vehicle for underwater inspection in a port area. In *Proceedings of the IEEE International Conference on Robotics and Automation* (pp. 188–193).

"Detection sonar data sheet". Searobotix.com. Retrieved December 27, 2024 form https://searobotix.com/oculus/.

Fitzpatrick, A., Mathews, R. P., Singhvi, A., & Arbabian, A. (2023). Multi-modal sensor fusion towards three-dimensional airborne sonar imaging in hydrodynamic conditions. *Communications Engineering, 2*, 16.

IEEE Std 1901.1.1-2020, IEEE Standard Test Procedures for IEEE Std 1901.1 for Medium Frequency (less than 15 MHz) Power Line Communications for Smart Grid Applications (2020). Retrieved December 30, 2024 from https://standards.ieee.org/standard/1901_1_1-2020.html.

"IEEE Std 802.3-2008, IEEE Standard for Ethernet" (2008). Retrieved December 30, 2024 from https://standards.ieee.org/standard/802_3-2008.html.

Jaffe, J. S., Franks, P. J. S., Roberts, P. L. D., Mirza, D., Schurgers, C., Kastner, R., & Boch, A. (2017). A swarm of autonomous miniature underwater robot drifters for exploring submesoscale ocean dynamics. *Nature Communications, 8*, 14189.

Khalid, A., Mushtaq, Z., Arif, S., Zeb, K., Khan, M. A., & Bakshi, S. (2023). Control schemes for quadrotor UAV: Taxonomy and survey. *ACM Computing Surveys, 56*, 124.

Lee, D., Balachandran, T., Cho, H.-W., & Haran, K. (2019). Exploring fully superconducting air-core machine topology for off-shore wind turbine applications. *IEEE Transactions on Magnetics, 55*, 1–6.

Lin, J. M., Fang, S. K., He, R. J., Tang, Q. S., Qu, F. Z., Wang, B. S., & Xu, W. (2024). Monitoring ocean currents during the passage of Typhoon Muifa using optical-fiber distributed acoustic sensing. *Nature Communications, 15*, 1111.

McNelly, B. P., Albert, C., & Appler, J. D. (2023). Prototyping an autonomous underwater vehicle to support port protection challenges. *OCEANS, 2023*, 1–7.

Mealey, R. M. (1963). A method for calculating error probabilities in a radar communication system. *IEEE Transactions on Space Electronics and Telemetry, 9*, 37–42.

New Data Reveals Risk of Maritime Incidents in Ports and Terminals, gcaptain.com, February 22, 2023. https://gcaptain.com/new-data-reveals-risk-of-maritime-incidents-in-ports-and-terminals/.

"Onshore camera data manual," LCSC Electronics (Shanghai) Co., LCSC.com. Retrieved June 6, 2024 from https://www.lcsc.com/datasheet/lcsc_datasheet_2409191732_HIKVISION-DS-2DE2Q140MY-T-GL2-2-8mm_C41359978.pdf.

Pendleton, S. D., Andersen, H., Du, X., Shen, X., Meghjani, M., Eng, Y. H., et al. (2017). Perception, planning, control, and coordination for autonomous vehicles. *Machines, 5*, 6.

Peng, X., Huang, L., Wu, L., Zhou, C., Wen, Y., Chen, H., & Xiao, C. (2021). Remote detection sulfur content in fuel oil used by ships in emission control areas: A case study of the Yantian model in Shenzhen. *Ocean Engineering,237*, Article 109652.

Ratnarajah, L., Abu-Alhaija, R., Atkinson, A., et al. (2023). Monitoring and modelling marine zooplankton in a changing climate. *Nature Communications, 14*, 564.

Selva, J., Lorito, S., Volpe, M., et al. (2021). Probabilistic tsunami forecasting for early warning. *Nature Communications, 12*, 5677.

Serra, M., Sathe, P., Rypina, I., Kirincich, A., Ross, S. D., Lermusiaux, P., Allen, A., Peacock, T., & Haller, G. (2020). Search and rescue at sea aided by hidden flow structures. *Nature Communications, 11*, 2525.

Shen, L., Hou, Y., Yang, Q., Lv, M., Dong, J.-X., Yang, Z., & Li, D. J. (2022). Synergistic path planning for ship-deployed multiple UAVs to monitor vessel pollution in ports. *Transportation Research Part D: Transport and Environment,110*, Article 103415.

Shlesinger, M. (2006). Search research. *Nature, 443*, 281–282.

Slade, P., Atkeson, C., Donelan, J. M., Houdijk, H., Ingraham, K. A., Kim, M., Kong, K., Poggensee, K. L., Riener, R., & Steinert, M. (2024). On human-in-the loop optimization of human robot interaction. *Nature, 633*, 8031.

Sladen, A., Rivet, R., Ampuero, J. P., Barros, L. D., Hello, Y., Calbris, G., & Lamare, P. (2019). Distributed sensing of earthquakes and ocean-solid Earth interactions on seafloor telecom cables. *Nature Communications, 10*, 5777.

Sun, G. B., Zhou, R., Ma, Z., Li, Y. Q., Grob, R., Chen, Z., & Zhao, S. Y. (2023). Mean-shift exploration in shape assembly of robot swarms. *Nature Communications, 14*, 3476.

Tran, T. A., Sridhar, S., Reece, S. T., Lunguya, O., Jacobs, J., Puyvelde, S. V., Marks, F., Dougan, G., Thomson, N. R., Nguyen, B. T., Bao, P. T., & Baker, S. (2024). Combining machine learning with high-content imaging to infer ciprofloxacin susceptibility in isolates of Salmonella Typhimurium. *Nature Communications, 15*, 5074.

Tsaimou, C. N., Sartampakos, P., & Tsoukala, V. K. (2023). UAV-driven approach for assisting structural health monitoring of port infrastructure. *Structure and Infrastructure Engineering, 19*, 1–20.

Xia, X., Fattah, S. M. M., & Babar, M. A. (2023). A survey on UAV-enabled edge computing: Resource management perspective. *ACM Computing Surveys, 56*, 78.

Xiao, Y., Lei, X. K., Zheng, Z. C., Xiang, Y. L., Liu, Y. Y., & Peng, X. G. (2024). Perception of motion salience shapes the emergence of collective motions. *Nature Communications, 15*, 4779.

Yang, K., Zhang, Z., Cui, R., & Yan, W. (2024). Acoustic-optic assisted multisensor navigation for autonomous underwater vehicles. *Ocean Engineering,297*, Article 117139.

Zhang, Z., Shi, Y., Zhang, Z., & Yan, W. (2019). New results on sliding-mode control for Takagi-Sugeno fuzzy multiagent systems. *IEEE Transactions on Cybernetics, 49*, 1592–1604.

Zhang, Z., Shi, Y., Zhang, Z., Zhu, M., Ji, L., & Wu, Y. (2024). Autonomous UAV safety oriented situation monitoring and evaluation system. *Drones, 8*, 308.

Zhao, Y., Luo, W., Wang, Z., et al. (2024). An oil and gas pipeline inspection UAV based on improved YOLOv7. *Measurement and Control, 57*, 1068–1086.

Zhuang, Y. M., Liu, J. M., Zhao, H. Y., Ma, L. Y., Fang, Z. R., Li, L., Wu, C. D., Cui, W., & Liu, Z. L. (2025). A deep learning framework based on structured space model for detecting small objects in complex underwater environments. *Communications Engineering, 4*, 24.

Chapter 8
Future Research Directions

Looking across the world today, the ocean is increasingly prominent as a critical domain for global resource development, economic growth, and strategic competition. The United States Unmanned Systems Integrated Roadmap (2021) proposed the construction of a cross-domain unmanned combat system. The European Union, through its European Defence Fund, prioritized support for the development of underwater autonomous systems. China is a major land and maritime nation, whose 14th Five-Year Plan also identifies unmanned systems as a development priority. Against this backdrop, MUSs, as a key component of cross-domain clusters, provide technical support for developing marine resources, and maintaining maritime security. Their capabilities include efficient networking in complex environments, precise detection, and intelligent control. However, constrained by the current technological development, MUSs still face multiple bottlenecks. These include communication reliability, detection accuracy, energy consumption optimization, and intelligent collaboration. In the future, in-depth research should target key technological shortcomings. Methodologies need continuous refinement. This chapter outlines several important yet unresolved issues in the detection-networking-control coordination of MUSs. It also suggests future directions worthy of further exploration.

8.1 Open Issues for Further Research

(1) Lack of a standardized framework
Current maritime cross-domain unmanned systems lack unified technical standards and operational protocols. This makes it difficult to integrate the equipments and systems from different domains efficiently. The problem is especially evident in international and multi-domain cooperation. There is a clear absence of unified top-level planning and technical interfaces. Creating a common framework for cross-domain

J. Yan and X. Guan, *Maritime Unmanned Systems*, Cognitive Intelligence
and Robotics 1, https://doi.org/10.1007/978-981-95-6907-6_8

cooperation is therefore a key bottleneck. Differences in technical standards, communication protocols, and data interfaces hinder coordination. Without unified cross-domain standards, system interoperability and efficiency remain limited. Furthermore, these systems operate across underwater, surface, aerial, and satellite spaces. This multi-layer structure increases the complexity of data fusion, task allocation, and cooperative optimization for MUSs. As a result, the existing theories and methods need significant advancement.

(2) Pressing need for better long-endurance MUSs

Long-duration, reliable MUSs are critical for cross-domain clusters, yet current technologies require significant advancements in endurance, energy management, and adaptability to extreme conditions. While conceptually viable, platforms designed for long-term maritime residence, such as offshore floating islands, need further validation of their engineering feasibility, cost-effectiveness, and operational practicality. The diverse requirements of underwater, surface, and aerial domains pose a major challenge for multi-mode integrated design, which must balance structural integrity, power systems, and communications. Furthermore, achieving long endurance demands innovations in energy density, efficiency management, and adaptive power systems. The immaturity of marine renewable energy technologies including fuel cells, wave, and solar power currently limits progress in this area. Finally, a key design challenge lies in ensuring operational reliability and mission continuity during dynamic task switching and multi-modal operations.

(3) Difficulty in deep-sea accurate detection

Deep-sea sensors lack precise detection capabilities. Signal attenuation is severe in complex deep-sea environments. Existing sensors need improvements in accuracy and anti-interference. Real-time performance also requires enhancement. Current equipment cannot meet refined detection requirements. Deep-sea unmanned systems show low flexibility and intelligence. They lack unified modular design standards. This hinders multi-task coordination. Extreme deep-sea conditions create major challenges. These include high pressure, low temperature, and strong corrosion. Miniaturized equipment requires better materials and sealing. Durability must be improved. Current technology cannot maintain both reliability and performance at small sizes. Intelligent systems need efficient perception and computing. However, miniaturization limits hardware integration. Communication delays and data transmission are also constrained. Energy supply remains a critical bottleneck. Effective energy management in limited space is still unsolved.

(4) Separation design of detection, communication, and control

Detection, communication, and control are often designed separately. This causes resource competition. System size, energy consumption, and stealth are limited. Integrated architecture needs environmental breakthroughs. Resource sharing must be improved. Collaboration efficiency should be enhanced. Hardware redundancy needs reduction. Cross-domain clusters face coordination challenges. Detection requires high-precision data. Communication needs high bandwidth and low latency.

Control demands rapid response and robustness. The maritime environment is highly complex. Electromagnetic interference and depth variations exist. Weather changes also cause problems. System interference and resource conflicts are common. Optimizing all three under limited resources remains difficult.

(5) Bottlenecks in cross-medium communication technology
Maritime cross-domain communication relies on multi-modal technologies such as acoustic, optical, and electromagnetic methods yet their integration faces multiple technical challenges. Issues such as inefficient photoelectric conversion, underdeveloped protocol adaptation, and immature frequency coordination contribute to the unreliability of relay-free cross-medium communication, thereby limiting seamless information exchange across multiple platforms. Establishing efficient and reliable links across different media-including water, air, and deep-sea environments-proves particularly difficult. Underwater acoustic communication, for instance, is constrained by signal attenuation, noise interference, and limited bandwidth, while air and surface-level links often suffer from transmission discontinuity and susceptibility to atmospheric conditions. Furthermore, achieving low power consumption without compromising communication efficiency remains a critical challenge for MUSs, making it difficult to meet the long-endurance requirements.

(6) Limitations in human-machine collaboration mechanisms
Fully autonomous modes have limited capability. Complex maritime missions are challenging. Weak communication scenarios are problematic. High-risk situations are especially difficult. Existing human-in-the-loop technologies need adaptation. Maritime environment compatibility is low. Human intelligence must better integrate with system autonomy. Situation awareness requires enhancement. Collaboration capability needs improvement. Human-machine collaboration frameworks face challenges. Efficient interaction between operators and systems is needed. Traditional manual operation is inadequate. Complex situations are hard to handle. AI systems must understand human decision intent. They need to make optimized decisions in dynamic environments. This remains a research difficulty.

(7) Challenges in AI technology adaptation to scenarios
While AI holds significant potential for enhancing system intelligence, cross-domain clusters impose stricter requirements, including low computational cost, high robustness, and real-time performance. Currently, there is a shortage of lightweight and reliable AI algorithms specifically designed for cluster characteristics. AI-driven optimization faces considerable challenges in complex maritime environments, where traditional methods often fail to address high-dimensional, nonlinear, and dynamically changing conditions. Multi-objective optimization under multiple constraints is particularly difficult, making global optimization a persistent challenge.

(8) High complexity and cost of simulation environments
Testing in marine environments is both costly and high-risk, while the current simulation methods often fail to adequately replicate real mission scenarios. In view of this, digital twin technology offers a promising solution by enabling the construction of dynamic virtual environments to support the cross-domain cluster optimization

and validation in a cost-effective manner. However, several challenges impede its reliability: achieving a high-fidelity match between virtual and real worlds requires not only accurate models built on extensive data but also detailed physical representations, which are difficult to obtain especially for deep-sea conditions. Furthermore, the computational demands are also substantial, as simulation systems must process massive data streams in real time while maintaining excellent responsiveness to support dynamic updates.

8.2 Directions for Future Research

(1) Cross-domain standards and collaboration framework
Future work should establish unified cross-domain and international standards. These standards must cover communication protocols and data interfaces. Operating procedures also need standardization. International organizations and industry bodies should collaborate. This will enhance system compatibility and adaptability. A universal framework for cross-domain cooperation is essential. It will promote open technology development and standardized processes. Multinational technical standards should be developed. These include task allocation rules and communication protocols. International cooperation and trust mechanisms must be strengthened. Joint experiments and open innovation platforms will help. Deeper collaboration in technology and data sharing should be pursued. Theoretical research should focus on cross-domain optimization algorithms. Adaptive control methods need development. These will enable efficient multi-domain operations. They also support sustainable development.

(2) Multi-mode long-endurance equipment design
Research should advance endurance technologies. Autonomous energy harvesting and management are key priorities. The “offshore floating island” concept requires engineering validation. Its cost-effectiveness and deployment flexibility need improvement. Deep-sea equipment development faces urgent challenges. Multi-mode integrated design requires innovation. New energy technologies should be explored. Intelligent energy scheduling algorithms require optimization. Energy efficiency and system endurance must be enhanced. Intelligent switching strategies for multi-modal tasks should be designed. Equipment should adapt to dynamic environments and missions. Digital twin technology and AI can optimize design processes. Multi-environment performance simulation should be implemented. Equipment design should evolve from passive adaptation to active collaboration.

(3) Miniaturization and intelligence of deep-sea equipment
In China, the 2025 Government Work Report has identified deep-sea technology as a national strategic priority. However, current technological capabilities fall short of balancing resource development with ecological protection. Existing deep-sea mining and detection equipment remains bulky, energy-intensive, and operationally inflexible. Future systems must achieve higher levels of mobility and intelligence to

adapt to complex terrains and dynamically changing tasks. Breakthroughs in new materials and micro-nano manufacturing are essential to overcome miniaturization barriers, with composite materials offering improved strength and corrosion resistance. Research should also advance AI-enabled embedded systems capable of integrating perception, computing, and decision-making. The exploration of new energy technologies such as deep-sea thermal gradient power generation and the integration of acoustic, optical, and wireless communication systems will be critical to enhance operational coordination in the deep-sea environment.

(4) Integrated design of detection, communication, and control
Research should prioritize the development of integrated detection-communication-control systems, where breakthroughs in key technologies such as dynamic resource scheduling, frequency band reuse, and stealth enhancement are essential. The design of modular and compact integrated equipment will further optimize cluster coordination efficiency. Future studies need to combine multi-objective optimization algorithms with adaptive control methods, supported by design frameworks capable of real-time parameter adjustment in dynamic operational environments. Leveraging advanced deep learning and reinforcement learning will enable systems to achieve self-adjustment and autonomous optimization, thereby improving both adaptability and robustness under complex conditions. Additionally, interdisciplinary design approaches must be strengthened to deepen the understanding of system synergy, while enhanced simulation of cooperative effects will ensure the seamless integration of detection, communication, and control functions.

(5) Efficient cross-medium communication systems
Multi-model communication integration needs development. Communication stability and speed must improve. Relay-free communication bottlenecks require solutions. Direct inter-platform connectivity should be enhanced. Photoelectric conversion technology needs research. Protocol adaptation requires further study. Future directions should integrate multiple technologies. These include acoustic, fiber optic, and satellite communication. Hybrid signal transmission mechanisms are designed. AI can optimize signal scheduling. This will improve bandwidth utilization and anti-interference capability. System stability will increase in complex environments.

(6) Human-machine integrated framework
A deeply integrated architecture should be established, effectively combining "human-in-the-loop" control with autonomous system capabilities. The introduction of multi-modal sensing technologies, coupled with enhanced human-machine interaction, will significantly improve system perception and decision-making efficiency. Future research should prioritize intelligent interaction by optimizing interface design and dynamic task allocation, as well as developing context-aware multimodal interaction systems. With AI effectively supporting operator decisions, capabilities in autonomous networking, target recognition, cooperative control, and decision-making will be substantially strengthened. Furthermore, environmental perception and prediction will be enhanced, leading to more reliable autonomous navigation

and obstacle avoidance. Advances in deep learning and reinforcement learning will further improve system adaptation to complex scenarios and increase the efficiency of human-machine collaboration.

(7) AI-driven autonomous optimization
MUSs are undergoing a rapid evolution, transitioning from standalone platforms to intelligent, collaborative clusters a shift largely driven by advances in artificial intelligence. Deep learning enhances underwater target recognition, while multi-agent reinforcement learning enables cross-domain cooperation, and brain-inspired computing supports higher-level autonomous reasoning. Moving forward, the development of lightweight machine learning algorithms is essential to meet real-time and low-power operational demands. Reinforcement learning will further improve system autonomy, though distributed collaboration capabilities must be strengthened in parallel. There is also a need for efficient reinforcement learning algorithms that can dynamically adapt to environmental changes. Integrating big data analytics with edge computing will allow systems to leverage real-time data streams and utilize historical data to anticipate future conditions. Ultimately, cross-disciplinary research will be crucial to fully realize AI's potential in this field.

(8) Digital twin-based simulation verification
Digital twin technology requires deeper integration across several fronts. Enhancements are needed in multi-sensor data fusion, the combination of virtual simulation with machine learning, and improvements in both model accuracy and real-time performance. The establishment of hardware-in-the-loop testing systems will be crucial to support performance evaluation and optimization. For instance, digital twins capable of mapping real ocean environments can provide reliable platforms for tracking control, enabling dynamic simulation under virtual-real integration. Blockchain technology could further ensure data security throughout this process. As simulation platforms become more closely linked with actual systems, digital twins must accurately reflect real-world performance. Finally, the integration of cloud and edge computing will enhance data processing capabilities, thereby increasing simulation efficiency and scalability.

8.3 Summary

The integrated detection-communication-control methodology for MUS represents a core technical pathway driving the intelligent transformation of maritime operations. This approach substantially enhances capabilities in maritime target detection, communication connectivity, and intelligent decision-making, while providing essential technical support for marine resource development, ecological protection, and maritime safety management. Based on this framework, the book systematically elaborates collaborative methodologies spanning the entire technology chain from networked communication and detection positioning to cooperative control. It provides a thorough analysis of current technical bottlenecks and challenges, reviews global research advances, and summarizes typical application scenarios. Future directions are also outlined in alignment with strategic ocean development needs. Moving forward, research should focus on advancing core technologies such as high-efficiency communication, precise detection, and intelligent collaboration methods. Strengthening multidisciplinary cooperation across relevant fields will be crucial to overcome existing technical constraints. These efforts will establish a solid foundation for building a fully functional, efficient, and broadly applicable collaborative system, ultimately supporting the long-term objective of sustainable and advanced maritime operations.

MIX
Papier aus verantwortungsvollen Quellen
Paper from responsible sources
FSC® C105338

If you have any concerns about our products,
you can contact us on
ProductSafety@springernature.com

In case Publisher is established outside the EU,
the EU authorized representative is:
Springer Nature Customer Service Center GmbH
Europaplatz 3, 69115 Heidelberg, Germany

Printed by Libri Plureos GmbH
in Hamburg, Germany